Advanced Joining Technologies

This book covers advances in fusion and solid-state welding processes, including basics, welding metallurgy, defect formation, and the effect of process parameters on mechanical properties. Details of the microstructural and mechanical behaviors of weldments are included. This book covers challenges encountered during dissimilar welding of metal by fusion and solid-state welding processes, including remedial solutions and hybrid processes to counter the same. Numerical and statistical simulation approaches used in the welding process for parameter optimization and material flow studies are described as well.

Features:

- Provides details related to the microstructural and mechanical behaviors of welded joints developed by different welding processes.
- Covers recent research content, metallurgical analysis, and simulation aspects.
- Discusses the joining of plastics and ceramics.
- Includes a dedicated chapter on machine learning and digital twin in welding.
- Explores difficulties associated with the joining of dissimilar metals and alloys.

This book is aimed at researchers and graduate students in material joining and characterization and welding.

Advanced Materials Processing and Manufacturing

Series Editor: Kapil Gupta

The CRC Press Series in Advanced Materials Processing and Manufacturing covers the complete spectrum of materials and manufacturing technology, including fundamental principles, theoretical background, and advancements. Considering the heightened significance of advancements in creating high-quality products for diverse applications, the titles within this series mirror the cutting-edge developments in comprehending and optimizing materials processing and manufacturing operations. Technological advancements for enhancement of product quality, process productivity, and sustainability are on special focus including processing for all materials and novel processes. This series aims to foster knowledge enrichment on conventional and modern machining processes. Micro-manufacturing technologies, such as micro-machining, micro-forming, and micro-joining, and hybrid manufacturing, such as additive manufacturing, near net shape manufacturing, and ultra-precision finishing techniques, are also covered.

Advanced Materials Characterization: Basic Principles, Novel Applications, and Future Directions
Ch Sateesh Kumar, M. Muralidhar Singh and Ram Krishna

Thin-Films for Machining Difficult-to-Cut Materials: Challenges, Applications, and Future Prospects
Ch Sateesh Kumar and Filipe Daniel Fernandes

Advanced Materials Processing and Manufacturing: Research, Technology, and Applications
Amogelang Sylvester Bolokang and Maria Ntsoaki Mathabathe

Advanced Joining Technologies
Edited by Manjaiah M, Shivraman Thapliyal and Adepu Kumar

For more information about this series, please visit: www.routledge.com/Advanced-Materials-Processing-and-Manufacturing/book-series/CRCAMPM

Advanced Joining Technologies

Edited by
Manjaiah M, Shivraman Thapliyal
and Adepu Kumar

CRC CRC Press
Taylor & Francis Group
Boca Raton London New York

CRC Press is an imprint of the
Taylor & Francis Group, an **informa** business

Designed cover image: © Shutterstock Images

First edition published 2024
by CRC Press
2385 NW Executive Center Drive, Suite 320, Boca Raton FL 33431

and by CRC Press
4 Park Square, Milton Park, Abingdon, Oxon, OX14 4RN

CRC Press is an imprint of Taylor & Francis Group, LLC

© 2024 selection and editorial matter, Manjaiah M, Shivraman Thapliyal and Adepu Kumar; individual chapters, the contributors

ISBN: 9781032356358 (hbk)
ISBN: 9781032356372 (pbk)
ISBN: 9781003327769 (ebk)

DOI: 10.1201/9781003327769

Typeset in Times
by codeMantra

Contents

Ajfarul Islam, Dipankar Bose, Dhiraj Kumar, and B. Acherjee

G. R. Joshi and Raghavendra Darji

Pankaj Kaushik, Ranjan Kumar, Manjaiah M., and Ajith G. Joshi

Chapter 8 Joining of Metallic Materials Using Microwave
            Hybrid Heating

*Kadapa Vijaya Bhaskar Reddy, K. V. Hari Shankar,
and Gudipadu Venkatesh*

Preface

This book comprises 13 chapters authored by professionals from different countries, all of whom possess extensive knowledge, substantial experience, and a background in welding methods. This book commences by elucidating the fundamental principles of explosive welding and their characterization. It then progresses to delve into cutting-edge welding technologies and the application of machine learning techniques in welding processes.

The welding of metallic materials can occur through various processes, utilizing diverse heat sources and under varying conditions. Conventionally, this is achieved by applying heat to metallic pieces, causing them to melt and subsequently re-solidify, resulting in their rejoining. However, this solidification leads to differing mechanical and chemical properties in the re-solidified piece, which can sometimes yield lower performance levels compared to the original pieces prior to welding. The utilization of high heat inputs during welding, particularly over larger areas, can impact the mechanical properties of both the welded and heat-affected zones. As a response to this challenge, recent decades have witnessed focused efforts toward discovering novel welding techniques that minimize heating and, consequently, reduce the extent of heat-affected areas. This pursuit has given rise to innovative welding methods, wherein pressure has been employed as an alternative to heating, as seen in diffusion and friction welding. Notably, advanced welding techniques like laser and electron beam welding have emerged, where the welded area is significantly smaller compared to traditional welding methods.

This book also encompasses advancements in fusion and solid-state welding processes. It thoroughly discusses the fundamentals of welding processes, welding metallurgy, defect formation, and the influence of process parameters on mechanical properties. The microstructural and mechanical behaviors of weldments are scrutinized from a research perspective, drawing upon the authors' published works. Moreover, this book addresses the challenges encountered during dissimilar metal welding through both fusion and solid-state welding processes, providing remedies and hybrid approaches to tackle these challenges. In conclusion, a portion of the book is dedicated to exploring numerical and statistical simulation methodologies employed in welding processes for parameter optimization and material flow analysis. This section also touches upon the utilization of machine learning and digital twin-based approaches in various welding processes.

Manjaiah M, Shivraman Thapliyal and Adepu Kumar

About the Editors

Manjaiah M. is an accomplished Assistant Professor in the Department of Mechanical Engineering at the National Institute of Technology (NIT), Warangal. With a strong background in Mechanical Engineering, he holds a B.E. degree and an M.Tech. in Manufacturing from the University B.D.T. College of Engineering, Karnataka, as well as a Ph.D. in Mechanical Engineering from the National Institute of Technology, Karnataka. Dr. Manjaiah has a wealth of experience and expertise in welding and additive manufacturing, having completed 3.5 years of postdoctoral experience at the University of Johannesburg, South Africa, and the Ecole Centrale de Nantes-CNRS lab, France.

Before joining the NIT, Warangal, Dr. Manjaiah served as an Assistant Professor at the Manipal Institute of Technology, Manipal Academy of Higher Education, where he supervised several M.Tech. and B.Tech. theses. He has made significant contributions to the field of Additive Manufacturing and has published 38 highly reputed journal articles and ten books. Dr. Manjaiah is a sought-after speaker and has presented 15 papers at international conferences across the globe. He has also conducted over ten training programs in Additive Manufacturing, sharing his expertise with professionals and students alike.

His contribution to the field of Additive Manufacturing is further evident in his reference book *Additive Manufacturing: A Tool for Industrial Revolution 4.0*. His expertise and accomplishments make him a valuable member of the academic community at NIT Warangal and a leading authority in the field of Additive Manufacturing.

Shivraman Thapliyal is an Assistant Professor in the Department of Mechanical Engineering, National Institute of Technology, Warangal, India. He received his undergraduate degree from the College of Engineering, Roorkee. He completed his Master's and Doctorate degrees from the Indian Institute of Technology, Roorkee, in 2012 and 2017, respectively. He works in areas of metal-based additive manufacturing, welding metallurgy, surface engineering, and AI/ML in welding and additive manufacturing. Most of his work has focused on the establishment of a structure–property correlation of welded and processed surfaces, keeping the view of industrial applications.

Adepu Kumar is a full-time Professor in the Department of Mechanical Engineering, National Institute of Technology, Warangal, India. He earned his PhD degree from Osmania University, Hyderabad, India, in 2001. His impressive academic journey includes a collection of over 200 publications, including about 90 journal articles, 80 conference proceedings, and an Indian patent. His publications have been cited more than 4000 times (Google Scholar 2022, h-index = 38). He has also successfully completed two international projects and two national projects. His research interests include dynamic recrystallization, microstructure evolution, and the intricate mechanics of plastic deformation during solid-state joining and processing of both similar and dissimilar metals. Additionally, his exploration extends to the domain of metal-based additive manufacturing of aluminum alloys.

List of Contributors

B. Acherjee
Department of Production and
  Industrial Engineering
Birla Institute of Technology
Ranchi, India

Arulmurugan, B.
Mechanical Engineering
KPR Institute of Engineering and
  Technology
Coimbatore, India

Arunkumar, P.
Mechanical Engineering
KPR Institute of Engineering and
  Technology
Coimbatore, India

Arunkumar Bongale
Robotics and Automation Department
Symbiosis Institute of Technology,
  Symbiosis International University
Pune, India

Dipankar Bose
Department of Mechanical Engineering
National Institute of Technical
  Teachers' Training and Research
Kolkata, India

Raghavendra Darji
Faculty of Technology and Engineering,
  Department of Metallurgy and
  Materials Engineering
The M S University
Baroda, India

Dharanikumar, S.
Mechanical Engineering
KPR Institute of Engineering and
  Technology
Coimbatore, India

K. V. Hari Shankar
Department of Mechanical Engineering
National Institute of Technology
  Warangal
Warangal, India

Kazuyuki Hokamoto
Institute for Industrial Nano Materials
Kumamoto University
Kumamoto, Japan

Daisuke Inao
Department of Advanced Mechanical
  Systems
Kumamoto University
Kumamoto, Japan

Ajfarul Islam
Department of Mechanical Engineering
National Institute of Technical
  Teachers' Training and Research
Kolkata, India

Ajith G. Joshi
Department of Mechanical Engineering
Madanapalle Institute of Technology
  and Science
Andhra Pradesh, India

G. R. Joshi
Department of Mechanical Engineering
Marwadi University
Rajkot, India
and
India Fusion Blanket Division
The Institute for Plasma Research
Gandhinagar, India

Pankaj Kaushik
Department of Mechanical Engineering
National Institute of Technology
Warangal, India

Dhiraj Kumar
Department of Production Engineering
Jadavpur University
Kolkata, India

Ranjan Kumar
Department of Mechanical Engineering
National Institute of Technology
    Warangal
Warangal, India

Satish Kumar
Robotics and Automation Department
Symbiosis Institute of Technology,
    Symbiosis International University
Pune, India

V. Kumar
Department of Mechanical Engineering
Indian Institute of Technology
    Bhubaneswar
Varanasi, India

Kambeyanda Rahul Machaiah
Institute for Industrial Nano Materials
Kumamoto University
Kumamoto, Japan

Manjaiah, M.
Department of Mechanical Engineering
National Institute of Technology
    Warangal
Warangal, India

P. M. Mashinini
Department of Mechanical and
    Industrial Engineering Technology
University of Johannesburg
Johannesburg, South Africa

J. P. Misra
Department of Mechanical Engineering
Indian Institute of Technology
    Bhubaneswar
Varanasi, India

Kallol Mondal
Department of Materials Science and
    Engineering
IIT Kanpur
Kanpur, India

Manidipto Mukherjee
Additive Manufacturing Group
CSIR-Central Mechanical Engineering
    Research Institute
Durgapur, India
and
Department of Engineering Sciences
Academy of Scientific and Innovative
    Research
Ghaziabad, India

Amrit Raj Paul
Department of Engineering Sciences
Academy of Scientific and Innovative
    Research
Ghaziabad, India

P. S. Samuel Ratna Kumar
Department of Mechanical and
    Industrial Engineering Technology
University of Johannesburg
Johannesburg, South Africa
and
Department of Mechanical Engineering
Kumaraguru College of Technology
Coimbatore, India

Satyanarayan
Department of Mechanical Engineering
Alva's Institute of Engineering and
    Technology
Moodbidri, India

Pragya Saxena
Mechanical Engineering
Symbiosis Institute of Technology,
    Symbiosis International University
Pune, India

Sameer Sayyad
Robotics and Automation Department
Symbiosis Institute of Technology,
    Symbiosis International University
Pune, India

Gaurav Sharma
Department of Mechanical Engineering
KIET Group of Institutions
Delhi-NCR, Ghaziabad, India

Sunil Sinhmar
Department of Materials Science and
    Engineering
IIT Kanpur
Kanpur, India

Suresh, R.
Department of Mechanical and
    Manufacturing Engineering
M S Ramaiah University of Applied
    Sciences
Bangalore, India

Shigeru Tanaka
Institute for Industrial Nano Materials
Kumamoto University
Kumamoto, Japan

Gudipadu Venkatesh
Department of Mechanical Engineering
National Institute of Technology
    Warangal
Warangal, India

Kadapa Vijaya Bhaskar Reddy
Department of Mechanical Engineering
National Institute of Technology
    Warangal
Warangal, India

Vivek Warke
Robotics and Automation Department
Symbiosis Institute of Technology,
    Symbiosis International University
Pune, India

1 Underwater Explosive Welding of Tin and Nickel Plates and Characterization of Their Interfaces

Kambeyanda Rahul Machaiah, Shigeru Tanaka, Daisuke Inao, Kazuyuki Hokamoto, and Satyanarayan

1.1 INTRODUCTION

Welding processes have become increasingly important in most of the manufacturing industries, such as automotive, aircraft, pipe lines, chemical and petrochemical industries, tanks and vessels, and railways (Kah and Martikainen 2012). Although many conventional techniques are available for welding, such as cold welding, forge welding, friction welding, roll welding, hot press welding, and ultrasonic welding, current developments in advanced/novel technologies are required for welding corrosion- and wear-resistant materials for industrial applications and electronic materials for electronic packaging applications (Mudali et al. 2003, Satyanarayan et al. 2017). The explosive welding (EXW) technique has been regarded as one of the best techniques to weld these materials because it is unique from other conventional joining methods as it does not depend on the melting of two metals to be fused or on plastic deformation of the surfaces in contact, as occurs with cold or hot pressure welding (Blazynski 2012, Vaidyanathan and Ramanathan 1992).

EXW or bonding is a solid-state process used for the welding of similar or dissimilar metals in which the joining/cladding is accomplished by accelerating one metal plate onto another at extremely high velocity through the use of chemical explosives to create a high-quality metallurgical bond (John and Edward 1993). The purpose of being called a solid-state welding process is because of the rapid

DOI: 10.1201/9781003327769-1

welding process, which occurs in a matter of a few microseconds, and during this short period of time, the chances of diffusion are very low. The EXW process has the capability to weld metals with extremely different properties, such as melting point, thermal coefficient, and other mechanical properties. Even EXW can be used to weld large surface areas because the high energy density of the explosives has a distributive characteristic over the weld surface. The welding process occurs due to the plastic deformation of the samples. The jetting phenomenon in the EXW process plays a significant role, which is responsible for the removal of the oxide layer present in the samples and ensures a smooth welding process. In fact, the formation of an oxide layer between the metals restricts the joining process from forming a strong metallurgical bond. The quality of the weld depends on the selection of several parameters, such as surface preparation of the sample, stand-off distance between the flyer plate and base plate, water distance, and detonation velocity.

The EXW process is usually performed in two basic configurations, namely the horizontal setup and the inclined setup. The representations of both configurations are shown in Figure 1.1a and b.

The clearance gap or parallel configuration is mainly used when the thickness of the metal plates/samples is large, whereas the angular configuration is used when the thickness of the metal plates/samples is very small; in other words, the inclined configuration is used in coating a base metal with a thin film of flyer plate. The buffer (cover) plate is placed on top of the flyer plate to protect it from high-velocity impact.

In a parallel configuration setup, the detonation velocity (the velocity at which the shock wave front travels through a detonated explosive) of the explosive must be less than the velocity of sound in the material to be joined. However, it is difficult to fulfill this condition with most of the explosives. The explosive with detonation velocities greater than 120% of the sonic velocity (the sound velocity of the internal fluid) of the metal should not be used, and it is desirable to use the explosive

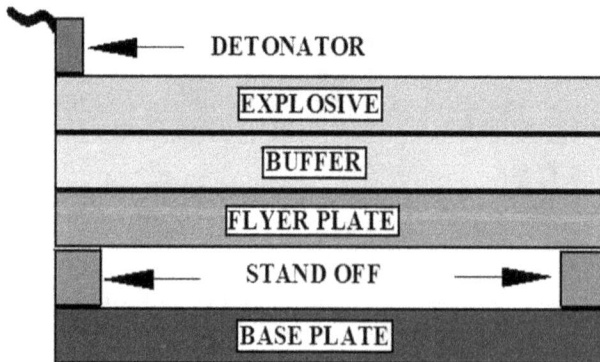

FIGURE 1.1A   Schematic illustration of the EXW process of a constant (parallel) interface clearance gap.

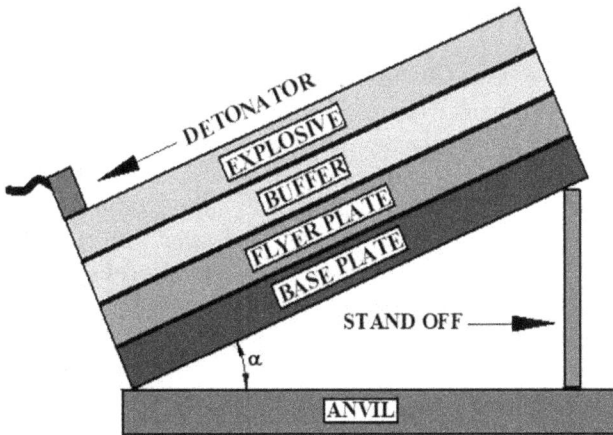

FIGURE 1.1B   Schematic illustration of the EXW process of an angular (inclined) interface clearance gap.

with detonation velocities ranging from 2000 to 6000 m/s (Colin 2006). Chances of localized melting at the interface in a parallel configuration are high. In order to overcome the above problems, an angular configuration is preferred. In the case of the inclined setup, the collision angle (α) plays a very important role because the angle determines the formation of the jet. The formation of straight and wavy interfaces completely depends on the collision angle. It is advisable to set the collision angle range to above 15° (Satyanarayan et al. 2019). The quality of the bond in EXW depends on careful control of process parameters such as collision point, material surface preparation, plate separation, explosive load, detonation velocity (V_d), and detonation energy (Colin 2006).

Literature suggests that Steel/Steel (Acarer et al. 2004), Al/Mg (Sherpa et al. 2015), Al/Cu (Paul et al. 2015), Cu/Ta (Skorokhod 2013), Cu/Stainless Steel (Durgutlu et al. 2005), and Titanium/Duplex Stainless Steel (Wang et al. 2022) combinations of materials were successfully welded using EXW. EXW is generally performed in an open atmosphere. However, it is stated that conventional EXW is found to be undesirable for the welding of thin metal plates (below 1 mm thickness) and brittle metals/materials, e.g., tungsten, amorphous ribbon, and ceramics (Iyama et al. 2001). This is because during a massive explosion, the high force and temperature created in the open air destroy the thin foils, as reported by Sun et al. (2014a, 2014b). Hence, a precise and controlled atmosphere is essential to maintain on metal plates to achieve successive bonding.

Kazuyuki Hokamoto et al. (2004) proposed a novel EXW method that uses underwater shockwaves to weld thin sheets, resulting in a significant decrease in kinetic energy loss at the interface of flyer plate and base plate. The reason for choosing the water is that it is more incompressible than air, due to which it cannot generate a high temperature during an explosion. The method states the importance of building an inclined setup. In their study, thin copper plates were explosively welded onto mild steel, and multilayered copper plates were also successfully welded. It is also reported that (Manikandan et al. 2011) underwater EXW is regarded as one of the best welding methods in which water acts as a transmitting medium and creates a uniform pressure at the point where it is essential.

Few combinations of similar and dissimilar materials are effectively fused using underwater explosive techniques such as Tungsten foil/Cu (Manikandan et al. 2011), Al/Mg (Manikandan et al. 2012), Sn/Cu (Satyanarayan et al. 2017, 2019), and Sn/Al (Satyanarayan et al. 2020). It is stated that Al/Steel, Al/Cu, and Cu/Stainless Steel combinations of materials are the most essential in electrical and electronics engineering applications (Paul et al. 2015). However, the problem with Al/Cu joints is that their mechanical properties can be reduced by the formation of brittle intermetallic compounds (IMCs) during welding, and these joints easily get oxidized, resulting in their resistance to the flow of electrical currents. A few problems with the Sn/Cu joint are the formation of needle-shaped Cu_6Sn_5 intermetallic at the interface (Satyanarayan and Prabhu 2012). The formation of cracks, voids, and delamination or spalling at the interfaces of Sn/Al joints are the few disadvantages of Sn/Al joints (Satyanarayan et al. 2020). It is reported that the nature of bond formation between metallic joints in electronic applications depends on the alloy composition and the type of metallic substrate used for soldering applications (Satyanarayan and Prabhu 2012). The selection of the substrate plays a vital role in electronic applications. Ni, Au, and Pd are the common substrate materials used in electronic applications other than Cu and Al (Satyanarayan and Prabhu 2011). Although numerous investigations on EXW of various metal combinations were conducted by the researchers Hokamoto and Hamashima (2021), Zlobin (2002), and Kahraman et al. (2007), the cladding of Sn and Ni plates using the underwater explosive technique has not been received much attention.

Hence, in the present chapter, discussion on an attempt to cladd Sn and Ni plates using the underwater EXW method is highlighted. The reason for choosing Sn is that it is used as a solder alloy (major constituent) in lead bases and lead-free solder alloys in electronic industries (Satyanarayan and Prabhu 2013). Pure tin is corrosion-resistant compared to other tin alloys. Moreover, solders are electrically connected with metallic substrates in electronic devices (Vignesh et al. 2012). The purpose of using Ni is that it has various advantages, such as a slow oxidation rate, corrosion resistance, high hardness, good ductility, being a fairly good conductor of heat and electricity. Ni is frequently used as a diffusion barrier material Under Bump Metallization in flip-chip and Ball–Grid–Array technology to prevent the rapid interfacial reaction between solders and Cu substrates (Chen et al. 2007). Moreover, Ni finds its applications in rechargeable batteries, electric guitar strings, and many other electronic devices. However, nickel allergy is one of the major problems that are being faced by many people; therefore, to reduce its effect by coating nickel with a layer of tin underwater EXW is preferred in the current chapter because the current method has less chances of melting metals. Therefore, chapter discussion is made on research carried out on the fusing of Sn/Ni, and further analysis of the effect of interfacial morphology on bending strength is highlighted for the welded Sn/Ni plates at different water distances.

1.2 PRINCIPLES OF UNDERWATER EXPLOSIVE WELDING

The underwater EXW method is a process in which the cladding occurs due to the high-pressure shockwaves that are produced after the detonation of the explosives. The current solid-state welding is significantly used for the welding process of thin metal sheets on the base plate, or it can also be implemented in the welding of multi-layer metal sheets (Hokamoto et al. 2004). The inclined configuration of the experimental setup of underwater EXW (schematic), as shown in Figure 1.2, is used to generate a uniform pressure along the surface of the samples. Moreover, the method is suitable to accelerate a thin plate with high enough velocity/speed over a short distance to satisfy the condition of EXW, and it is possible to control the pressurizing condition on plates by modifying the assembly.

The reflector and the spacers that can be used are polymethyl methacrylate (PMMA) or other plastic materials of required dimensions that are consumed during the welding process, which does not affect the welding quality of the samples. The properties of samples used in underwater EXW are industrial pure tin and nickel. Table 1.1 indicates that even though the properties of both Sn and Ni are different, they could be successfully welded by the method of EXW using underwater shockwaves. Especially the melting point of both metals has to be emphasized because of the large difference.

1.3 DESIGN CONSIDERATIONS FOR EXPERIMENTATION OF UNDERWATER EXPLOSIVE WELDING

It is very important to maintain a uniform pressure on the flyer plate during the welding process. Therefore, a simple design criterion is taken into consideration, in which the ratio between the explosive thickness and the water distance at the initial point of

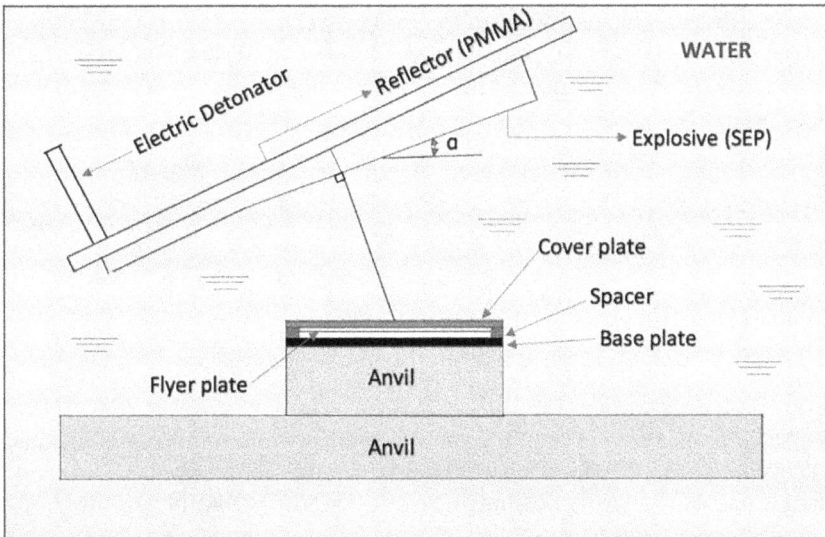

FIGURE 1.2 Inclined setup of underwater explosive welding.

TABLE 1.1

General Properties of Pure Nickel and Tin

Properties	Nickel	Tin
Density	8.89 g/cm³	7.3 g/cm³
Melting point	1435°C–1446°C	232°C
Specific heat	456 J/kg °C	228 J/kg °C
Young's modulus	190–220 GPa	41.6–50 GPa
Electrical conductivity	1.43×10^7 S/m	9.17×10^6 S/m

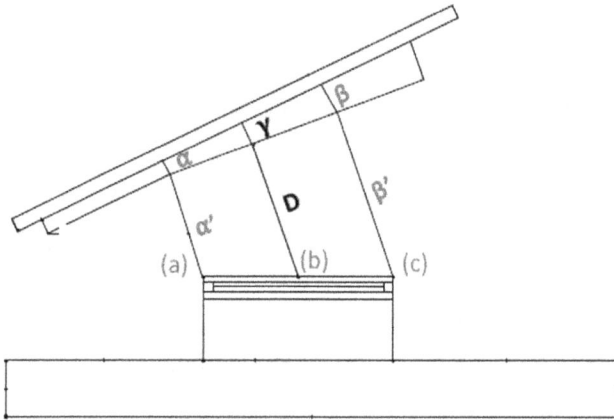

FIGURE 1.3   Design considerations for the experimentation.

the sample is approximately equal to the ratio between the explosive thickness and the water distance at the final point. Figure 1.3 shows an illustration of the criterion mentioned above. It can be mathematically represented as

$$\frac{\alpha}{\alpha'} \approx \frac{\gamma}{D} \approx \frac{\beta}{\beta'}$$

At points (a), (b), and (c), respectively.

1.4   METHODOLOGY ADOPTED TO WELD AND CHARACTERIZE METAL PLATES

The work samples used in the current chapter were pure tin plate (Sn, 50 mm× 50 mm×0.3 mm), nickel plate (Ni, 70 mm×70 mm×1 mm, l×b×t), and a stainless steel cover plate (SUS 304, 70 mm×70 mm×0.1 mm). The cover plate was sprayed with a coating paint to avoid welding the cover plate and the flyer plate. PMMA spacers of thickness 0.5 mm were used. The stand–off distance between the flyer plate and base plate was maintained at 0.2 mm. The complete setup was arranged on top of a mild steel anvil, and insulation tapes were used in order to waterproof the samples.

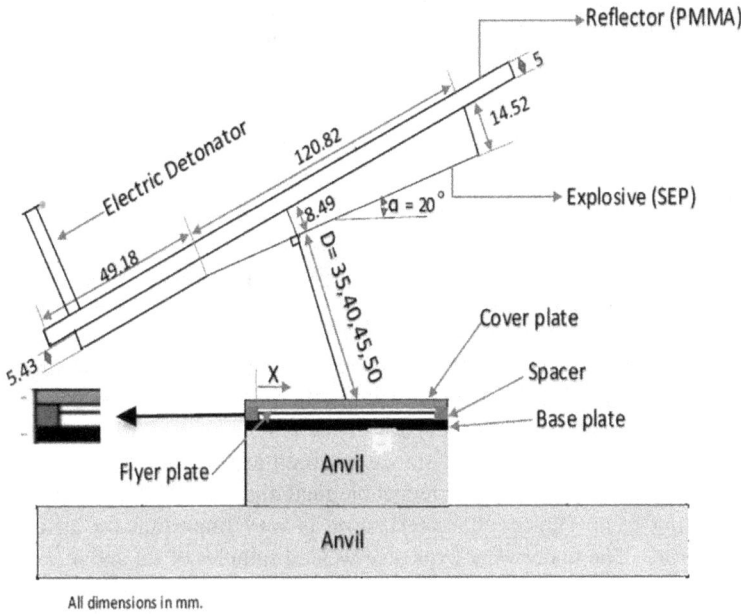

FIGURE 1.4 Experimental setup.

The explosive used to conduct the experiment is SEP, with a detonation velocity of 7 km/s and a density of around $1300 \, kg/m^3$. The SEP explosive was procured from Kayaku Japan Co., Ltd., Japan. The explosive was placed in an inclined position to obtain uniform pressure. It was supported by 5 mm PMMA sheets. PMMA sheets were precisely cut with the help of the laser cutter by using the designs prepared. The distances between the explosive and the center of the sample (D), i.e., the water distances maintained, were $D = 35$, 40, 45, and 50 mm, respectively. The entire setup was placed inside a cardboard box with a vinyl cover to fill with water. Figure 1.4 shows the schematic diagram of the underwater shockwave explosion welding technique with weldable conditions.

Underwater explosion-bonded Sn/Ni plates were sectioned along the direction of wave propagation using a shear cutting machine (Aizawa, AST-612). The samples were prepared for polishing by inserting them into resin molds, and later the resin around the samples was hardened. The samples were polished with silicon carbide papers of grit sizes (400–1500). The final polishing was carried out using a diamond polisher (Velnus, Asahikasemake) with a size of 1 μm. The final polishing was carried out on a Struers Labpol-1 disc polisher using silica liquid lubricant. The Sn/Ni-bonded interfacial regions were micro-examined using an optical microscope (Nikon LM 2) and an analytical scanning electron microscope (SEM, JEOL JSM 6510A). Samples for optical microcopy and SEM were cut at the center parallel to the detonation direction. Vickers hardness tests (using Akashi MVK-E 40716) were performed on the samples along the welded interface of the sample Sn/Ni. The load used for the measurement of hardness at the nickel section was 50 gf for 10 seconds, whereas at the tin section, it was 2 gf for 10 seconds. The results were individually cross-checked with the formula $HV = 1.854$ (F/d^2), where F is the force and d is the mean diagonal length of the trace. The three-point bending tests were performed by cutting the samples into strips that were bent to an angle of 90° to examine the bonding strength. Table 1.2 represents the thickness of the Ni, Sn, and cover plate, which were constant with a stand distance of 0.2 mm and a varying water distance value (D).

1.5 DISCUSSION ON WELDED PLATES AND THEIR CHARACTERIZATION

EXW occurred upon a high-velocity oblique collision of metal plates under different water distances. Therefore, the question of whether or not bonding is possible or, in case it is possible, whether the properties of the resulting weld are strong by varying water distance. Generally the distance between explosive and the center of the sample has significant effect on the morphological and metallurgical characteristics of the resulting weld. Hence, characterization is very important for all explosive-welded materials. The underwater explosive-welded samples of tin and nickel (Sn/Ni) plates were performed. The surface of the tin plate welded over the nickel plate is shown in Figure 1.5.

The electric detonation of the explosives converts the high energy of the explosives into high-pressure shockwaves, which results in the welding of the Sn and Ni plates with each other. The explosively welded samples and their interfaces are generally characterized using several methods, including optical microstructures for better bondability, SEM images in order to identify the welding defects, the formation of intermediate phases, and the distribution of elements at the interfaces. Even mechanical properties such as Vickers Hardness Number (VHN) and bond strength are determined for the explosively welded samples. The VHN indicates the distribution of hardness at the welded interface, and the bending test provides information about the maximum bending/bond strength of the welded samples. In the current chapter, the above-cited methods and tests are used for the Sn/Ni-welded plates.

TABLE 1.2
Thickness of Samples and Parameters

Flayer Plate (Thickness, mm)	Base Plate (Thickness, mm)	Cover Plate (Thickness, mm)	S.O.D. (mm)	Water Distance (mm)
Sn (0.3 mm)	Ni (1 mm)	SUS 304 (0.2 mm)	0.2	35
Sn (0.3 mm)	Ni (1 mm)	SUS 304 (0.2 mm)	0.2	40
Sn (0.3 mm)	Ni (1 mm)	SUS 304 (0.2 mm)	0.2	45
Sn (0.3 mm)	Ni (1 mm)	SUS 304 (0.2 mm)	0.2	50

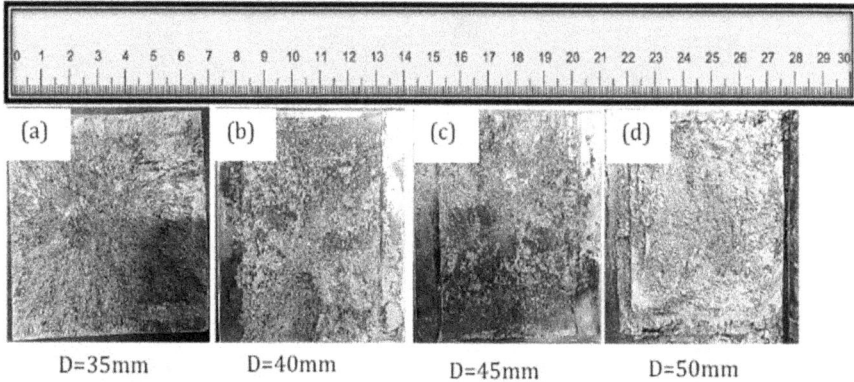

FIGURE 1.5 Surface of the recovered samples welded at (a) $D=35$ mm, (b) $D=40$ mm, (c) $D=45$ mm, and (d) $D=50$ mm.

The optical microstructures of the interfacial regions of the welded Sn/Ni plates are shown in Figure 1.6, which clearly indicates that the combination of Sn and Ni plates can be successfully welded by using the underwater shockwave method. A strong metallurgical bond between Sn and Ni was observed for all the samples due to the higher rate of collision between the plates. The interface of the samples welded at a water distance of 35 mm exhibited a thick continuous interface (Figure 1.6a), clearly indicating the formation of a molten layer at the interface. The reason for the formation of the molten layer is the high temperature, which was caused by the increased pressure of the underwater shockwave and also due to shorter water distance. Even samples welded at $D=40$ mm also showed similar results by the formation of the molten layer but thinner than the sample welded at $D=35$ mm. The reason is that as the water distance increased, the chances of the formation of a molten zone also decreased. However, a molten intermediate layer was discovered when carefully examined under the microscope.

A transition from smooth interface to wavy interface was observed at $D=40$ mm. For the samples welded at $D=45$ mm and $D=50$ mm, the interfaces between Sn/Ni plates demonstrated a pronounced wavy interface, as shown in Figure 1.6c and 1.6d. Zamani and Liaghat (2012) implemented the explosive technique to prepare coaxial pipes using stainless steel and carbon steel. They stated that the transition from a wavy interface to a smooth interface occurred due to the decrease in explosive load. Manikandan et al. (2011) suspected the reason of the formation of a wavy interface to be collision angle, flyer velocity, stand-off distance, loading ratio, and thickness of the base plate. SEM images of interfacial regions of explosively welded Sn/Ni joints at varied water distances are shown in Figure 1.7. Various types of wavy interfaces were observed, such as regular wavy interfaces, low-amplitude wavy interfaces, and irregular wavy interfaces, for the sample welded at $D=40$ mm (Figure 1.7b), and almost similar interfaces were observed in samples $D=45$ mm and $D=50$ mm, as shown in Figure 1.7c and d.

Across the interface of Sn/Ni plates welded at $D=35$ mm other than smooth interface, vortices, intermetallic, molten zones, shear cracks, and voids were formed

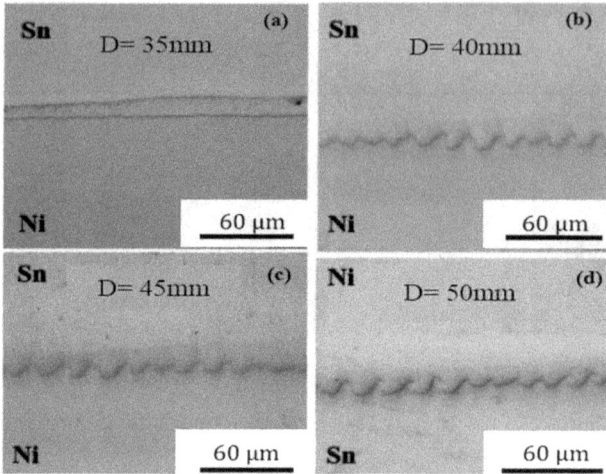

FIGURE 1.6 Welding interface for sample welded at (a) $D=35$ mm, (b) $D=40$ mm, (c) $D=45$ mm, and (d) $D=50$ mm.

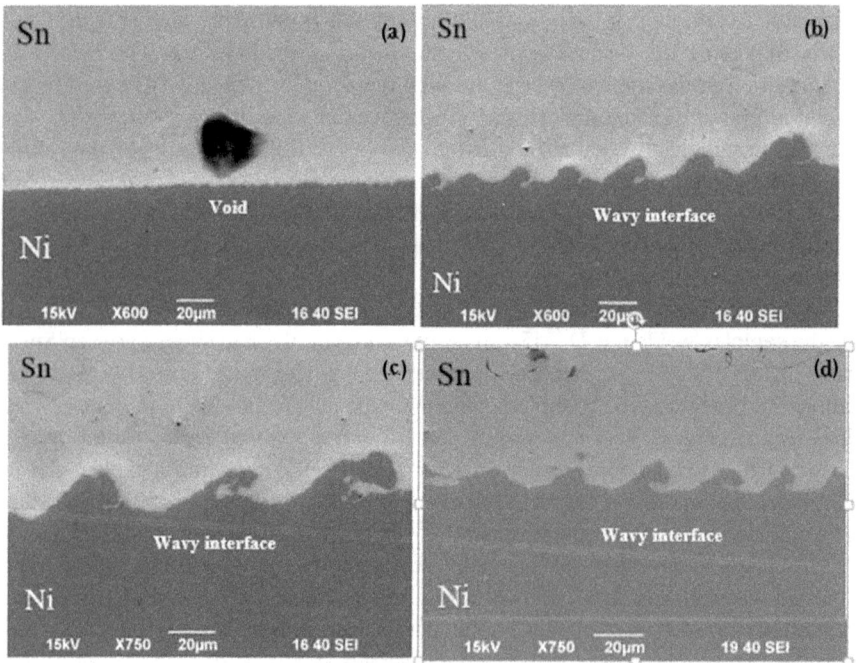

FIGURE 1.7 SEM images: (a) voids at $D=35$ mm, (b) irregular wavy interface at $D=40$ mm, (c) wavy interface at $D=45$ mm, and (d) wavy interface at $D=50$ mm.

(Figure 1.7a) because the energy shockwave was greater and the loss of kinetic energy of the flyer plate was higher compared to $D=40$, 45, and 50 mm. Moreover, the joining metals got fluidized at the interface. These defects (voids, molten zones) are caused when the gases during the welding process could not escape before the solidification of the molten metal, as shown in Figure 1.7a. Another possible reason for the formation of the voids is the low melting point of tin. A wavy interface was observed for the samples $D=40$, 45, and 50 mm. However, the waves formed were non-uniform at a few locations (Figure 1.7b–d). An almost similar trend was observed by the researchers Satyanarayan et al. (2017). However, the studies carried out by Kahraman et al. (2007) reported that there were no joining defects (such as oxide remains, melting cavities, etc.) observed on welded samples. This could be due to the different types and thicknesses of materials and explosives used in the current study.

The mechanical properties of welded Sn/Ni joints were evaluated using the Vickers indentation hardness test. Vickers hardness was measured for four samples, which were welded at water distances of 35, 40, 45, and 50 mm. It was observed that the hardness near the interface exhibited almost similar values in the tin section, whereas it was observed that the hardness slightly decreased near the interface in the nickel section. Higher hardness was measured in the intermediate layer than in the area of Sn. The reason behind the increase in hardness at the intermediate layer is due to the solidification of the molten layer at a very high rate of cooling. As a result, the layer was brittle, and naturally, the hardness increased. The measured thickness of the intermediate layer was found to be 17–19 μm since it was slightly non–uniform. The VHN of the intermediate layer was around 18 VHN, as shown in Figure 1.8. However, Satyanarayan et al. (2017, 2019) reported that the hardness increased near the interface.

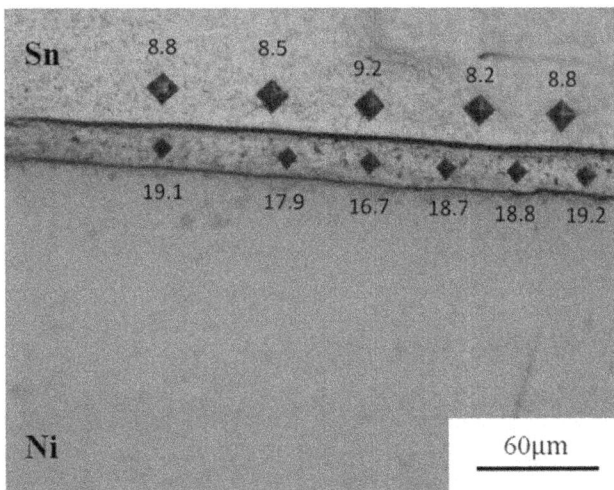

FIGURE 1.8 Trace of the Vickers hardness measurements in Sn and the intermediate layer.

SEM images with elemental mapping and distribution across the welded interface of the samples with a lower water distance were recorded and shown in Figure 1.9. It was observed that the intermediate layer was present in both samples and consisted of a small amount of dispersed intermetallic compound. The concentration of Ni across the interface and in the matrix of Sn for the samples welded at $D=35$ mm (Figure 1.9) was found to be higher, which indicates that the distribution of Sn and Ni atoms is not uniform. This clearly indicated that precipitation of Ni–Sn IMC particles occurred across the interface. The dispersed intermetallic compound was confirmed to be Ni_3Sn_4 by referring to the phase diagram of Ni–Sn (Okamoto 2006).

The elemental mapping along with the line analysis of the sample of $D=35$ mm exhibited the presence of a minute amount of nickel and a large amount of tin, as shown in Figure 1.9. A similar trend was observed for sample $D=40$ mm (Figure 1.10). These intermediate layers are formed mainly due to the maximum effect of heat at water distances $D=35$ mm and $D=40$ mm.

The excessive formation of the IMCs is undesirable as the bonding strength considerably reduces.

The mechanical properties of welded Sn/Ni joints have been assessed using the bending test for all the samples. Since there is no general testing method for measuring the bonding strength of such samples, three-point bending test was performed, which is identical to the method employed by Inao et al. (2020). The samples were bent up to an angle of 90°. The length between the supports (L) was considered to be 28.6 mm, the thickness of the sample (t) was 1.3 mm, the force applied to the sample (F) was measured to be 205.1 N, and the width of the sample (w) was considered to be 10 mm for all the samples.

Figure 1.11 shows the schematic diagram of the method adopted for the bending test. The interface of welded Sn/Ni joints at a water distance of 35 mm was fractured when bent to an angle of 90° (Figure 1.12a). Indicating that it has exhibited a lower bonding strength compared to the remaining samples.

At a water distance of 40 mm, a minute gap between the Sn/Ni joint was observed (Figure 1.12b), whereas at a water distance of 45 and 50 mm, the welded interface was observed to be intact (Figure 1.12c and d).

For the interface Sn/Ni sample welded at $D=35$ mm, the crack initiated slightly above the intermediate layer, and the separation increased as it moved toward the center. Finally, the tin layer was completely fractured at the center of the sample. Figure 1.12a shows that this is due to the brittle nature of the molten layer with the intermetallic compound and also to defects that appeared at the interface. Bending tests conducted on welded samples with different diameters indicated that the interfaces of the bonded samples have no defect (Durgutlu et al. 2005). Bending tests of samples show that the welding strength increases with the increase in water distance, up to an extent. It also clearly indicates that the interfaces exhibiting wavy morphology have better bending strength compared to continuous interfaces; this could be due to the better interlock property of the wavy interface.

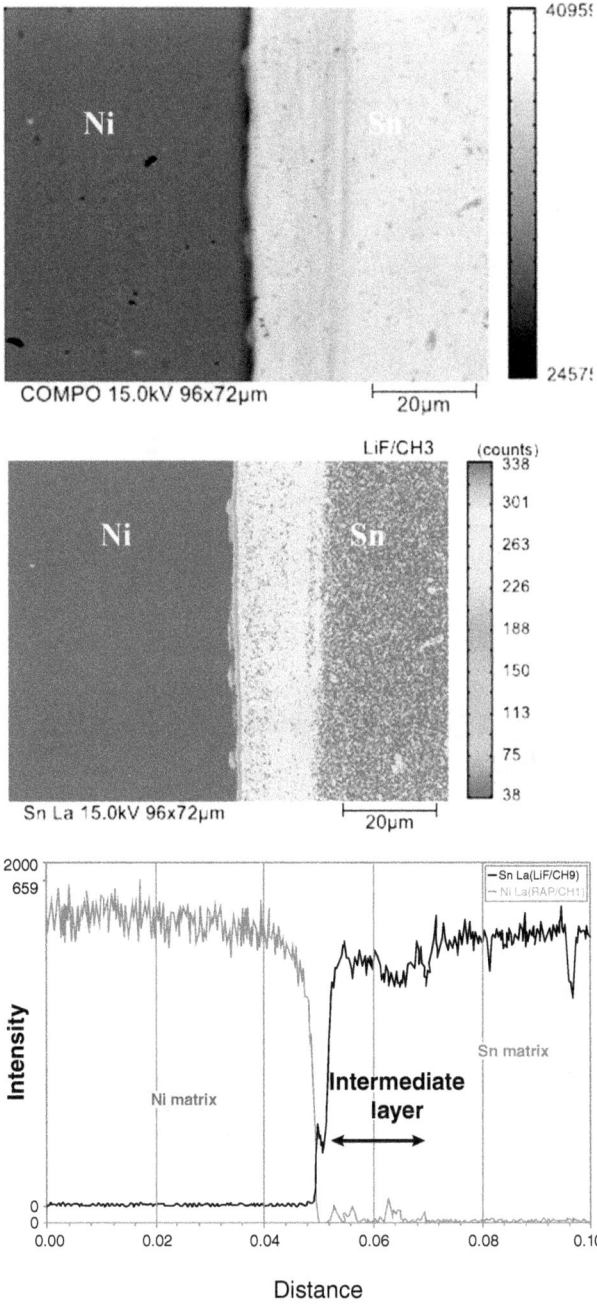

FIGURE 1.9 EPMA line analysis of an interfacial joint welded at $D = 35$ mm.

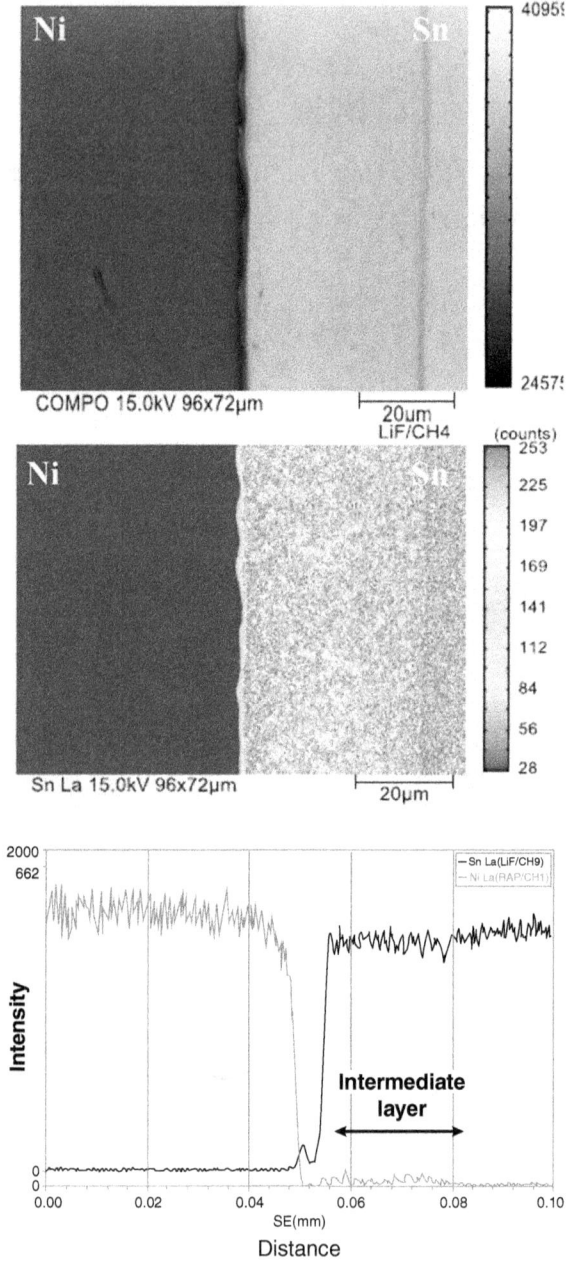

FIGURE 1.10   EPMA line analysis of an interfacial joint welded at $D = 40\,mm$.

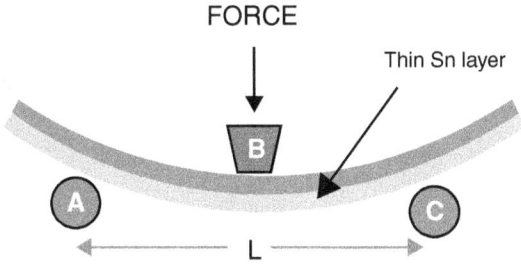

FIGURE 1.11   The schematic diagram of the method adopted for the bending test.

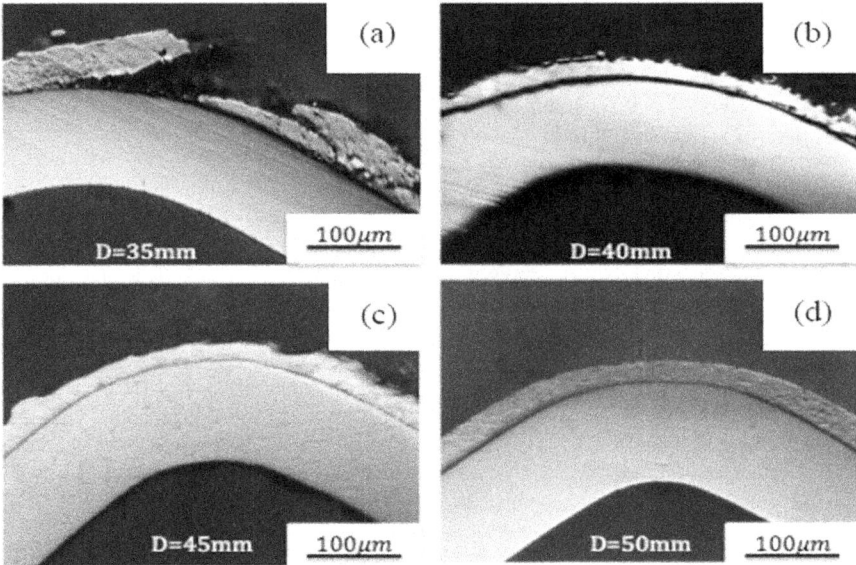

FIGURE 1.12   The results of three-point bending tests for the sample at (a) $D=35$ mm, (b) $D=40$ mm, (c) $D=45$ mm, and (d) $D=50$ mm.

1.6   CONCLUSION

In this chapter, the effect of water distance between the explosive and the center of the sample ($D=35$, 40, $D=45$ and 50 mm) on underwater explosive-welded Sn/Ni interfaces was discussed. Based on the discussion made in this chapter, the following conclusions are drawn:

- Commercially available pure tin and nickel plates can be successfully welded by EXW using underwater shockwaves.
- The interface of samples welded at $D=35$ mm is associated with IMCs, voids, and molten zones.

- The microhardness of the material was observed to decrease near the interface.
- Increase in the shockwave pressure increases the effect of heat at the interface.
- The IMCs formed at the interfaces of Sn/Ni joined at $D = 35\,mm$ reduced the bond strength of the samples.
- Bending tests of samples showed that the welding strength increased with the increase in water distance to some extent.
- The wavy morphology at interfaces exhibited better bending strength compared to continuous interfaces.

REFERENCES

Acarer, M., Gulenc, B., Findik, F., 2004. "The influence of some factors on Steel/Steel bonding quality on their characteristics of explosive welding joints". *Journal of Materials Science*, 39 (21): 6457–6466.

Blazynski, T.Z., 2012. *Explosive Welding, Forming and Compaction*, Applied Science Publisher, UK.

Chen, S.W., Wang, C.H., Lin, S.K., Chiu, C.N., Chen, C.C., 2007. "Phase transformation and microstructural evolution in Solder Joints". *Journal of Metals*, 59 (2007): 39–43.

Colin, M., 2006. "The fundamentals of explosion welding". *Welding Journal*, 85 (7): 27–29.

Durgutlu, A., Gulenc, B., Findik, F., 2005. "Examination of copper/stainless steel joints formed by explosive welding". *Material and Design*, 26 (6): 497–507.

Hokamoto, K., Ujimoto, Y., Fujita, M. 2004. "Basic characteristics of the explosive welding technique using underwater shock wave and its possibilities". *Materials Transactions*, 45 (9): 2897–2901.

Hokamoto, K., Hamashima, H., 2021. Chapter 1: Basic issues in explosion and other high-rate processing of materials", In *Explosion, Shock-Wave and High-Strain-Rate Phenomena of Advanced Materials*, Multiphysics: Advances and Applications, edited by Kazuyuki Hokamoto, pp. 1–16, Academic Press: An imprint of Elsevier, London. ISBN: 978-0-12-821665-1

Inao, D., Mori, A., Tanaka, S., Hokamoto, K., 2020. "Explosive welding of thin aluminum plate onto magnesium alloy plate using a gelatin layer as a pressure-transmitting medium". *Metals*, 10 (1): 106.

Iyama, H., Kira, A., Fujita, M., Kubota, S., Hokamoto, K., Itoh, S., 2001. "Aninvestigation on underwater explosive bonding process". *Journal of Pressure Vessel Technology*, 123 (4): 486–492.

John, G.B., Edward, G.R., 1993. "Explosive welding, welding, brazing and soldering". *ASM Hand Book*, 6, 304–305.

Kah, P., Martikainen, J., 2012. "Current trends in welding processes and materials: improve in effectiveness." *Reviews on Advanced Material Science*, 30: 189–200.

Kahraman, N., Gulenc, B., Findik, F., 2007. "Corrosion and mechanical-microstructural aspects of dissimilar joints of Ti–6Al–4V and Al plates". *International Journal of Impact Engineering*, 34 (8): 1423–1432.

Manikandan, P., Lee, J.O, Mizumachi, K., Mori, A., Raghukandan, K., Hokamoto, K., 2011. "Underwater explosive welding of thin tungsten foils and copper". *Journal of Nuclear Materials*, 418 (1–3): 281–285.

Manikandan, P., Lee, J.O., Mizumachi, K., Ghaderi, S.H., Mori, A., Hokamoto, K., 2012. "Transition joints of aluminum and magnesium alloy made by underwater explosive welding technique". *Material Science Forum*, 706–709, 757–762.

Mudali, U.K., Ananda Rao, B.M., Shanmugam, K., Natarajana, R., Raja, Baldev. 2003. "Corrosion and microstructural aspects of dissimilar joints of titanium and type 304L stainless steel". *Journal of Nuclear Materials*, 321 (1): 40–48.

Okamoto, H., 2006. "Ni–Sn (Nickel-Tin)". *Journal of Phase Equilibria and Diffusion*, 27: 315.

Paul, K., Cyril, V., Jukka, M., Raimo, S., 2015. "Factors influencing Al–Cu weld properties by intermetallic compound formation". *International Journal of Mechanical and Mechatronics Engineering (IJMME)*, 10, 10.

Satyanarayan, Prabhu, K.N., 2011. "Wetting behaviour and interfacial microstructure of Sn-Ag-Zn". *Materials Science and Technology*, 27 (7): 1157.

Satyanarayan, Prabhu, K.N., 2012. "Effect of temperature and substrate surface texture on wettability and morphology of IMCs between Sn–0.7Cu solder alloy and copper substrate". *Journal of Materials Science: Materials in Electronics*, 23: 1–9.

Satyanarayan, Prabhu, K.N., 2013. "Spreading behavior and evolution of IMCs during reactive wetting of SAC solders on smooth and rough copper substrates". *Journal of Electronic Materials*, 42: 2696–2707.

Satyanarayan, Tanaka, S., Mori, A., Hokamoto, K., 2017. "Welding of Sn and Cu plates using controlled underwater shock wave". *Journal of Pressure Vessel Technology*, 245: 300–308.

Satyanarayan, Mori, A., Nishi, M., Hokamoto, K., 2019. "Underwater shock wave weldability window for Sn-Cu plates". *Journal of Materials Processing Technology*, 267: 152–158.

Satyanarayan, Tanaka, S., Mori, A., Hokamoto, K., 2020. *Explosive Welding of Sn and Al Plates Using Controlled Underwater Shock Wave*, Nova Science Publishers, New York, USA.

Sherpa, B.B., Kumar, P.D., Batra, U., Upadhyay, A., Agarwal, A., 2015. "Study of the explosive welding process and applications", In *Advances in Applied Physical and Chemical Sciences- A Sustainable Approach*, edited by Mishra, Govind Chandra and Braj B. Singh, pp. 33–39, Excellent Publishing House, New Delhi. ISBN: 978-93-83083-72-5, 2015

Skorokhod, K.A., 2013. "Structure and microhardness of Cu–Ta joints produced by explosive welding". *Scientific World Journal*, 7: 1–7, Article ID 256758.

Sun, W., Li, X., Hokamoto, K., 2014a. "Numerical simulation of underwater explosive welding process". *Material Science Forum*, 767: 120–125.

Sun, W., Li, X., Yan, H., Hokamoto, K., 2014b. "Effect of initial hardness on interfacialfeatures in underwater explosive welding of tool steel SKS3". *Journal of Materials Engineering and Performance (JMEP)*, 23: 421–428.

Vaidyanathan, P.V., Ramanathan, A.R., 1992. "Design for quality explosive welding". *Journal of Materials Processing Technology*, 32 (1–2): 439–448.

Vignesh U.N., Prabhu, K.N., Stanford, N., Satyanarayan, 2012. "Wetting behavior and evolution of microstructure of Sn–3.5Ag solder alloy on electroplated 304 stainless steel substrates". *Transactions of the Indian Institute of Metals*, 65(6): 713–717.

Wang, K., Kuroda, M., Chen, X., Hokamoto, K., Li, X., Zeng, X., Nie, S., Wang, Y., 2022. "Mechanical properties of explosion-welded titanium/duplex stainless steel under different energetic conditions". *Metals*, 12: 1354.

Zamani, E., Liaghat, G.H., 2012. "Explosive welding of stainless steel-coaxial pipes". *Journal of Materials Science*, 47 (2): 685–695.

Zlobin, B.S., 2002. "Explosion welding of steel with aluminum". *Explosion and Shock Waves*, 38: 374

2 Advances in Gas Tungsten and Gas Metal Arc Welding – A Concise Review

Arulmurugan B., Arunkumar P., and Dharanikumar S.

2.1 INTRODUCTION

In Gas Tungsten Arc Welding (GTAW) or Tungsten Inert Gas (TIG) Welding, electrodes are employed to generate the arc that forms the weld. Tungsten's 3410°C melting point makes it a good electrode material. GTAW is ideally suited for welding thin sections of different alloys. Compared to other welding methods, GTAW allows the operator to exercise greater control over the welding process, resulting in welds with greater strength and integrity. When a welding arc is struck, it generates a high amount of heat that melts the base metal and filler material. However, as the metal cools, atmospheric gases such as oxygen, nitrogen, and water have the potential to react with the weld during the GTAW process, leading to undesirable effects such as oxidation, porosity, and brittleness in the weld. To prevent this, a shielding gas is used to create an inert atmosphere around the Heat-Affected Zone (HAZ) and electrode. Shielding gases have low ionisation potential, which makes them ideal for welding processes that require high voltages. The selection of shielding gas is critical, as it can significantly affect the quality and characteristics of the weld. It is influenced by various parameters like the substrate type, process, and preferred characteristics of the melt pool (Rathod 2021). The process can be done in autogenous mode or with the addition of filler metal. GTAW can be done using both direct current and alternating current. GTAW has the drawback of being more complicated and taking more time than many other welding methods because it requires a more precise and controlled welding technique. In GTAW, the welder must hold the torch steadily and consistently to produce a high-quality weld. This precision takes time and can make GTAW slower than Gas Metal Arc Welding (GMAW). However, GTAW produces welds with a superior aesthetic appearance due to its controlled arc, which results in minimal spatter and a smooth, clean bead appearance. In contrast, GMAW uses a consumable wire electrode that can produce a lot of spatter, which can result in an unsightly weld that requires more post-weld finishing processes, such as grinding

DOI: 10.1201/9781003327769-2

or cleaning, to achieve a smooth appearance. GTAW is often preferred for welding materials that require high-quality and aesthetically pleasing welds, such as stainless steel, aluminium, and other non-ferrous metals. The ability to produce high-quality welds without spatter makes GTAW ideal for applications where post-weld finishing processes must be minimised (Mallick 2021). In this section, different kinds of GTAW are discussed. Figure 2.1 shows variants of GTAW. These will be detailed in the subsequent section.

2.2 ADVANCEMENTS IN GAS TUNGSTEN ARC WELDING

2.2.1 COLD WIRE GAS TUNGSTEN ARC WELDING PROCESS

Welding thin materials manually or automatically with 250 A welding currents is ideal for GTAW. It is frequently utilised for connecting different and similar weld joints due to its high-quality welds. The high heat supply requirement and speed of the process remain significant disadvantages that can be offset by the soundness of the welds (Rathod 2021; Mallick 2021). Due to the aforementioned benefits, the GTAW method is utilised in a variety of industrial applications, including overlay

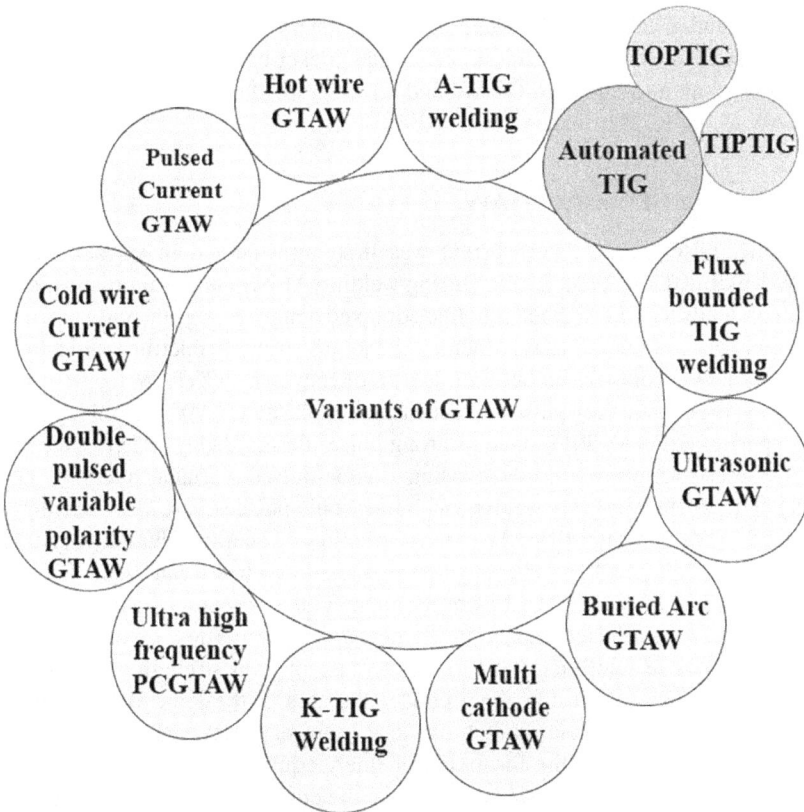

FIGURE 2.1 Variants of GTAW.

applications. The disadvantages of GTAW include slower welding speed, higher cost, concentration variation in shielding gas, intense UV radiation, and reduced deposition rates. Some of its variations are capable of resolving these issues (Arulmurugan and Manikandan 2017).

2.2.2   PULSED-CURRENT GAS TUNGSTEN ARC WELDING (PCGTAW)

Pulsed-current gas tungsten arc welding (PCGTAW) differed from GTAW in terms of heat supply. In GTAW, the supply of heat to the metal remains constant throughout the welding cycle, so it disturbs the surrounding area, resulting in a wider HAZ, an extended solidification time, and reduced weld quality. PCGTAW provides energy in two cycles: it primarily melts and welds the base material (Arulmurugan and Manikandan 2018). The absence of energy during the base current helps to solidify the weld metal. PCGTAW periodically supplies heat energy, allowing the material to cool effectively and decreasing the time required for solidification, thereby enhancing the quality of the weld. PCGTA welding consistently manifests its impact on the development of the fine microstructure. Pulse current (I_p) is regulated from its highest to lowest level at a predetermined frequency, which minimises thermal gradients in the fusion zone. Which also improves fluidity and accelerates cooling rates (M. Manikandan et al. 2014). Numerous researchers have examined the strength characteristics of various stainless steels and nickel-based super alloys utilising PCGTAW as an alternative to GTAW and have obtained superior results compared to GTAW (S. G. K. Manikandan et al. 2014).

2.2.3   VARIABLE-POLARITY GTAW

Ultrasonic pulse square wave current was first implemented in variable-polarity GTAW welding (Cong et al. 2009). During welding Al-2219 aluminium alloy, coarse columnar dendrites transformed into fine equiaxed dendrites, significantly narrowing the fusion zone and thereby strengthening the joint's strength, ductility, and hardness distribution. In a comparison between low-frequency and PCGTAW for welding Al 2219 alloy, high-frequency pulse current variable-polarity GTAW showed higher tensile properties (LI et al. 2017). Due to strong pulse stirring, Alloy Al2219's mechanical characteristics were improved using double-pulsed variable-polarity GTAW (Wang et al. 2018). VP-GTA welding with ultrasonic pulse square wave current was introduced (Wang et al. 2019). The transformation of coarse columnar crystals in the 2219 aluminium alloy weld into fine equiaxed dendrites can result in several improvements in the characteristics of the joint. This is because the microstructure of a weld has a significant impact on its mechanical properties. Firstly, the reduction in fusion zone width can lead to an increase in joint strength as the HAZ is minimised. A narrower fusion zone reduces the area of the joint that is subject to heat input and thermal cycles, which can lead to fewer defects and a more uniform microstructure. Secondly, the formation of finely equiaxed dendrites can result in an increase in joint plasticity. This is because the equiaxed dendritic structure produces a more evenly balanced distribution of grains and reduces the occurrence of preferential paths for crack propagation. The fine dendritic structure also promotes a more isotropic deformation response, which improves the joint's ability to withstand stresses from different directions. Finally, the transformation of the coarse columnar crystals into fine equiaxed dendrites can also improve the distribution of hardness in the weld, as the dendritic structure provides a more uniform distribution of strengthening elements and reduces the occurrence of large, hard grains that can lead to stress concentration and cracking.

2.2.4 Ultra High Frequency of Pulsed Gas Tungsten Arc Welding (UHFP-GTAW)

Welding penetration is greatly enhanced by improvements in the pulse frequency as compared to previous methods; this process is known as ultra-high-frequency pulse GTAW (Z. Yang et al. 2015). Narrower weld beads are the result of arc constriction during UHFP-GTAW and considerably reduce the root radius of the arc. For deeper welds, a greater arc force was necessary, as it depressed the surface by a greater amount. UHFP-GTAW produces excellent weld quality in titanium alloy welding (Morisada et al. 2013). The weld quality and molten metal solidification behaviour are examined in alloy Ti-6Al-4V weld joints made by UHFP-GTAW (M. Yang et al. 2015). Results showed that UHFP-GTAW significantly increased the melt pool pressure and radius of curvature while decreasing the filtration angle and viscosity of the molten liquid. This resulted in a high escape speed and simple detachment of the weld gas pore (Yang, Yang, and Qi 2015). The results showed that the UHFP-molten GTAW's pool was significantly affected by the arc plasma in terms of both heat and force. More so, the arc's heat would have less of a chance to penetrate the substrate because the HAZ temperature gradient would be lower.

2.2.5 Hot-Wire Gas Tungsten Arc Welding (HW-GTAW) Process

In 1964, Manz and Saenger were the ones who came up with the idea for the GTAW hot-wire welding process. The GTAW process can be altered to create the hot-wire (HW) gas tungsten arc welding (HW-GTAW) process to improve the deposition efficiency and thus productivity. In this process, prior to being added to the HAZ, the filler is heated by resistance until it reaches its melting point (Padmanaban, Neelakandan, and Kandasamy 2016). HW processes are typically operated at faster welding speeds than cold wire systems to increase the deposition rate. As a result of its superior deposition rates, the HW process is typically employed in the flat weld condition (Padmanaban, Neelakandan, and Kandasamy 2016). These benefits of HW-GTAW have the potential to be optimally utilised in more developed applications of this method, such as the welding of high-temperature materials (Pai et al. 2020).

Trailing gas shields deliver shielding gas and prevent solidifying weld metal at greater deposition rates and speeds. Many researchers have found that increasing both the welding current and the HW current results in a significant increase in the deposition rate (Ungethüm et al. 2020). Fabrication of the Fast Breeder Nuclear Reactor at Kalpakkam was accomplished with the HW-GTAW process, and modified 9Cr-1Mo was utilised as the candidate material. High weld deposition rates, a superior weld profile, and a decrease in dilution and porosities are just some of the benefits of the HW-GTAW system in this project (Dinesh et al. 2018). The stability of the weld morphology of electrolytic tough pitch copper of 12 mm thick fabricated by the HW-GTAW technique was investigated using CuNi filler wire (Darji et al. 2022). The results revealed enhanced weld bead geometry, refined microstructure, and improved hardness.

2.2.6   TWIN TIG

The conventional TIG welding process is subject to a significant constraint on its ability to increase welding current beyond a specific point because of the restricted current-carrying capacity of the electrode and the greater arc pressure it produces (Shah and Agrawal 2019). For that reason, its efficiency is reduced. Twin TIG (T-TIG) welding, in which two tungsten electrodes are used in a single torch, was developed as a countermeasure (Dinaharan et al. 2022). Electrodes are connected to two synchronised power sources. Hot filler wire reduces arc plasma heat and increases metal deposition. Electrodes are connected to two synchronised power sources. Pulsed power source-assisted welding is an additional innovation that has been incorporated into the HW-GTAW process. This allows for greater control over the weld bead profile and weld morphology, along with the joining of dissimilar materials (Silwal and Santangelo 2016).

2.2.7   KEYHOLE – GTAW WELDING

Australia's CSIRO invented the keyhole GTAW (K-GTAW) process for heavy-section welding of different alloys. Using already established knowledge of the GTAW process, the method is based on a standard arc welding device and a high-current GTAW torch. K-TIG welding's keyhole generation mechanism is different from that of LBW and EBW (Liu et al. 2017). It is derived from conventional TIG welding and involves generating a stronger arc force and more heat input by improving the current and the cooling of the gun. The molten metal rapidly vaporises and repels the liquid metal. High welding current (I) creates Lorentz-induced arc jet pressure. The superposition of arc force and metal vapour force on liquid metal creates small holes in the molten pool, improving welding. The process parameters create a keyhole in the plate's root surface. When the keyhole is opened, the pool surface will automatically anchor itself to the root face of the weldment, thereby producing an extremely stable structure. The keyhole opening at the base also facilitates the smooth flow of arc gases, which helps to reduce arc turbulence and instability. When compared to GTAW, laser welding, and other methods for joining plates with a mid-thickness, it is more efficient, cheaper, and has deeper penetration (Cui et al. 2017). The AISI 316L stainless steel plates of 10 mm thick are effectively joined using autogenous K-TIG welding (Feng et al. 2015). The weld had better corrosion resistance than the substrate and almost the same tensile strength and impact toughness. The commercially pure titanium plate weld joints were fabricated using K-TIG welding without any joint preparation or filler metal (Lathabai, Jarvis, and Barton 2001). The authors found the mechanical characteristics of the joints were the same as those of the base material. Given its much higher productivity, ease of use, and use of tried-and-true technology, it is determined that keyhole GTAW can be efficaciously utilised in the joining of heavy sections. A 12 mm Ti6Al4V titanium in a single pass is joined using K-TIG welding (Gao et al. 2021). The authors found refined weld microstructure and improved mechanical properties in this study. Similar results were observed in K-TIG welding of electrolytic tough pitch copper (Darji et al. 2020).

2.2.8 MULTICATHODE GTAW

Multicathode GTAW has been attempted for a long time to increase both the efficiency of the welding and the properties of the melt pool. Multicathode GTAW increases the speed of the process and prevents undercuts by lengthening the melt pool (Norrish 2006). By increasing cathodes, noticeable benefits have been observed. However, the arc deflection issue in this method can be effectively prevented by using high-frequency pulsing or magnetic arc stabilisation. To weld, a torch with two electrodes is used, with each electrode being powered by its own separate power source and electrically isolated from the other. HW feeding to the weld pool improves weld deposition further (Egerland et al. 2015).

2.2.9 ACTIVE GTAW (A-TIG)

Active flux TIG (A-TIG) is a novel modified version of the GTAW process that has attracted the attention of many researchers (Vidyarthy, Dwivedi, and Vasudevan 2017). An active flux is applied to the juncture's surface, which increases the local surface tension. Because of this, the Marangoni flow is altered, allowing for deeper penetration. The primary physical mechanisms have been proposed to explain the phenomenon (Singh and Khanna 2021). The first reason is that the flux decreases the surface tension of the HAZ, which in turn causes an increased depression of the surface of the HAZ. This increased depression provides a greater curvature radius on the surface of the HAZ, which aids in supporting the arc pressure and keyhole mode. The arc constriction that is caused by the vaporised flux molecules is the second mechanism. Third, when the surface active element in the HAZ is greater than the critical value, the surface tension temperature coefficient flips signs, inducing Marangoni convection to reverse. Different alloys with high thickness have been welded in single passes using A-TIG welding successfully (Singh and Khanna 2021; Vidyarthy, Dwivedi, and Vasudevan 2017; Garg et al. 2019).

2.2.10 ADVANCED A-TIG

AA-TIG, a new GTAW mode, uses two concentric shielding gas streams instead of one. The weld pool and tungsten electrode are protected by the typical inert gas. An oxidative gas layer around this gas stream reacts with the molten pool for ultra-deep penetration (Fujii et al. 2008). The use of He and gas that contained a small amount of oxidising gas was done in an effort to tame the Marangoni convection that the surface tension differential imposed on HAZ. Bead-on-plate TIG welding of SUS304 stainless steel was investigated to determine the impact of O_2 and CO_2 concentrations on the resulting weld bead profile (Zou, Ueji, and Fujii 2014). The authors noticed that when oxygen is introduced to the HAZ, the outward-to-inward Marangoni convection on the surface of the pool can be controlled effectively (Zou, Ueji, and Fujii 2014).

2.2.11   FLUX-BOUNDED TIG WELDING (FBTIG)

While A-TIG welding has been proven to work in academic research, it has not yet been used in industry. This is primarily because of its significant drawbacks, such as the necessity of welding automation and the challenge posed by maintaining arc control during the welding process. In addition to this, flux coating on the surface has a high probability of causing inclusions to occur in the fusion zone, which in turn will cause the joints to function poorly. FBTIG is a subset of the more common A-TIG technique. Flux was coated on both sides of a small gap positioned at the centre of the plate during the FBTIG process. In 2002, Sire and Marya first proposed the method now known as FBTIG welding (Jayakrishnan and Chakravarthy 2017). It was discovered that using this method led to greater penetration for the similar value of energy introduced into the HAZ. Compared to flux-assisted TIG welding, the FBTIG welding process has greater penetrant depth due to its supplementary effects. The tensile strength of FBTIG welds made from the aluminium alloy AA2219-T87 has been investigated (Babu et al. 2014). The fine grain structure of FBTIG welds was found to increase their tensile strength, making them superior to conventional TIG welds.

2.2.12   BURIED ARC GTAW

The electrode is buried beneath the base material's surface in the Buried Arc GTAW method, where the plasma force of the arc compresses the HAZ. The arc operates by improving its thermal efficiency and allowing the welder to work more quickly. The substrate of steel grade 80 with a thickness of 19 mm was investigated for its potential to be welded using the buried GTAW process (Lukens 1987). When compared to high-thickness welding techniques, this method reduces the heat flux of the finished weldment, the pulse cycle, and the filler requirement.

2.2.13   ULTRASONIC GTAW (U-TIG)

Ultrasonic energy is utilised in U-TIG welding to increase weld depth and cause a constant high-frequency mechanical vibrations in the arc discharge (Hua et al. 2017). The morphology of pure Al GMA joints produced in ultrasonic fields is examined (He et al. 2006). The researchers found that the utilisation of ultrasound caused in a finer and more uniform grain structure in the HAZ. This is because of the mechanical agitation provided by the ultrasonic waves, which promotes the breakdown of large dendritic grains and the formation of smaller equiaxed grains. The refinement of the weld grain structure has several advantages. It improves the mechanical properties of the weld, including its strength and ductility. It reduces the likelihood of defects such as porosity and cracking, which can be caused by large dendritic grains. The microstructure and mechanical properties of pulsed and unpulsed ultrasonic-assisted GTA-welded Q235 joints are examined (Chen et al. 2019). PU-GTAW refined the grain size and improved microstructure more efficiently than U-GTAW (He et al. 2006).

2.2.14   TOPTIG

TOPTIG is an automated form of GTAW welding with a non-consumable electrode and a separate wire feeder to provide filler material. This method increased deposition rates compared to conventional automated TIG welding. The primary technological component of this method is the welding torch itself, which is designed to feed the wire at a specific angle (20°). TOPTIG's high degree of control and consistency allows for increased productivity of up to 300% by cutting down on material waste, welding errors, and unnecessary rework while simultaneously increasing travel speeds (Silva, Schwedersky, and Rosa 2020). This new robotic welding process from ALW Welding, Inc. has the potential to provide high-quality welds in a shorter amount of time, leading to increased productivity and cost savings for manufacturers. In addition to the benefits of substantially reduced dimensions and improved accessibility of the weld torch, this arrangement also facilitates robotic welding of intricate geometries. The need for manually positioning and guiding the weld wire in relation to the weld torch and the joint being welded has been eliminated, rendering it unnecessary in the new welding process. This means the robot's sixth axis can move independently (Opderbecke and Guiheux 2009).

2.2.15   TIPTIG

TIPTIG is a welding process that combines the features of GTAW and HW Welding. It involves using a wire feeder that preheats the wire before it enters the weld puddle. The preheating of the wire helps to decrease the surface tension of the fusion metal, allowing for a more controlled and consistent weld pool. The TIPTIG process uses a high-frequency current to create an electromagnetic field around the wire, which further preheats and shapes the wire as it enters the HAZ. This leads to a smoother weld bead, which can help minimise distortion and improve weld quality. One of the main advantages of the TIPTIG process is its ability to increase welding speed while still maintaining high-quality welds. Additionally, the preheating of the wire can help reduce the amount of filler material required, leading to cost savings and reduced waste. Inside the TIP TIG welding machine is a secondary power source that generates current. The oscillation is transferred into the weld due to the mechanical action of the filler material. This causes the molten pool to become more agitated, which then breaks up the surface tension. Benefits of TIPTIG welding include increased travel speed, increased fluidity of the weld pool, reduced heat input, and cleaner welds (Wilson 2007).

2.3   ADVANCEMENTS IN GAS METAL ARC WELDING

GMAW is an electric arc welding technique and the most popular industrial welding technology because of its versatility, speed, and automation. MIG and MAG are other names for GMAW. In the 1940s, GMAW was established for the welding of non-ferrous metals. However, GMAW was quickly used on steels because it required less welding time than previous techniques. In GMAW, the wire electrode used in GMAW welding is typically made of a spool of thin, continuously fed wire that is made of a similar material as the substrate. The welding gun is equipped with a trigger that controls the flow of the wire and the shielding gas, allowing the operator to control the welding process (Rodrigues et al. 2022). By collaborating, the power source and wire feed speed unit in GMAW welding can regulate the arc length automatically, ensuring consistency throughout the welding process without requiring manual intervention from the operator. This feature makes GMAW a more efficient and reliable welding process (Le, Mai, and Hoang 2020). Average current increases metal droplet transfer from short-circuiting to globular to spray transfer modes. Figure 2.2 shows the modes of droplet transfer in the GMAW process (Kah, Suoranta, and Martikainen 2013).

Figure 2.3 shows variants of MIG welding. These variants originated in the ideation of little adjustments to the basic procedure that produce superior tailored outcomes than the MIG process. Thus, the impact of changes in some factors on the geometry of the HAZ, the degree of dilution, the weld penetration, the rate of deposition, and the morphology will be demonstrated.

2.3.1 Pulsed-Current Gas Metal Arc Welding

Continuous wire feed and high welding currents make the GMAW process a fast method that produces good-quality welds; nevertheless, GMA welding's vulnerability to porosity and fusion faults restricts its employment to situations where weld quality is important. In addition, issues with distortion, extreme heat input, and burn-through are preventing the GMAW method from being widely employed for attaching thin sheets of metal (Palani and Murugan 2006). However, the SP-GMAW method is

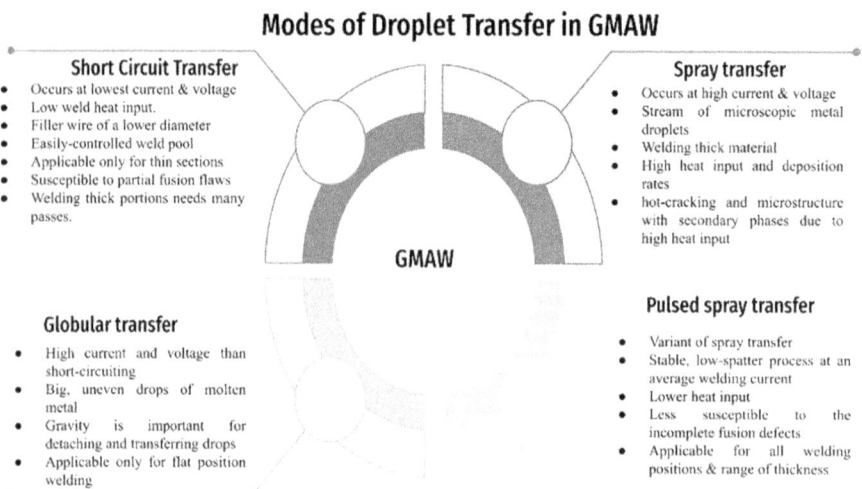

Modes of Droplet Transfer in GMAW

Short Circuit Transfer
- Occurs at lowest current & voltage
- Low weld heat input.
- Filler wire of a lower diameter
- Easily-controlled weld pool
- Applicable only for thin sections
- Susceptible to partial fusion flaws
- Welding thick portions needs many passes.

Spray transfer
- Occurs at high current & voltage
- Stream of microscopic metal droplets
- Welding thick material
- High heat input and deposition rates
- hot-cracking and microstructure with secondary phases due to high heat input

GMAW

Globular transfer
- High current and voltage than short-circuiting
- Big, uneven drops of molten metal
- Gravity is important for detaching and transferring drops
- Applicable only for flat position welding

Pulsed spray transfer
- Variant of spray transfer
- Stable, low-spatter process at an average welding current
- Lower heat input
- Less susceptible to the incomplete fusion defects
- Applicable for all welding positions & range of thickness

FIGURE 2.2 Modes of metal droplet transfer in GMAW.

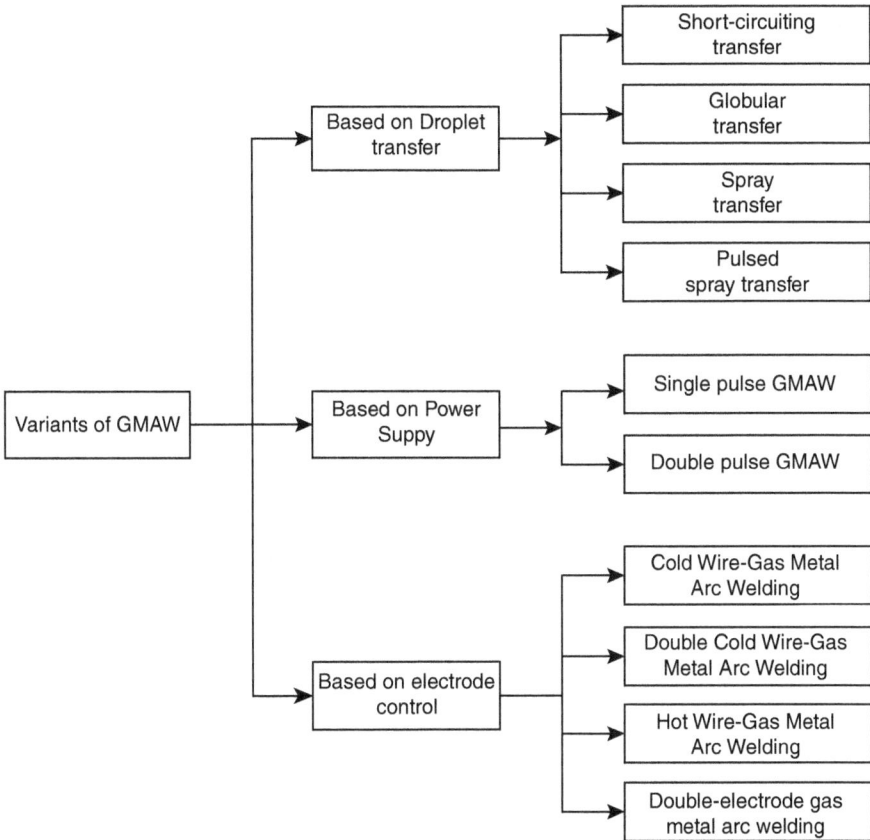

FIGURE 2.3   Variants of MIG welding.

producing superior results because it uses pulsed current rather than continuous current. Pulsed-current GMAW is a regulated spray transfer procedure in which the arc current is enough to allow spray transfer droplet mode. Pulsed-Current Gas Metal Arc Welding supplies power in two cycles: peak current (I_p) and background current (I_b). With pulsed GMAW, the average current needed for a regulated spray transfer is much smaller than the transition current needed when using constant voltage. One droplet per pulse is the target of pulsed GMAW's controlled metal transfer, which necessitates a certain set of process parameters (Sánchez-Cruz et al. 2023). As a result, the process of droplet detachment occurs at the rate of one droplet every second. The arc is maintained by lowering the current once the droplet is transferred. Low current periods lower the average arc current for positional welding, whereas fast current pulses allow metal transfer in spray mode. This current pulsing helps in microstructure refinement and in regulating the segregation of alloying components.

The reduced mean current of this pulsed current (I_p) lowers heat supply, which in turn leads to improved spatter control, greater penetration, faster deposition rates, and a higher welding speed. Moreover, SP-GMAW lessens defects like porosity (Praveen, Yarlagadda, and Kang 2005). SP-GMAW involves rapid switching between high peak current and low background current at rates of 30–400 kHz/s. At this instant, a small bit of wire is pinched off by the peak current and propelled toward the weld joint. Also, the weld pool can freeze slightly, which helps to prevent melt-through because the I_b keeps the arc going but doesn't generate enough heat to cause metal transfer (Sathishkumar et al. 2021). This is in contrast to the conventional spray transfer process, which constantly transfers tiny droplets of molten metal into the weld joint.

2.4   DP-GMAW

After examining the differences between DP-GMAW and SP-GMAW, the outcomes of using DP-GMAW are even more favourable. In this approach, high-energy and low-energy thermal pulse cycles are broken up into many smaller pulses and peaks. When compared to SP-GMAW, this further decreases the average current supply. DP-GMAW reduces crack sensitivity, allows all-position welding, reduces heat input, improves joint efficiency, and has a wider root gap design than GMAW and SP-GMAW (Sen et al. 2018). Researchers (Wang et al. 2018, 2017) looked into the DP-GMAW method, and they found that the finer grain structure observed at the greater cooling rates resulting from the lesser heat supply. The DP-GMAW technique was used to study the properties of AA5754 Al-alloy (Liu, Tang, and Lu 2013). Thermal pulses, which involve a higher rate of droplet transfer than thermal bases, are more effective at reducing grain size and enhancing mechanical properties. Single and double pulse welding were tested for their effects on austenitic stainless steel 310 (Mathivanan, Senthilkumar, and Devakumaran 2015). From the results, it is evident that the DP-GMAW method increases the quality of the fusion zone and the resulting mechanical characteristics. Reasons for this include less heat supply and heat reduction in the SP-GMAW method, which is used in most other countries.

2.5   HIGH-FREQUENCY PULSED GAS METAL ARC WELDING

Pulsed gas metal arc welding is a technique that uses a nearly square current profile at high frequencies to accomplish both streaming transfer and short arc lengths. Thanks to technological advancements in electronics and the introduction of high-speed inverters, numerous welding power supply manufacturers now provide a wide variety of current pulse waveforms. By using a high pulse frequency and a short arc length, the process can achieve a stable stream of molten metal transfer from the electrode to the workpiece, known as streaming transfer. This results in a more precise and controlled weld bead and reduces the likelihood of defects such as spatter or porosity. Because of this, an efficient welding method for welding heat-sensitive materials is introduced (Kuang, Qi, and Zheng 2022).

2.6   ULTRA-HIGH-FREQUENCY PULSE METAL-INERT GAS WELDING (UFP-MIG)

In order to enhance the microstructure and properties of various aluminium alloy weld joints, the use of ultra-high-frequency pulse (UFP) Metal-Inert Gas Welding was evaluated. Various modes and frequencies of UFP-MIG welding were studied for their effects on pore reduction efficiency, microstructure, and mechanical properties. The results showed that welding porosity in aluminium alloy joints could be nearly wiped out using the UFP-MIG process. An increase in the ultrasonic frequency used in the UFP-MIG welding process causes a gradual decrease in the crystal size of the welding zone. This is in contrast to the CP-MIG welding process. Grain size might be decreased by 14.85% in the HAZ close to the welds, which is similar to the effect seen elsewhere. By combining conventional pulsed MIG with UFP current, it may be possible to increase the ductility and strength of joints concurrently. The key explanations for the enhancement in the tensile characteristics of UFP-MIG joints were the welding porosity, grain refinement, and lesser segregation of Cu elements in the eutectic (Kuang, Qi, and Zheng 2022).

2.6.1 COLD METAL TRANSFER-GMAW

Developed in 2004 by Fronius of Austria, Cold Metal Transfer (CMT) is a welding process that uses GMAW with a controlled short-circuit transfer. A consumable electrode wire is fed through a welding gun, which is also equipped with a shielding gas nozzle. The CMT process is designed to minimise heat input and reduce the risk of distortion or burn-through in thin materials. During the high-current phase, the wire is pushed into the joint, causing a short circuit that melts a small amount of wire and creates a droplet. When the current drops to a low level, the droplet separates and falls into the joint. The short-circuiting phase helps with material transport into the welding pool and liquid fracture. The large amount of spatter produced during droplet detachment using electromagnetic force is its main downside. Using the CMT welding method, this restriction can be managed. During the CC-GMAW process, the wire keeps moving until a short circuit happens. During a short circuit, the welding current goes up quickly. This lets the short circuit open up again, which starts the arc again (Ramaswamy, Malarvizhi, and Balasubramanian 2020). The CMT-GMAW process is designed to address the issues of high heat input and spatter that are associated with traditional GMAW. The process achieves this by using a controlled short-circuit transfer, which limits the amount of heat supply and reduces the spatter quantity produced. The backward movement helps to separate the droplet from the wire, which reduces the amount of spatter produced. When it hits the HAZ, the arc is extinguished by decreasing the welding current. This controlled extinguishing of the arc helps to reduce the amount of spatter produced and ensures that the weld pool is not overheated (Ramaswamy, Malarvizhi, and Balasubramanian 2019). With the CMT-GMAW process, it can weld faster and with less distortion than with the Pulsed-Current Gas Metal Arc Welding process.

2.6.2 PCMT-GMAW

When the arc's ignition time exceeds the pulse peak period, the wire electrode melts, resulting in pulsed cold metal transfer (PCMT-GMAW). The arc's intensity goes down to zero when the welding current is at its lowest, and droplet detachment takes place during the pulse period. During the pulsing phase, the current is increased to a higher level to create a droplet on the end of the wire. The droplet then separates from the wire and falls into the HAZ during the base phase, when the welding current is decreased to a lower level. The droplet continues to detach into the HAZ as the wire is retracted, and the arc is reignited at the very end. Afterwards, the procedure cycle keeps going and going. The PCMT-GMAW method allowed for consistent droplet transfer with minimal spatter. One droplet at a time is moved over the arc in a regulated manner in this process, as opposed to the CMT cycle's uncontrolled droplet detachment (Ribeiro et al. 2019).

2.6.3 CW-GMAW

The GMAW process has undergone several variations and improvements. One of the variations is Continuous Wave GMAW (CW-GMAW), which is also known as conventional GMAW. The name "cold wire" comes from the fact that an unpowered wire is introduced into the HAZ to speed up the deposition of material through an electric arc. It can be employed in manual, semi-automated, or fully-automated modes, and there are several ways in which reducing the number of cycles and optimising the welding process can help lessen the energy burden on a component. By minimising the number of passes required to fill the chamfer, less energy is needed to complete the welding process. Fewer passes mean less time spent welding and less heat input into the material, which can lead to lower energy consumption. Lowering dilution can also help reduce the energy burden on a component. Lower dilution means less filler material is required, leading to less energy consumption during the welding process. By shrinking the HAZ, where the base metal has been subjected to heat but not melted, less energy is needed to weld the material. This is because a smaller HAZ means that less material needs to be heated and cooled, resulting in lower energy consumption. Applying coatings to surfaces can also help lessen the energy burden on a component. Coatings can provide thermal insulation, reducing heat loss during the welding process and leading to lower energy consumption. Coatings can also help prevent oxidation and other forms of degradation, which can reduce the energy required to repair or replace components that have deteriorated due to exposure to harsh conditions (Assunção et al. 2019). Multiple studies show that the CW-GMAW variation has a wider range of potential applications and industries because it can be used for both straight and narrow gap chamfers, as well as for the application of coatings, general or otherwise. For splices, the variation with the lowest electrode wire feed speed and the lowest percentage of excess wire is preferred since linear penetration is unaffected. The penetration, dilution, and width-to-height ratio of the reinforcement are all reduced when the wire feed speed is increased and the fraction of cold wire is increased, making the reinforcement taller and more suitable for use in coatings. This possibility grows when the wires employed can be mixed and matched to produce a wide variety of chemical compositions in the deposited metal (Assuncão 2013).

2.6.4 DCW-GMAW

DCW-GMAW is more restricted for use in coating scenarios, and hence its behaviour varies with respect to the possibility of usage for various industries. Weld metal is less likely to be diluted by the cold wire's weight, which favours low w/h values.

Additionally, there is a likelihood of reducing residual stresses when using this form of welding; however, it has only been experienced so far in the CW-GMAW variant. DCW-double cold wire GMAW is similar to CW-GMAW in that it employs an energised electrode wire, but it differs in that it also incorporates two non-energised wires (colds). The concept was first proposed in (Assunção et al. 2020), where the author evaluated various addition percentages of cold wire to electrode wire (up to 100%) with successful outcomes in terms of the shape of the weld zone (width and reinforcing), minimal dilution, and the lack of discontinuities. However, there is an alternative to the addition of wires. The authors of this study also found that, when comparing the identical metal to that used in GMAW with coplanar feeding, the weld metal loses about 15% of its hardness qualities. However, the mechanical qualities of hardness are improved over the standard GMAW process.

2.6.5 Hot-Wire GMAW

Fast melting and depositing rates, design flexibility, and cheap operating costs are just a few of the benefits of the HW-GMAW method, made possible by the lower heat supply required to melt the warmed wire in the melt pool. According to research (RIBEIRO et al. 2020), HW-GMAW is a promising choice for welding in small gaps or depositing corrosion-resistant coatings. Preheating the filler material in HW-GMAW reduces the heat supply needed for the electrode and the fusing of the sample piece. By heating the filler material before it enters the welding pool, less energy is required to melt the material, which can reduce the heat supply and increase the efficiency of the process. The cooler weld pool in HW-GMAW can also offer several benefits. A cooler weld pool can reduce the risk of burn-through or warping, which can be especially useful in welding applications with small gaps or complex geometries. Additionally, a cooler weld pool can lead to improved control over the fusion zone shape and penetration depth, resulting in a higher-quality weld. The reduced main arc energy in HW-GMAW can also be beneficial for depositing corrosion-resistant coatings. By using a lower-energy welding process, there is less risk of damaging the coating material or causing other forms of degradation, resulting in a more reliable and durable coating. These authors claim that this variation can become more common because of recent technological advancements in inverter welding sources, including the sophisticated manipulation of waveforms. Since more wire may be preheated with HW, the amount of energy needed to melt it is decreased, which is one of the main reasons for using HW. This variant's weld bead profile is distinguished by low dilution and tiny w/h values, which are profoundly affected by the polarity of the current supplied to the HW (Kuang, Qi, and Zheng 2022). All recent publications dealt with enhancing TIG welding with UFP current, which has been shown to significantly enhance both welding quality and productivity. UFP current (10–100 kHz) can enhance the stability of the electric arc. This is because the high frequency can help reduce spatter and improve the arc stability, resulting in a smoother and more consistent welding process. The improved arc stability can lead to improved joint quality. A more stable arc can lead to better control over the heat input, resulting in a higher-quality weld with improved penetration, fusion, and consistency. UFP current can also offer several advantages over conventional pulsed MIG welding. By using a higher frequency, it is possible to achieve greater precision and control over the welding process, resulting in a more consistent and reliable weld. Additionally, the high frequency can help reduce heat input and distortion, leading to improved joint quality and reduced post-welding processing (Rodrigues et al. 2022; Lu et al. 2014).

2.7 DOUBLE-ELECTRODE GAS METAL ARC WELDING (DE-GMAW)

DE-GMAW is a unique welding technology in which a second electrode is used to bypass a portion of the wire current. In consumable DE-GMAW, the bypass current speeds up the deposition rate, while in non-consumable DE-GMAW, it reduces the heat supply needed. This allows for more precise regulation of standard GMAW and its variants by altering the previously unchangeable relationship between heat input and deposition. New welding processes have emerged in response to rising productivity and reliability standards in the welding industry. These methods often result in higher deposition rates, but a larger HAZ could compromise their mechanical qualities. In this article, we take a look at DCW-GMAW, a promising technique that promises increased productivity. Adjustable waveforms are a significant plus for pulsed spray transfer, and adaptive pulse is a cutting-edge power source that enables synergistic control. These cutting-edge innovations have made possible the application of pulsed spray transfer, wherein the control system incorporates and is tied to the wire feed speed as well as pulse characteristics including pulse current, pulse length, background current, and pulse frequency.

2.8 CONCLUSION

According to the literature, sophisticated GTAW technologies may successfully weld thick sectional materials in one pass. GTAW technologies such as pulse PCGTAW, A-TIG, K-TIG, and high-frequency GTAW increase weld penetration depth with minimal heat input, improving production efficiency and product quality. The fabrication industry's use of GTA welding technology has increased, and research into weld joint quality, especially for sophisticated materials, has increased as well. On the other hand, recent advances in controlled transfer have improved GMAW's efficiency, which has improved production and quality. There has been a substantial reduction in the HAZ, enhanced weld bead profile, and refined microstructure by implementing these advancements, which gives support to the inference that there has been a reduction in the number of defects. These advancements have also been the subject of intensive research and have been a solution to a number of existing problems.

REFERENCES

Arulmurugan, B., and M. Manikandan. 2017. "Development of Welding Technology for Improving the Metallurgical and Mechanical Properties of 21st Century Nickel Based Superalloy 686." *Materials Science and Engineering: A* 691 (April): 126–40. doi:10.1016/j.msea.2017.03.042.
Arulmurugan, B., and M. Manikandan. 2018. "Improvement of Metallurgical and Mechanical Properties of Gas Tungsten Arc Weldments of Alloy 686 by Current Pulsing." *Transactions of the Indian Institute of Metals* 71 (12): 2953–70. doi:10.1007/s12666-018-1395-8.

Assuncão, P. D. C. 2013. Estudo da Viabilidade do Processo de Soldagem GMAW-DCW (double cold wire). Universidade Federal do Pará, Belém.

Assunção, P. D. C., R. A. Ribeiro, P. M. G. P. Moreira, E. M. Braga, and A. P. Gerlich. 2020. "A Preliminary Study on the Double Cold Wire Gas Metal Arc Welding Process." *The International Journal of Advanced Manufacturing Technology* 106 (11–12): 5393–405. doi:10.1007/s00170-020-05005-6.

Assunção, P. D. C., R. A. Ribeiro, E. B. F. dos Santos, E. M. Braga, and A. P. Gerlich. 2019. "Comparing CW-GMAW in Direct Current Electrode Positive (DCEP) and Direct Current Electrode Negative (DCEN)." *The International Journal of Advanced Manufacturing Technology* 104 (5–8): 2899–910. doi:10.1007/s00170-019-04175-2.

Babu, A. V. S., P. K. Giridharan, P. Ramesh Narayanan, S. V. S. Narayana Murty, and V. M. J. Sharma. 2014. "Experimental Investigations on Tensile Strength of Flux Bounded TIG Welds of AA2219-T87 Aluminum Alloy." *Journal of Advanced Manufacturing Systems* 13 (02): 103–12. doi:10.1142/S0219686714500073.

Chen, C., C. Fan, X. Cai, Z. Liu, S. Lin, and C. Yang. 2019. "Arc Characteristics and Weld Appearance in Pulsed Ultrasonic Assisted GTAW Process." *Results in Physics* 15 (December): 102692. doi:10.1016/j.rinp.2019.102692.

Cong, B. Q., B. J. Qi, X. G. Zhou, and J. Luo. 2009. "Influences of Ultrasonic Pulse Square-wave Current Parameters on Microstructures and Mechanical Properties of 2219 Aluminum Alloy Weld Joints." *Acta Metall Sin* 45 (9): 1057–62.

Cui, S. L., Z. M. Liu, Y. X. Fang, Z. Luo, S. M. Manladan, and S. Yi. 2017. "Keyhole Process in K-TIG Welding on 4mm Thick 304 Stainless Steel." *Journal of Materials Processing Technology* 243 (May): 217–28. doi:10.1016/j.jmatprotec.2016.12.027.

Darji, R., V. Badheka, K. Mehta, J. Joshi, and A. Yadav. 2020. "Processing of Copper by Keyhole Gas Tungsten Arc Welding for Uniformity of Weld Bead Geometry." *Materials and Manufacturing Processes* 35 (15): 1707–16. doi:10.1080/10426914.2020.1784932.

Darji, R., V. Badheka, K. Mehta, J. Joshi, A. Yadav, and A. K. Chakraborty. 2022. "Investigation on Stability of Weld Morphology, Microstructure of Processed Zones, and Weld Quality Assessment for Hot Wire Gas Tungsten Arc Welding of Electrolytic Tough Pitch Copper." *Materials and Manufacturing Processes* 37 (8): 908–20. doi:10.1080/10426914.2021.1981931.

Dinaharan, I., R. Palanivel, H. M Alswat, and M. A. Rasheed. 2022. "Effect of Hot Wire Feed Rate on Microstructural Evolution and Mechanical Strength of Pure Nickel Tubes Joined Using Gas Tungsten Arc Welding." *Proceedings of the Institution of Mechanical Engineers, Part B: Journal of Engineering Manufacture.* doi:10.1177/09544054221136530.

Dinesh, K., B. Anandavel, and K. Devakumaran. 2018. "Visualization of Hot Wire Gas Tungsten Arc Welding Process." *International Research Journal of Engineering and Technology* 5 (5): 2512–17.

Egerland, S., J. Zimmer, R. Brunmaier, R. Nussbaumer, G. Posch, and B. Rutzinger. 2015. "Advanced Gas Tungsten Arc Weld Surfacing Current Status and Application." *Soldagem & Inspeção* 20 (3): 300–14. doi:10.1590/0104-9224/SI2003.05.

Feng, Y., Z. Luo, Z. Liu, Y. Li, Y. Luo, and Y. Huang. 2015. "Keyhole Gas Tungsten Arc Welding of AISI 316L Stainless Steel." *Materials & Design* 85 (November): 24–31. doi:10.1016/j.matdes.2015.07.011.

Fujii, H., T. Sato, S. Lu, and K. Nogi. 2008. "Development of an Advanced A-TIG (AA-TIG) Welding Method by Control of Marangoni Convection." *Materials Science and Engineering: A* 495 (1–2): 296–303. doi:10.1016/j.msea.2007.10.116.

Gao, F., Y. Cui, Y. Lv, W. Yu, and P. Jiang. 2021. "Microstructure and Properties of Ti-6Al-4V Alloy Welded Joint by Keyhole Gas Tungsten Arc Welding." *Materials Science and Engineering: A* 827 (October): 142024. doi:10.1016/j.msea.2021.142024.

Garg, H., K. Sehgal, R. Lamba, and G. Kajal. 2019. "A Systematic Review: Effect of TIG and A-TIG Welding on Austenitic Stainless Steel." In *Advances in Industrial and Production Engineering: Select Proceedings of FLAME 2018*, 375–85. doi:10.1007/978-981-13-6412-9_36.

He, L.-B., L.-M. Li, H.-W. Hao, M.-S. Wu, and R.-L. Zhou. 2006. "Grain Refinement and High Performance of Titanium Alloy Joint Using Arc-Ultrasonic Gas Tungsten Arc Welding." *Science and Technology of Welding and Joining* 11 (1): 72–4. doi:10.1179/174329306X77083.

Hua, C., H. Lu, C. Yu, J.-M. Chen, X. Wei, and J.-J. Xu. 2017. "Reduction of Ductility-Dip Cracking Susceptibility by Ultrasonic-Assisted GTAW." *Journal of Materials Processing Technology* 239 (January): 240–50. doi:10.1016/j.jmatprotec.2016.08.018.

Xingguo, CONG Baoqiang, Q. I. Bojin Zhou, and L. U. O. Jun. 2009. "Influences of Ultrasonic Pulse Square-Wave Current Parameters on Microstructures And Mechanical Properties of 2219 Aluminum Alloy Weld Joints." *Acta Metall Sin* 45 (9): 1057–62. https://www.researchgate.net/publication/301554782.

Jayakrishnan, S., and P. Chakravarthy. 2017. "Flux Bounded Tungsten Inert Gas Welding for Enhanced Weld Performance-A Review." *Journal of Manufacturing Processes* 28 (August): 116–30. doi:10.1016/j.jmapro.2017.05.023.

Kah, P., R. Suoranta, and J. Martikainen. 2013. "Advanced Gas Metal Arc Welding Processes." *The International Journal of Advanced Manufacturing Technology* 67 (1–4): 655–74. doi:10.1007/s00170-012-4513-5.

Kuang, X., B. Qi, and H. Zheng. 2022. "Effect of Pulse Mode and Frequency on Microstructure and Properties of 2219 Aluminum Alloy by Ultrahigh-Frequency Pulse Metal-Inert Gas Welding." *Journal of Materials Research and Technology* 20 (September): 3391–407. doi:10.1016/j.jmrt.2022.08.094.

Lathabai, S., B. L. Jarvis, and K. J. Barton. 2001. "Comparison of Keyhole and Conventional Gas Tungsten Arc Welds in Commercially Pure Titanium." *Materials Science and Engineering: A* 299 (1–2): 81–93. doi:10.1016/S0921-5093(00)01408-8.

Le, V. T., D. S. Mai, and Q. H. Hoang. 2020. "A Study on Wire and Arc Additive Manufacturing of Low-Carbon Steel Components: Process Stability, Microstructural and Mechanical Properties." *Journal of the Brazilian Society of Mechanical Sciences and Engineering* 42 (9): 480. doi:10.1007/s40430-020-02567-0.

LI, Q., A.-p. WU, Y.-j. LI, G.-q. WANG, B.-j. QI, D.-y. YAN, and L.-y. XIONG. 2017. "Segregation in Fusion Weld of 2219 Aluminum Alloy and Its Influence on Mechanical Properties of Weld." *Transactions of Nonferrous Metals Society of China* 27 (2): 258–71. doi:10.1016/S1003-6326(17)60030-X.

Liu, A., X. Tang, and F. Lu. 2013. "Study on Welding Process and Prosperities of AA5754 Al-Alloy Welded by Double Pulsed Gas Metal Arc Welding." *Materials & Design* 50 (September): 149–55. doi:10.1016/j.matdes.2013.02.087.

Liu, Z. M., Y. X. Fang, S. L. Cui, S. Yi, J. Y. Qiu, Q. Jiang, W. D. Liu, and Z. Luo. 2017. "Keyhole Thermal Behavior in GTAW Welding Process." *International Journal of Thermal Sciences* 114 (April): 352–62. doi:10.1016/j.ijthermalsci.2017.01.005.

Lukens, W. E. 1987. "Mechanical Properties of HSLA Steel Buried Gas Tungsten Arc Weldments." *Weld. J.* 66 (7): 215.

Lu, Y., S. J. Chen, Y. Shi, X. Li, J. Chen, L. Kvidahl, and Y. M. Zhang. 2014. "Double-Electrode Arc Welding Process: Principle, Variants, Control and Developments." *Journal of Manufacturing Processes* 16 (1): 93–108. doi:10.1016/j.jmapro.2013.08.003.

Mallick, P. K. 2021. "Joining for Lightweight Vehicles." In *Materials, Design and Manufacturing for Lightweight Vehicles*, 321–71. Elsevier. doi:10.1016/B978-0-12-818712-8.00008-2.

Manikandan, M., N. Arivazhagan, M. Nageswara Rao, and G. M. Reddy. 2014. "Microstructure and Mechanical Properties of Alloy C-276 Weldments Fabricated by Continuous and Pulsed Current Gas Tungsten Arc Welding Techniques." *Journal of Manufacturing Processes* 16 (4): 563–72. doi:10.1016/j.jmapro.2014.08.002.

Manikandan, S. G. K., D. Sivakumar, K. Prasad Rao, and M. Kamaraj. 2014. "Effect of Weld Cooling Rate on Laves Phase Formation in Inconel 718 Fusion Zone." *Journal of Materials Processing Technology* 214 (2): 358–64. doi:10.1016/j.jmatprotec.2013.09.006.

Mathivanan, A., A. Senthilkumar, and K. Devakumaran. 2015. "Pulsed Current and Dual Pulse Gas Metal Arc Welding of Grade AISI: 310S Austenitic Stainless Steel." *Defence Technology* 11 (3): 269–74. doi:10.1016/j.dt.2015.05.006.

Morisada, Y., H. Fujii, F. Inagaki, and M. Kamai. 2013. "Development of High Frequency Tungsten Inert Gas Welding Method." *Materials & Design* 44 (February): 12–6. doi:10.1016/j.matdes.2012.07.054.

Norrish, J. 2006. *Advanced Welding Processes*. Woodhead Publishers, Cambridge.

Opderbecke, T., and S. Guiheux. 2009. "TOPTIG: Robotic TIG Welding with Integrated Wire Feeder." *Welding International* 23 (7): 523–9. doi:10.1080/09507110802543146.

Padmanaban, M. R., A., B. Neelakandan, and D. Kandasamy. 2016. "A Study on Process Characteristics and Performance of Hot Wire Gas Tungsten Arc Welding Process for High Temperature Materials." *Materials Research* 20 (1): 76–87. doi:10.1590/1980-5373-mr-2016-0321.

Pai, A., I. Sogalad, S. Basavarajappa, and P. Kumar. 2020. "Assessment of Impact Strength of Welds Produced by Cold Wire and Hot Wire Gas Tungsten Arc Welding (GTAW) Processes." *Materials Today: Proceedings* 24: 983–94. doi:10.1016/j.matpr.2020.04.411.

Palani, P. K., and N. Murugan. 2006. "Selection of Parameters of Pulsed Current Gas Metal Arc Welding." *Journal of Materials Processing Technology* 172 (1): 1–10. doi:10.1016/j.jmatprotec.2005.07.013.

Praveen, P., P. K. D. V. Yarlagadda, and M. J. Kang. 2005. "Advancements in Pulse Gas Metal Arc Welding." *Journal of Materials Processing Technology* 164–165 (May): 1113–9. doi:10.1016/j.jmatprotec.2005.02.100.

Ramaswamy, A., S. Malarvizhi, and V. Balasubramanian. 2019. "Influence of Post Weld Heat Treatment on Tensile Properties of Cold Metal Transfer (CMT) Arc Welded AA6061-T6 Aluminium Alloy Joints." *Journal of the Mechanical Behavior of Materials* 28 (1): 135–45. doi:10.1515/jmbm-2019-0015.

Ramaswamy, A., S. Malarvizhi, and V. Balasubramanian. 2020. "Effect of Variants of Gas Metal Arc Welding Process on Tensile Properties of AA6061-T6 Aluminium Alloy Joints." *The International Journal of Advanced Manufacturing Technology* 108 (9–10): 2967–83. doi:10.1007/s00170-020-05602-5.

Rathod, D. W. 2021. "Comprehensive Analysis of Gas Tungsten Arc Welding Technique for Ni-Base Weld Overlay." In *Advanced Welding and Deforming*, 105–26. Elsevier. doi:10.1016/B978-0-12-822049-8.00004-9.

RIBEIRO, P. P. G., P. D. C. ASSUNÇÃO, E. M. BRAGA, R. A. RIBEIRO, and A. P. GERLICH. 2020. "Metal Transfer Mechanisms in Hot-Wire Gas Metal Arc Welding." *Welding Journal* 99 (11): 281s–94s. doi:10.29391/2020.99.026.

Ribeiro, R., P. Assunção, E. dos Santos, A. Filho, E. Braga, and A. Gerlich. 2019. "Application of Cold Wire Gas Metal Arc Welding for Narrow Gap Welding (NGW) of High Strength Low Alloy Steel." *Materials* 12 (3): 335. doi:10.3390/ma12030335.

Rodrigues, L. A. S., P. P.G. Ribeiro, E. da S. Costa, T. dos S. Cabral, and E. de M. Braga. 2022. "A Brief Study of Unconventional Variants of GMAW Welding: Parameters, Weld Bead, and Microstructures." In *Engineering Principles - Welding and Residual Stresses*. IntechOpen. doi:10.5772/intechopen.104525.

Sánchez-Cruz, T. del N. J., F. F. Curiel-López, V. H. López-Morelos, J. A. González-Sánchez, A. Ruiz, and E. Carrillo. 2023. "Optimization of Macro and Microstructural Characteristics of 316L/2205 Dissimilar Welds Obtained by the GMAW-Pulsed Process." *Materials Today Communications* 34 (March): 105401. doi:10.1016/j.mtcomm.2023.105401.

Sathishkumar, M., Y. J. Bhakat, K. Gokul Kumar, S. Giribaskar, R. Oyyaravelu, N. Arivazhagan, and M. Manikandan. 2021. "Investigation of Double-Pulsed Gas Metal Arc Welding Technique to Preclude Carbide Precipitates in Aerospace Grade Hastelloy X." *Journal of Materials Engineering and Performance* 30 (1): 661–84. doi:10.1007/s11665-020-05360-1.

Sen, M., M. Mukherjee, S. K. Singh, and T. K. Pal. 2018. "Effect of Double-Pulsed Gas Metal Arc Welding (DP-GMAW) Process Variables on Microstructural Constituents and Hardness of Low Carbon Steel Weld Deposits." *Journal of Manufacturing Processes* 31 (January): 424–39. doi:10.1016/j.jmapro.2017.12.003.

Shah, P., and C. Agrawal. 2019. "A Review on Twin Tungsten Inert Gas Welding Process Accompanied by Hot Wire Pulsed Power Source." *Journal of Welding and Joining* 37 (2): 41–51. doi:10.5781/JWJ.2019.37.2.7.

Silva, R. H. G. e, M. B. Schwedersky, and Á. F. da Rosa. 2020. "Evaluation of Toptig Technology Applied to Robotic Orbital Welding of 304L Pipes." *International Journal of Pressure Vessels and Piping* 188 (December): 104229. doi:10.1016/j.ijpvp.2020.104229.

Silwal, B., and M. Santangelo. 2016. "Vibration Assisted Hot-Wire Gas-Tungsten Arc Welding of Duplex Stainless Steel 2205." In *Volume 2: Advanced Manufacturing*. American Society of Mechanical Engineers. doi:10.1115/IMECE2016-67665.

Singh, S. R., and P. Khanna. 2021. "A-TIG (Activated Flux Tungsten Inert Gas) Welding: - A Review." *Materials Today: Proceedings* 44: 808–20. doi:10.1016/j.matpr.2020.10.712.

Ungethüm, T., E. Spaniol, M. Hertel, and U. Füssel. 2020. "Analysis of Metal Transfer and Weld Geometry in Hot-Wire GTAW with Indirect Resistive Heating." *Welding in the World* 64 (12): 2109–17. doi:10.1007/s40194-020-00986-0.

Vidyarthy, R. S., D. K. Dwivedi, and M. Vasudevan. 2017. "Influence of M-TIG and A-TIG Welding Process on Microstructure and Mechanical Behavior of 409 Ferritic Stainless Steel." *Journal of Materials Engineering and Performance* 26 (3): 1391–403. doi:10.1007/s11665-017-2538-5.

Wang, L. L., H. L. Wei, J. X. Xue, and T. DebRoy. 2017. "A Pathway to Microstructural Refinement through Double Pulsed Gas Metal Arc Welding." *Scripta Materialia* 134 (June): 61–5. doi:10.1016/j.scriptamat.2017.02.034.

—. 2018. "Special Features of Double Pulsed Gas Metal Arc Welding." *Journal of Materials Processing Technology* 251: 369–75.

Wang, Y., B. Cong, B. Qi, M. Yang, and S. Lin. 2019. "Process Characteristics and Properties of AA2219 Aluminum Alloy Welded by Double Pulsed VPTIG Welding." *Journal of Materials Processing Technology* 266 (April): 255–63. doi:10.1016/j.jmatprotec.2018.11.015.

Wang, Y., B. Qi, B. Cong, M. Zhu, and S. Lin. 2018. "Keyhole Welding of AA2219 Aluminum Alloy with Double-Pulsed Variable Polarity Gas Tungsten Arc Welding." *Journal of Manufacturing Processes* 34 (August): 179–86. doi:10.1016/j.jmapro.2018.06.006.

Wilson, M. 2007. "TIP TIG: New Technology for Welding." *Industrial Robot: An International Journal* 34 (6): 462–6. doi:10.1108/01439910710832057.

Yang, M., Z. Yang, B. Cong, and B. Qi. 2015. "How Ultra High Frequency of Pulsed Gas Tungsten Arc Welding Affects Weld Porosity of Ti-6Al-4V Alloy." *The International Journal of Advanced Manufacturing Technology* 76 (5–8): 955–60. doi:10.1007/s00170-014-6324-3.

Yang, Z., B. Qi, B. Cong, F. Liu, and M. Yang. 2015. "Microstructure, Tensile Properties of Ti-6Al-4V by Ultra High Pulse Frequency GTAW with Low Duty Cycle." *Journal of Materials Processing Technology* 216 (February): 37–47. doi:10.1016/j.jmatprotec.2014.08.026.

Yang, Z., M. Yang, and B. Qi. 2015. "Fluid and Arc Behavior with Ultra High Frequency Pulsed GTAW." *Quarterly Journal of the Japan Welding Society* 33 (2): 11s–4s. doi:10.2207/qjjws.33.11s.

Zou, Y., R. Ueji, and H. Fujii. 2014. "Effect of Oxygen on Weld Shape and Crystallographic Orientation of Duplex Stainless Steel Weld Using Advanced A-TIG (AA-TIG) Welding Method." *Materials Characterization* 91 (May): 42–9. doi:10.1016/j.matchar.2014.02.006.

3 Welding of AISI 304 Steel Using TIG and Pulse TIG
Weld Deposition and Relative Joint Strength Comparisons

Ajfarul Islam, Dipankar Bose,
Dhiraj Kumar, and B. Acherjee

3.1 INTRODUCTION

Tungsten inert gas (TIG) welding, also referred to as gas tungsten arc welding (GTAW), is an arc welding procedure that melts and welds metal pieces by creating an arc between a non-consumable tungsten electrode and the workpiece. Because of tungsten's exceptionally high melting point, the arc can be produced without melting the electrode. The high temperatures produced by the process may cause oxidation of the electrode and the workpiece if they are not properly shielded from the environment. Shielding gases such as argon or helium are routinely employed in TIG welding (Ogundimu et al., 2019). TIG welding is utilized practically in all industrial sectors; however, it is mostly employed in applications that need high-quality welds (Khanna, 2008). TIG welding with the pulse function is frequently used for welding thin metals because it provides better regulation of the heat input to the weld. In pulse TIG (pTIG) welding, the welding current rapidly switches between two levels: high (pulse current) and low (background current). The weld region is melted and fused because of the pulse current application period. During the period when the current level drops to the background current, the weld region cools and solidifies. As heat input is well controlled in pTIG welding, it has metallurgical benefits such as fusion zone grain refinement, reduced HAZ, lesser deformation, and minimal residual stresses. This is used to join metals like stainless steel that melt and flow freely. Pulsing is employed to achieve desired penetration while minimizing heat buildup. The power sources offer a square waveform for the pulse cycle. Because the welder must maintain the arc length short, significant caution and experience are necessary to avoid electrode contact with the workpiece (Gao et al., 2013). TIG welding has various process factors that must be adjusted to achieve the required weld quality, including electrode size, welding current, arc voltage, gas flow rate, welding speed, and electrode sticking out. Additional factors, viz.,

DOI: 10.1201/9781003327769-3

base current, peak current, pulse frequency, pulse duty, etc., must be adjusted in pTIG welding. Because it creates higher-quality welds, has less electrode waste, has greater control, and is more versatile than TIG welding, pTIG welding has grown in popularity. TIG welding offers the following advantages: no flux is required, hence no slag or slag inclusion issues; it can be operated in both automated and manual modes; it creates high-quality and strong welds with a minimum quantity of flames and sparks; and it produces non-corrosive and ductile joints. pTIG welding has several advantages over TIG welding, including pulse action that agitates and stirs the weld pool, providing resistance to lack of fusion, reduced heat input and spatter, and a decrease in distortion, porosity, and inclusions. TIG and pTIG welding, on the other hand, have a few limitations: slower than other arc welding processes such as MIG and MMAW; lower filler deposition rate; not appropriate for thicker metal sheets. pTIG welding is more costly and slower than TIG welding, and it requires more careful maintenance.

The potential of TIG and pTIG welding in diverse application areas not only entices industries to enthusiastically apply this welding process to a variety of application areas, but it also entices several researchers to study the process and examine its application to a variety of materials with the goal of improving several aspects of metal joining technology, including weld quality enhancement and cost reduction. Raveendra and Kumar (2013) used TIG welding to assess the impact of pulsed and non-pulsed current at various frequencies on the hardness, tensile strength, and microstructure of stainless steel 304. The observations for various welding settings in TIG employing non-pulsed and pulsed current are described and contrasted. Chennaiah et al. (2015) examined the effect of pTIG welding factors on microhardness and microstructures by performing pTIG welding on aluminum alloys. pTIG welding appears to give a finer weld grain structure as compared to TIG welding, as well as less HAZ. Subhasmita and Mohanty (2016) studied the effect of TIG welding variables on aluminum alloy using welding current, voltage, and gas flow rate as process variables and tensile strength as a weld quality measure. Ugla (2016) studied the impact of welding variables on weld bead shape and mechanical properties using TIG welding with non-pulse and pulse currents. Because it breaks the dendritic arms during welding, the pulse frequency has a considerable influence on tensile strength. The travel speed has the greatest influence on the bead shape and aspect ratio. Susmitha et al. (2016) examined TIG welding of aluminum alloy and observed that the mechanical properties of the weld joint, such as tensile strength and impact strength, vary linearly with the welding current increase. Deepak et al. (2016) optimized the welding variables for pulsed bead-on-plate TIG welding of maraging steel C300 by considering input factors including peak current, base current, pulse frequency, and weld bead depth as process responses. Ramkumar et al. (2016) studied the outcome of post-weld heat treatment on the microstructure and mechanical characteristics of Inconel X750 active flux TIG welding. Welding efficiency of 60.7% and 94.07% were attained in as-welded and post-weld heat treatment situations, respectively. Ramkumar et al. (2017) studied the impact of welding current on weld joint aspect ratio by TIG welding of dissimilar metals without the use of activated flux. Full penetration is obtained in multi-pass welding, and failure occurs at the parent material during tensile testing of the welded specimen.

Yan et al. (2017) investigated the influence of post-weld heat treatment on Ti6Al4V alloy TIG welding and compared the mechanical properties to the as-welded state. It is observed that post-weld heat treatment improves mechanical characteristics and reduces residual stress. Kulkarni et al. (2018) investigated the mechanism of activated flux TIG welding of P91 steel to P22 steel utilizing fluxes such as SiO_2, TiO_2, Cr_2O_3, MoO_3, and CuO. After post-weld heat treatment, impact toughness is seen to increase.

The present study examines the effect of welding parameters on desired weld performance using TIG and pTIG welding of AISI 304 stainless steel. The primary goal of the present study is to compare TIG and pTIG welding of AISI 304 to assess the degree of efficacy of pTIG welding over TIG welding. Weld deposition and relative joint strength are used as response metrics to compare TIG and pTIG welding performance. Besides that, the impact of welding parameters on weld deposition and relative joint strength is examined. Response surface methodology (RSM) is employed for experimental design and the development of empirical models to determine the correlation between process factors and responses. The parametric trends are examined using various graphs created using the developed models.

3.2 MATERIALS AND METHODS

AISI 304 stainless steel is chosen as the work material since it is widely used in structural applications. The mechanical characteristics and chemical compositions of AISI 304 are shown in Tables 3.1 and 3.2. The work specimen is cut from a rolled plate into strips with dimensions of 100 mm × 75 mm × 2 mm using a shearing machine. A hand grinder is used to smoothen the edges of the specimens. The square butt joint configuration is used for welding the steel strips. TIG and pTIG welding are used to weld steel specimens and investigate the relative performance of both welding processes. The welding system consists of a TIG welding power source, a welding torch, a shielding gas supply unit, and a welding table with a speed variation attachment. A three-phase DC generator is utilized as the power supply for TIG welding, and the welding current alternates between two levels, one greater, known as peak current, and one lower, known as background current. The torch may be operated manually or automatically, and its primary components are the collet, tungsten electrode, screwed cap, ceramic cap, and handle. A profile gas-cutting machine with adjustable speed settings is used to attach the welding torch and realize a consistent welding speed. The TIG welding torch mounted on the profile gas-cutting machine moves at a constant speed along a desired path. The same setup is employed for pTIG welding

TABLE 3.1
Mechanical Characteristics of Stainless Steel AISI 304

Workpiece Material	Tensile Strength (MPa)	Melting Temp. (°C)	Density (g/cm³)	Hardness (HRB)
Stainless Steel 304	540–750	1400–1450	8.00	85–92

TABLE 3.2
Chemical Compositions of Stainless Steel AISI 304 (wt. %)

Elements	C	Mn	Si	P	S	Cr	Ni	Fe	N
Percentage	0.06	1.40	0.38	0.023	0.018	18.94	9.05	70.02	0.11

since the pulsing parameters can be adjusted using the same power source as for TIG welding. Figure 3.1 shows a graphical depiction of the TIG/pTIG welding setup utilized in this experimental investigation. During welding, a welding fixture is utilized to properly clamp the work sample. Argon gas is employed as a shielding gas, while ER-316L filler material with a diameter of 1 mm is used for welding.

Process parameters in TIG welding include welding speed, welding current, and gas flow rate, whereas the pTIG welding process also includes pulse frequency. During a pTIG welding process, the pulse frequency is fixed at 85 Hz. Through a series of experimental trials, the range of welding parameters for producing an acceptable weld is determined. The experiments are designed using a face-cantered cubic design of the central composite with three levels. The central composite design is the most prominent form of second-order design used to estimate second-order polynomials for response variables using RSM. The adoption of a face-cantered cubic design resulted in twenty experimental runs. The experimental design used has a significant influence on the accuracy of the approximation and the response surface. RSM is used not only for experimental design but also to develop empirical models to establish the interrelationship between process variables and response parameters (Acherjee et al., 2017). Table 3.3 shows the welding parameters and their notation, units, and levels.

FIGURE 3.1 (a) TIG/pTIG welding power source, (b) welding torch setup with profile cutter machine for torch movement, and (c) welding fixture setup.

TABLE 3.3

Welding Parameters, Their Notation, Units, and Levels

Welding Parameters	Notations	Units	Levels		
			−1	0	+1
Welding speed	WS	mm/min	55.80	67.20	72.60
Welding current	WC	A	80	90	100
Gas flow rate	GFR	L/min	8	9	10

Weld deposition and relative joint strength are considered response parameters to compare TIG and pTIG welding performance. The effect of welding parameters on weld deposition and relative joint strength is also investigated. A digital weighing machine (Figure 3.2) is utilized for measuring the weight of the material before and after welding. The two plates to be welded are weighed together before welding and again after welding, and the weight difference represents the quantity of weld deposition. The weld deposition percentage is calculated as:

$$\text{Weld deposition } (\%) = \frac{\text{Weight after welding} - \text{Weight before welding}}{\text{Weight after welding}} \times 100\%$$

(3.1)

TIG and pTIG weld samples are cut for tensile tests according to the ASTM-E-646-98 standard. A universal testing machine is utilized to evaluate the ultimate tensile strength (UTS) of the welded stainless steel samples and a base material sample. The UTS of the base material (748 MPa) is used to calculate the relative joint strength of the weld using the following equation:

$$\text{Relative Joint Strength } (\%) = \frac{\text{UTS of weld sample}}{\text{UTS of base sample}} \times 100\% \qquad (3.2)$$

FIGURE 3.2 A welded specimen is weighed on a digital weighing machine.

3.3   RESULTS AND DISCUSSION

The experiments are carried out in accordance with the experimental design, and the results are obtained. RSM is used to construct empirical models for each of the responses using the experimental results. Analysis of variance (ANOVA) is performed to verify the adequacy of the empirical models created. The parametric trends are analyzed using the developed models and various graphs. The selected responses are then used to compare the performance of TIG and pTIG welding processes.

3.3.1   WELD DEPOSITION

Weld Deposition (%) – TIG welding: The fit summary analysis for "Sequential Model Sum of Squares" suggests a quadratic model for Weld Deposition (%) – TIG. Table 3.4 shows the ANOVA findings for Weld Deposition (%) – TIG. The model has an F-value of 241.37, indicating that it is significant. There is just a 0.01% probability that a high "Model F-value" may be caused by noise. When the "p-value" (probability value) is less than 0.05, model terms are significant. Model terms with probability values greater than 0.10 are considered insignificant. Model reduction improves model accuracy by eliminating insignificant model terms (except those essential to preserving hierarchy) when there are several insignificant model terms (Acherjee et al., 2011). Significant model terms for Weld Deposition (%) – TIG are A, B, C, AC, and A^2, while insignificant model terms are excluded by applying the backward elimination method. The "LOF" (lack of fit) F-value of 1.56 indicates that the lack of fit is not significant when compared to "PE" (pure error). A "LOF" of this magnitude has a 32.52% probability of arising due to noise. An insignificant "LOF" is excellent since the model fits. The "Pred. R^2" (predicted R^2) of 0.98 agrees reasonably with the "R^2" of 0.99, which is near unity and indicates excellent data fitting to a polynomial curve. "Adeq. Prcn." (adequate precision) is used to assess the signal-to-noise ratio,

TABLE 3.4
ANOVA Results for Weld Deposition (%) – TIG

Source	SS	DF	MS	F-value	p-Value	
Model	0.04375	5	0.00875	241.37	< 0.0001	significant
A-WS	0.02401	1	0.02401	662.34	< 0.0001	
B-WC	0.01089	1	0.01089	300.41	< 0.0001	
C-GFR	0.00169	1	0.00169	46.62	< 0.0001	
AC	0.00031	1	0.00031	8.62	0.0108	
A^2	0.00685	1	0.00685	188.83	< 0.0001	
Residual	0.00051	14	0.00004			
LOF	0.00037	9	0.00004	1.56	0.3252	insignificant
PE	0.00013	5	0.00003			
Corrected Total	0.04426	19				
R^2	0.99	Pred. R^2	0.98	Adeq. Prcn. 57.62		

and a value greater than 4 is desired. The "Adeq. Prcn." of 57.62 suggests that the signal is adequate. According to the results of the statistical tests, the constructed model is adequate. The quadrature model of Weld Deposition (%) – TIG in actual parameter values is given below.

$$\text{Weld Deposition } (\%)_{\text{TIG}} = 3.3257 - 0.0798A + 3.3000E - 003B - 0.0608C$$
$$+ 7.4405E - 004AC + 5.2438E - 004A^2$$

(3.3)

The perturbation plot of Weld Deposition (%) – TIG in Figure 3.3a shows the influence of all selected welding parameters on weld deposition (%) for TIG welding experiments in a single plot. Figure 3.3a shows that the weld deposition (%) declines as welding speed increases but increases with increasing welding current. More heat is delivered into the weld zone when the welding current is increased, resulting in more filler material melting and deposition. Increasing welding speed reduces the duration of travel and interaction, resulting in less time available for deposition. The use of shielding gas protects the weld pool and surrounding zone against contamination from the environment. Argon as a shielding gas in TIG welding not only shielded the weld zone and the tungsten electrode from contamination, but it also resulted in arc stability and decreased spatter. The trend of gas flow rate on weld deposition (%) with the chosen parametric range is negative, which might be attributed to increased convection loss of heat from filler wire. Figure 3.3b depicts a cube plot of Weld Deposition (%) for TIG welding, which is a graphical representation of the weld deposition (%) for TIG welding at various input variable combinations.

Weld Deposition (%) – pTIG welding: For Weld Deposition (%) – pTIG, the fit summary analysis for "Sequential Model Sum of Squares" suggests a quadratic model. Table 3.5 shows the ANOVA findings for Weld Deposition (%) – pTIG. The model has an F-value of 224.01, indicating that it is significant. For the Weld Deposition (%) – pTIG

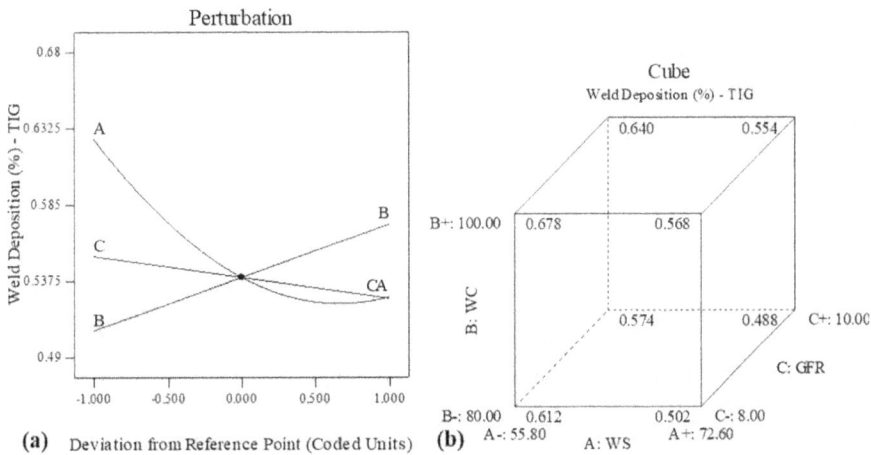

FIGURE 3.3   (a) Perturbation plot and (b) cube pot of Weld Deposition (%) – TIG.

TABLE 3.5
ANOVA Results for Weld Deposition (%) – pTIG

Source	SS	DF	MS	F-value	p-Value	
Model	0.04756	6	0.00793	224.01	< 0.0001	significant
A-WS	0.02401	1	0.02401	678.54	< 0.0001	
B-WC	0.01369	1	0.01369	386.89	< 0.0001	
C-GFR	0.00121	1	0.00121	34.20	< 0.0001	
AC	0.00045	1	0.00045	12.72	0.0034	
BC	0.00020	1	0.00020	5.65	0.0335	
A^2	0.00800	1	0.00800	226.09	< 0.0001	
Residual	0.00046	13	0.00004			
LOF	0.00018	8	0.00002	0.39	0.8865	insignificant
PE	0.00028	5	0.00006			
Corrected Total	0.04802	19				
R^2	0.99	Pred. R^2	0.98	Adeq. Prcn. 55.98		

model, the model terms A, B, C, AC, BC, and A^2 are significant, while insignificant terms other than those necessary to sustain the hierarchy are eliminated. The "LOF" of 0.39 shows that the lack of fit is not significant, which is exactly what is needed. The "Pred. R^2" of 0.98s is in good accordance with the "R^2" of 0.99, indicating excellent data fit. The "Adeq. Prcn." of 55.98 suggests that the signal is adequate. The developed model of Weld Deposition (%) – pTIG in actual parameter values is given below:

$$\text{Weld Deposition}(\%)_{\text{pTIG}} = 3.0809 - 0.0867A + 8.2000E - 003B - 0.0233C + 8.9286E$$

$$-004AC - 5.0000E - 004BC + 5.6689E - 004A^2$$

$$(3.4)$$

The perturbation plot of Weld Deposition (%) – pTIG in Figure 3.4a shows similar patterns as TIG welding, with weld deposition (%) decreasing with increasing welding speed while increasing with increasing welding current. The gas flow rate has a small negative influence on pTIG welding weld deposition (%). According to the ANOVA results and perturbation plots for weld deposition (%) for TIG and pTIG, welding speed has the greatest impact on weld deposition (%), trailed by welding current and gas flow rate within their selected ranges. Figure 3.4b shows a cube plot of Weld Deposition (%) – pTIG with different input variable combinations.

Figure 3.5 compares weld deposition (%) for TIG and pTIG welding throughout all twenty experimental runs. TIG-welded samples have more weld deposition than pTIG in all twenty experimental runs. Although the degree of weld deposition achieved by TIG and pTIG varies, they follow roughly the same pattern in rise and fall as acquired with the same set of parameters for TIG and pTIG. Additional parameters, such as pulse frequency, are employed for pTIG but are maintained constant across all experimental runs.

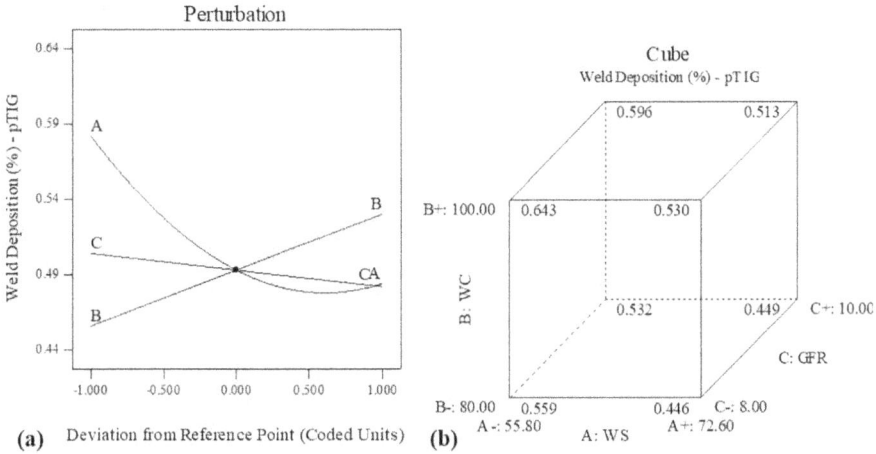

FIGURE 3.4   (a) Perturbation plot and (b) cube pot of Weld Deposition (%) – pTIG.

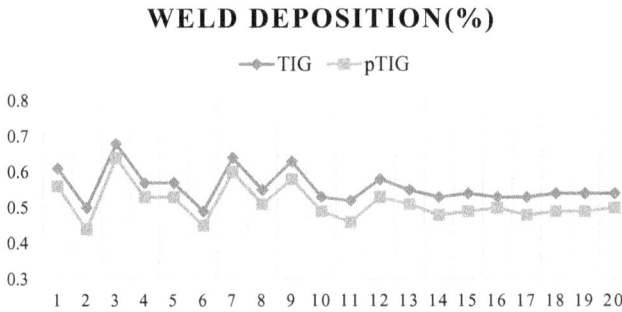

FIGURE 3.5   Weld Deposition (%) comparison for TIG and pTIG welding.

3.3.2   RELATIVE JOINT STRENGTH

Relative Joint Strength (%) – TIG: The "Sequential Model Sum of Squares" suggests a quadratic model on the basis of the fit summary analysis of Relative Joint Strength (%) – TIG results. Table 3.6 presents the ANOVA findings for Relative Joint Strength (%) – TIG. The model has an F-value of 71.09, indicating that it is significant. With "p-values" less than 0.05, model terms A, B, C, BC, A^2, B^2, and C^2 are significant for Relative Joint Strength (%) – TIG, but terms with "p-values" larger than 0.10 are insignificant and eliminated. The lack of fit is insignificant, as indicated by the "LOF" of 2.49, indicating that the model fits well. The "Pred. R²" of 0.93 approaches unity and corresponds well with the "R²" of 0.98, signifying adequate data fitting. A signal with an "Adeq. Prcn." of 30.74 is considered sufficient. The model's adequacy is justified by the ANOVA findings. The quadrature model of Relative Joint Strength (%) – TIG in actual parameter values is given below:

TABLE 3.6
ANOVA Results for Relative Joint Strength (%) – TIG

Source	SS	DF	MS	F-value	p-Value	
Model	398.39	7	56.91	71.09	< 0.0001	significant
A-WS	40.00	1	40.00	49.97	< 0.0001	
B-WC	119.30	1	119.30	149.02	< 0.0001	
C-GFR	10.02	1	10.02	12.52	0.0041	
BC	7.28	1	7.28	9.09	0.0108	
A^2	48.11	1	48.11	60.10	< 0.0001	
B^2	5.84	1	5.84	7.29	0.0193	
C^2	49.85	1	49.85	62.27	< 0.0001	
Residual	9.61	12	0.80			
LOF	7.46	7	1.07	2.49	0.1666	insignificant
PE	2.14	5	0.43			
Corrected Total	407.99	19				
R^2	0.98	Pred. R^2	0.93	Adeq. Prcn. 30.74		

$$\text{Relative Joint Strength}(\%)_{\text{TIG}} = -333.30 + 7.37A - 3.1360B + 69.06C$$
$$+ 0.10BC - 0.06A^2 + 0.02B^2 - 4.26C^2 \tag{3.5}$$

The perturbation plot of Relative Joint Strength (%) – TIG in Figure 3.6a depicts the impact of all specified welding parameters on relative joint strength (%) for TIG welding. It is observed that welding speed influences relative joint strength (%) in such a way that it has a favorable influence at lower speeds but a negative influence at higher speeds. The limiting value lies between a low speed that produces excessive interaction time between the arc and the base metal, resulting in a broader and deeper HAZ, and a high speed that results in less interaction time and reduced melt deposition. The graph depicts how welding current increases relative joint strength (%). As the welding current increases, more heat is introduced into the weld zone, resulting in more base and filler material melting and a sound weld. The pattern for the gas flow rate is similar to that of welding speed, where relative joint strength (%) initially increases and then falls with the gas flow rate. Low flow rates might result in weld defects as the weld pool is not sufficiently shielded. When the flow rate is increased to a significant level, the flow becomes more turbulent, maximizing the risk of contamination of the weld. Figure 3.6b depicts a cube plot of Relative Joint Strength (%) – TIG for TIG welding at various input variable combinations.

Relative Joint Strength (%) – pTIG: Based on the fit summary analysis using "Sequential Model Sum of Squares", the quadratic model is chosen as the highest-order polynomial with significant model terms for Relative Joint Strength (%) – pTIG. Table 3.7 shows the ANOVA findings for Relative Joint Strength (%) – pTIG. The model has an F-value of 50.98, indicating that it is significant. The model terms A, B, C, BC, A^2, B^2, and C^2 are significant for the Relative Joint Strength (%) – pTIG model.

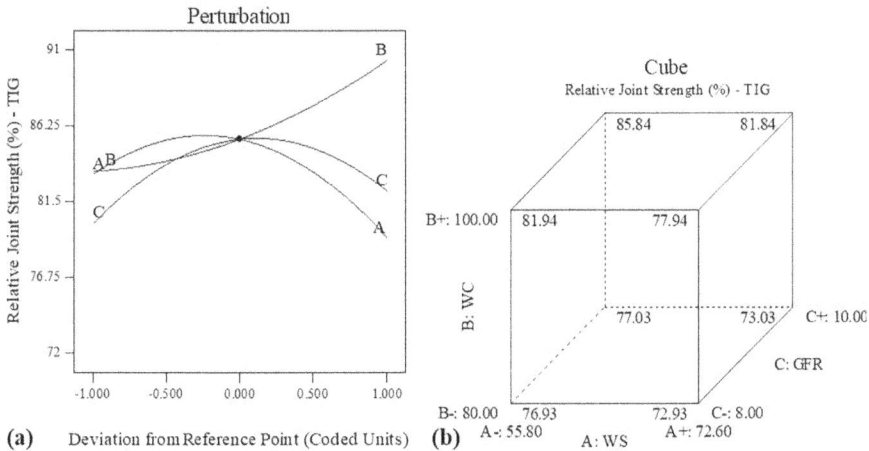

FIGURE 3.6 (a) Perturbation plot and (b) cube pot of Relative Joint Strength (%) – TIG.

TABLE 3.7
ANOVA Results for Relative Joint Strength (%) – pTIG

Source	SS	DF	MS	F-value	p-Value	
Model	381.93	7	54.56	50.98	< 0.0001	significant
A-WS	47.65	1	47.65	44.53	< 0.0001	
B-WC	126.81	1	126.81	118.49	< 0.0001	
C-GFR	14.76	1	14.76	13.79	0.0030	
BC	7.09	1	7.09	6.62	0.0244	
A^2	21.72	1	21.72	20.30	0.0007	
B^2	3.45	1	3.45	3.22	0.0979	
C^2	61.02	1	61.02	57.02	< 0.0001	
Residual	12.84	12	1.07			
LOF	7.34	7	1.05	0.95	0.5404	insignificant
PE	5.50	5	1.10			
Corrected Total	394.77	19				
R^2	0.97	Pred. R^2	0.95	Adeq. Prcn.	26.14	

The "LOF" of 0.95 signifies that the lack of fit is not significant, which is desired. The "Pred. R^2" of 0.95 is in good accordance with the "R^2" of 0.97 and near unity, indicating appropriate data fitting. An "Adeq. Prcn." of 26.14 shows that the signal is adequate. The derived quadratic model of Relative Joint Strength (%) – pTIG in actual parameter values is given below:

$$\text{Relative Joint Strength}(\%)_{\text{pTIG}} = -310.80 + 4.85A - 2.51B + 77.53C$$

$$+ 0.09BC - 0.04A^2 + 0.01B^2 - 4.71C^2$$

(3.6)

The Relative Joint Strength (%) – pTIG perturbation plot in Figure 3.7a displays the influence of all specified welding parameters on relative joint strength (%) for pTIG welding. Because of the comparable phenomena associated with pTIG welding for those parameters, all of the parameters follow the same pattern as TIG welding. The ANOVA results and perturbation plots for TIG and pTIG relative joint strength (%) show that welding current has the greatest influence, followed by welding speed and gas flow rate within their respective ranges. Figure 3.7b depicts a cube plot of Relative Joint Strength (%) – pTIG with different input variable combinations.

The relative joint strength (%) for TIG and pTIG welding is shown in Figure 3.8. For all twenty experimental runs, the relative joint strength of pTIG-welded samples is greater than TIG. Although the relative joint strength attained by pTIG and TIG differs in magnitude, they follow nearly the same pattern in rise and fall as acquired with the same set of parameters for pTIG and TIG, except for a constant value of pulse frequency for pTIG. In general, weld strength is proportional to weld

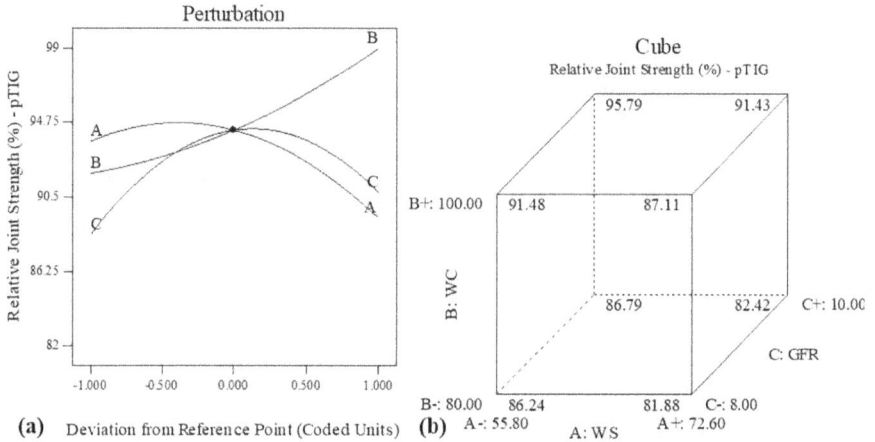

FIGURE 3.7   (a) Perturbation plot and (b) cube pot of Relative Joint Strength (%) – pTIG.

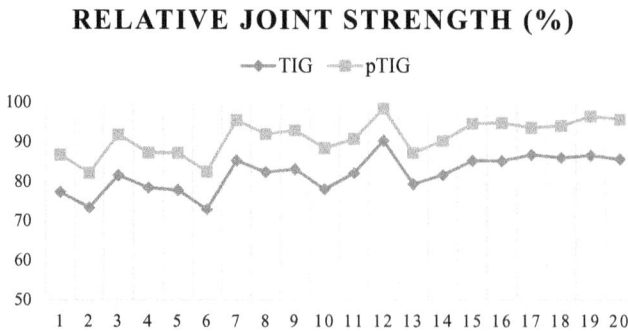

FIGURE 3.8   Relative Joint Strength (%) comparison for TIG and pTIG welding.

deposition. However, in this investigation, it was found that pTIG resulted in lesser deposition than TIG for the same set parameter values but produced greater weld strength than TIG. The pulsed thermal cycle and refined heat input in pTIG may have had an impact on the mechanical characteristics of the HAZ. Because all the welded samples failed from HAZ, heat input may have had a substantial influence in determining joint strength. Because HAZ reduces weld strength, the pTIG welding method is used to lower HAZ.

3.4 CONCLUSION

1. Within the conditions adopted in this work, the weld deposition is higher for TIG welding than for pTIG welding.
2. The relative joint strength of pTIG welding is found to be greater than that of TIG welding, which is achieved up to 97% within the chosen design space.
3. Welding current has a favorable impact on weld deposition, but welding speed and gas flow rate have a negative impact on weld deposition.
4. Welding current has a positive influence on relative joint strength; however, welding speed and gas flow rate have a negative impact on weld deposition at lower values and a positive impact at higher values.
5. Welding speed has the largest influence on weld deposition, followed by welding current and gas flow rate within their respective limits.
6. Welding current has the largest influence on relative joint strength, followed by welding speed and gas flow rate within their respective ranges.
7. The second-order polynomial equations established for weld deposition and relative joint strength for TIG and pTIG welding are checked for adequacy using ANOVA and found to be adequate in all cases.

REFERENCES

Acherjee, B., D. Misra, D. Bose, S. Acharyya. 2011. "Optimal process design for laser transmission welding of acrylics using desirability function analysis and overlay contour plots". *International Journal of Manufacturing Research* 6 (1), 49–61, doi: 10.1504/IJMR.2011.037913.

Acherjee, B., D. Maity, and A.S. Kuar. 2017. "Parameter optimization of transmission laser welding of dissimilar plastics using RSM and flower pollination algorithm integrated approach". *International Journal of Mathematical Modelling and Numerical Optimisation* 8 (1), 1–22, doi: 10.1504/IJMMNO.2017.083656.

Chennaiah, M.B., P.N. Kumar, and K.P. Rao. 2015. "Effect of pulsed TIG welding parameters on the microstructure and micro-hardness of AA6061 joints". *Journal of Material Sciences and Engineering* 4, 1000182, doi: 10.4172/2169-0022.1000182.

Deepak, P., M.J. Jualeash, J. Jishnu, P. Srinivasan, M. Arivarasu, R. Padmanaban, and S. Thirumalini., 2016. "Optimization of process parameters of pulsed TIG welded maraging steel C300". *IOP Conference Series: Materials Science and Engineering* 149, 012007, doi: 10.1088/1757-899X/149/1/012007.

Gao, X.L., L.J. Zhang, J. Liu, and J.X. Zhang. 2013. "A comparative study of pulsed Nd:YAG laser welding and TIG welding of thin Ti6Al4V titanium alloy plate". *Materials Science and Engineering A* 559, 14–21, doi: 10.1016/j.msea.2012.06.016.

Khanna, O.P. 2008. *"A Textbook of Welding Technology"*. Dhanpat Rai Publications (P) Ltd., New Delhi, ISBN: 9788189928360.

Kulkarni, A., D.K. Dwivedi, and M. Vasudevan. 2018. "Study of mechanism, microstructure and mechanical properties of activated flux TIG welded P91 Steel-P22 steel dissimilar metal joint". *Materials Science and Engineering A* 731, 309–323, doi: 10.1016/j.msea.2018.06.054.

Ogundimu, E.O., E.T. Akinlabi, and M.F. Erinosho. 2019. "Comparative study between tig and mig welding processes". *Journal of Physics: Conference Series* 1378 (2), 022074. doi: 10.1088/1742-6596/1378/2/022074.

Ramkumar, K.D., R. Ramanand, A. Ameer, K.A. Simon, and N. Arivazhagan. 2016. "Effect of post weld heat treatment on the microstructure and tensile properties of activated flux TIG welds of Inconel X750". *Materials Science and Engineering A* 658, 326–338, doi: 10.1016/j.msea.2016.02.022.

Ramkumar, T., M. Selvakumar, P. Narayanasamy, A.A. Begam, P. Mathavan, and A.A. Raj. 2017. "Studies on the structural property, mechanical relationships and corrosion behaviour of Inconel 718 and SS 316L dissimilar joints by TIG welding without using activated flux". *Journal of Manufacturing Processes* 30, 290–298, doi: 10.1016/j.jmapro.2017.09.028.

Raveendra, A., and B.R. Kumar. 2013. "Experimental study on pulsed and non-pulsed current TIG welding of stainless steel sheet (SS304)". *International Journal of Innovative Research in Science, Engineering and Technology* 2(6), 2337–2344.

Subhasmita, M. and A.M. Mohanty. 2016. "Performance analysis of TIG welding on Al alloy by using Taguchi method", *International Journal of Multidisciplinary Research and Development* 3, 05–08.

Susmitha, J.D., V.M. Ram, and K. Srinivas. 2016. "Analysis of TIG welding process on mechanical properties and microstructure of AA6063 aluminum alloy joints". *International Journal on Recent and Innovation Trends in Computing and Communication* 4(4), 623–626, doi: 10.17762/ijritcc.v4i4.2088.

Ugla, A.A. 2016. "A Comparative study of pulsed and non-pulsed current on aspect ratio of weld bead and microstructure characteristics of AISI 304L stainless steel". *Innovative Systems Design and Engineering*, 7(4), 88–98.

Yan, G., M.J. Tan, A. Crivoi, F. Li, S. Kumar, and C.H.N. Chia. 2017. "Improving the mechanical properties of TIG welding Ti-6Al-4V by post weld heat treatment". *Procedia Engineering* 207, 633–638, doi: 10.1016/j.proeng.2017.10.1033.

4 Processing of Bimetallic Steel–Copper Joint by Beam Welding

G. R. Joshi and Raghavendra Darji

4.1 INTRODUCTION

Welding of the bimetallic material system is challenging due to its diversified metallurgical, mechanical, thermal, and chemical properties. Even though industry and academia are interested in this area, it is attributed to the techno-economic advantages of dissimilar material joints [1]. Therefore, bimetallic material joints are rapidly employed in different industrial sectors [2]. Bimetallic joints of copper (Cu) and stainless steel (SS) are constrained by diversified properties such as melting point, thermal conductivity, strength, and electrical conductivity. Hence, an amalgamation of these properties is essential for the structural component where service conditions demand high-temperature resistance and strength. SS can withstand repetitive thermal stress, wherein Cu reduces the thermal energy concentration. These reduce or eliminate the adverse effects of repetitive heating and cooling on the structural component. Therefore, the efficient joining of Cu with SS is suitable as a primary component material for the first wall and divertor [3] in the international thermonuclear fusion experimental reactor (commercial energy production unit).

On the other hand, Cu and SS joints are also applicable in the ultra-high vacuum system, large hadron collider, plan wave transformer linac structure, and heat exchangers [4–9]. Apart from particle accelerators and fusion technology, the Cu to SS joint can be useful in electronics [10]. Hence, a sustainable joining solution is necessary to enhance the joint integrity of Cu and SS.

Cu and Fe's limited solubility is the main challenge while joining them [11–14]. It results in an individual solid solution upon solidification after melting. The formation of a separate solid solution affects the weld joint integrity, but it depends on the joint area's compositional gradient. It means a sound joint can be achieved through an adequate degree of intermixing of both materials. The widespread diffusion of both Cu and Fe without presenting defects or discontinuities must join both materials. However, limited mutual solubility (of Cu and Fe) is a contributing

DOI: 10.1201/9781003327769-4

factor for defects or discontinuities. Prolonged solidification time promotes the longer liquid-to-solid interaction of limited soluble materials (Cu and Fe). It will create a heterogeneous distribution of thermal stress, leading to a higher chance of cracks and pore defects. The high heat intensity of the beam forms efficient turbulent flow, leading to widespread material within the joint area, and the increased cooling rate reduces the liquid–solid interaction time. Hence, Laser Beam Welding (LBW) and Electron Beam Welding (EBW) are the right candidates for joining Cu/SS bimetallic materials. On the other hand, these processes can serve the qualitative joint at high speed. Additionally, thicker material can be welded in a single-pass configuration. Similarly, surface oxidation can be avoided with EBW due to the vacuum weld environment. Referring to advantages mentioned above and the commercial interest of the joint, researchers have studied the joint formation of Cu/SS dissimilar material while beam welding [15–41] (see Table 4.1). The detailed review of the work done so far on LBW and EBW is discussed separately in subsequent sections.

4.2   LASER BEAM WELDING OF SS-COPPER

The study of LBW for the Cu/SS bimetallic system (see Table 4.1) can be divided into two sections: (1) articles focused on the microstructure formation [15, 16, 25, 35] and (2) articles focused on identifying the effect of the parameters to the mechanical properties [26, 27, 29, 30, 36]. Shuhai Chen et al. [35] have reported the detailed microstructure formation mechanism. The LBW of Cu/SS has been categorized into two distinct modes: welding-brazing and fusion welding [35]. The terms are associated with the laser beam position [25–27, 29, 30, 35, 36]. However, the role of laser heat is also vital. In the laser brazing mode, Cu remains solid, and the melting of SS is obtained by laser heating. On the other hand, in fusion welding mode, melting of both materials is experienced. It indicates that the welding mode depends on the laser power, laser beam diameter, laser beam offset, and focusing parameters. The mentioned parameters govern the welding heat. It means that the interaction of Cu and laser beams, considering laser power and beam focusing points, determines the mode of welding. The conductive heat magnitude was a critical aspect when the laser beam shifted completely toward the SS side while choosing welding mode. Hence, the study of heat input and welding temperature is necessary to categorize the mode. The available literature (see Table 4.1) is about the effect of the parameters on the mechanical properties and formation of microstructures.

Shuhai Chen et al. have reported the detailed microstructure formation mechanism [35]. The effect of both welding-brazing and fusion welding modes on the microstructure formation has been explained in detail. While in the welding-brazing mode, the Cu base material was not affected much and did not show any melting signs. However, grain coarsening presented the recrystallization of the region at Cu HAZ (Figure 4.1). The liquid SS is also solidified rapidly at the Cu/Fusion Zone (FZ) interface, i.e., attributed to Cu's high thermal conductivity. Hence, the rough interface is observed at the Cu side (Figure 4.1). The efficient turbulent flow imposes the deformation at the interface and the conduction of heat toward Cu, i.e., the rough interface. Similar results are also published elsewhere [15, 16, 25–27, 29, 30, 35, 36]. The Cu interface and melted SS's direct interaction can be confirmed through the

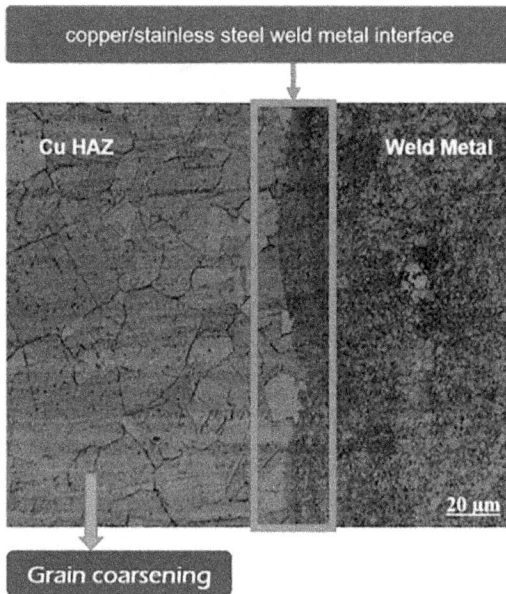

FIGURE 4.1   Micrograph of a laser beam-welded copper to stainless steel joint [34].

EDS analysis [35]. The author [35] reports that the mentioned interface and limited diffusion of Cu into FZ are the reasons for higher mechanical properties. It indicated a strong metallurgical bond between both materials; however, the article did not report the mechanical properties.

On the other hand, the fusion welding mode presented a heterogeneous composition within the weld area. It is the effect of both diffusion and melting. No doubt the degree of melting is higher in the SS than that of Cu, i.e., the higher thermal conductivity of Cu.

In the fusion welding zone, the quantity of mixed Cu in the weld zone is higher. In addition to that, the formation of spherical particles from Cu and SS is observed. It is the sign of liquid separation, and the mentioned article explained it in two sections: (1) primary and (2) secondary (Figure 4.2). The two immiscible liquids, namely, Cu and Fe, separated upon cooling in the primary separation. On the other hand, spherical particle formation indicates the secondary separation, which is solid Fe and liquid Cu. This phenomenon is reported as supersaturation under the action of rapid cooling. The supersaturation will end up with fully solidified immiscible materials (Cu, Fe). The solidification is in the form of matrices and spheres. The Cu spheres within the SS matrix at FZ and the SS spheres within the Cu matrix interface at the interface of Cu/weld metal can be achieved through primary separation (Figure 4.2). However, the quantity of diffused materials plays a vital role in either spheres or matrix formation. The primary separated liquid was further separated due to the supercooling within the boundaries of the miscibility gap. Hence, the area with a larger Cu volume in the primary separated liquid promotes the spherical particles of the SS. This kind of behavior of materials (Cu, SS) is in line with the lever rule: SS spheres form when

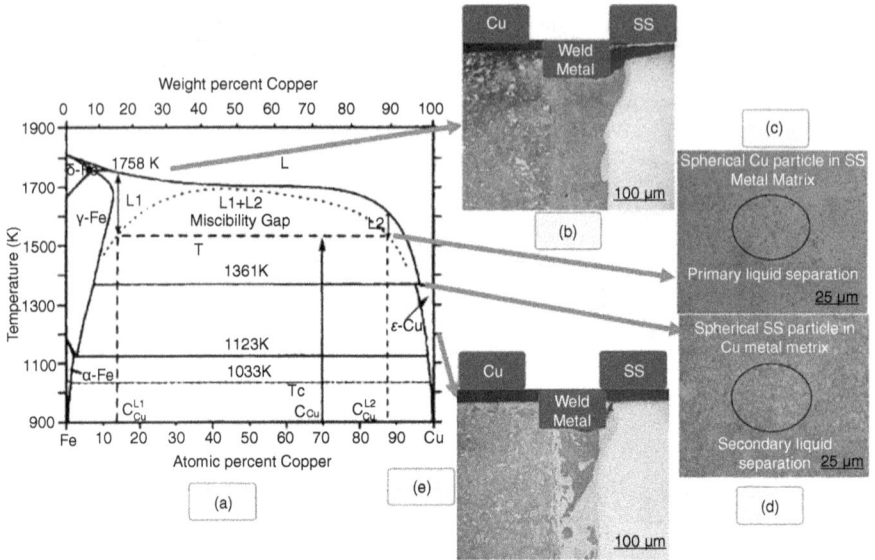

FIGURE 4.2   (a) Fe-Cu phase diagram [42] and liquid separation phenomenon (b–e) [34] of laser beam-welded copper/stainless steel bimetallic joint.

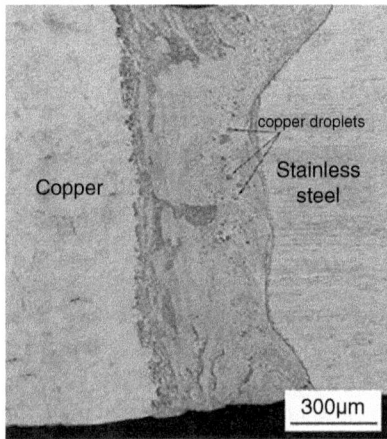

FIGURE 4.3   Micrograph of a Cu/SS joint welded by LBW in fusion welding mode [35].

the Cu participation in solution is more than 50%, and Cu spheres form when the Cu participation is less than 50%. No doubt, when observing the microimage (Figure 4.3), it is clear that the formation of spheres from SS is at the adjacent Cu/weld interface, where the materials experience rapid cooling action. Moreover, the secondary separation is more dominant adjacent to that of the Cu/weld metal interface, and primary separation is being observed at the FZ. The above discussion directly concludes that the cooling rate governs: (1) the separation mechanism during the miscibility gap, (2) atomic diffusion of the immiscibility material, and (3) weld interface shape.

Based on the discussion mentioned above, it is safe to say that the microstructure formation suggests metallurgical bonding despite limited solubility. Herein, the welding mode is responsible for the degree of Cu and SS intermixing. The mode of welding has been exploited up until now by shifting the laser beam toward either side of the joint line [25–27, 29, 30, 35, 36]. The literature concludes that restricted mixing of Cu and Fe improve joint integrity. The suppression of the Cu material within the weld area is especially claimed [25–27, 30, 35, 36] to obtain a better joint.

In contrast, the laser beam offset toward the Cu side reported good joint quality [25, 27, 29]. However, Chengwu Yao et al. [36] observed the significant suppression of Cu melting after increases in laser beam offset toward the SS side from 0.5 to 1 mm, as shown in Figure 4.4. It is attributed to the thickness of the material being welded (see Table 4.1). There is no doubt that the presented mechanical properties align with the work done in the area (see Table 4.1). Additionally, the weld area decreases from face to root, indicating the thermal gradient across the central thickness. The intervention of SS into the Cu interface confirms the turbulent busting. It concludes the more extended liquid stage of Fe compared to Cu. Such a liquid phase of Fe eliminates the possibility of a Cu attack on its grain boundaries at the HAZ of SS. It eventually prevents solidification cracking.

The phenomenon mentioned above of primary and secondary liquid separation and its formation of spherical particles of Cu and SS is limited [15].

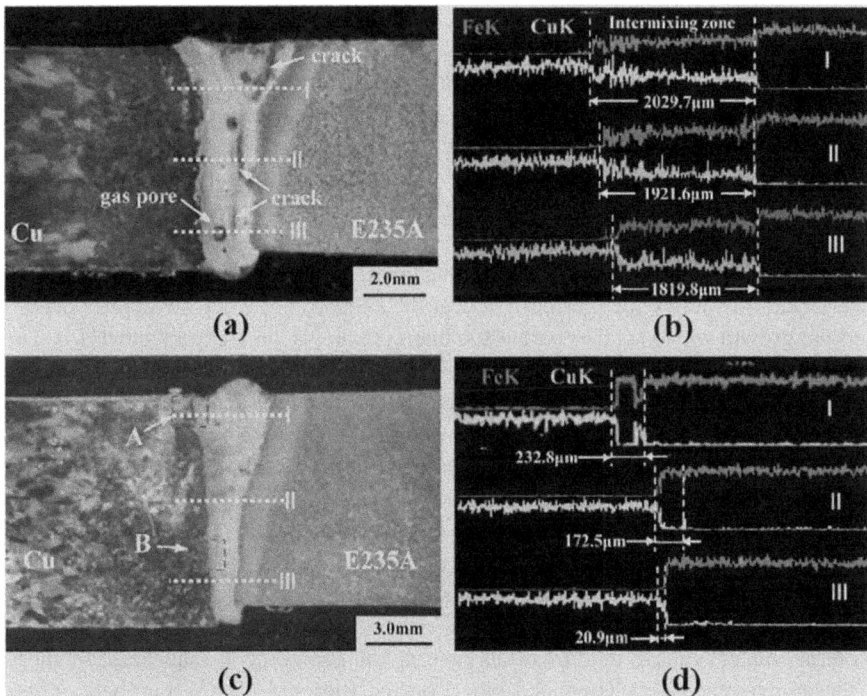

FIGURE 4.4 Macrographs of material welded by LBW at offset toward the SS side: (a) 0.5 mm, (b) 1 mm, and corresponding EDS analyses (c) and (d) [36].

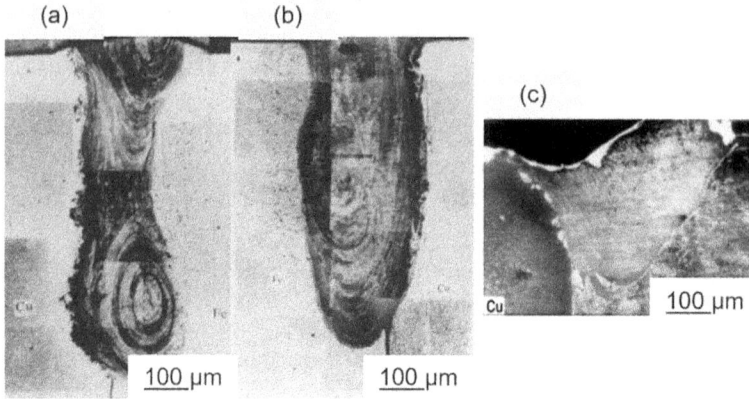

FIGURE 4.5 LBW of Cu/SS at keyhole mode (a and b) and conduction mode (c) [15].

Gandham Phanikumar et al. [15] attempted to make the microstructure of the joint free from welding modes such as welding-brazing and fusion welding. Several experiments ranging from the keyhole mode to the conduction mode of LBW have been attempted. They have welded and presented LBW results for Cu/SS at three specific parametric conditions (see Table 4.1). The weld's depth is significantly increased when moving from the conduction to the keyhole mode of welding (Figure 4.5). However, the presented microstructural results talk only about conduction modes. The discussion on keyhole mode is not shown in the mentioned article. The microstructural band or compositional gradient is significant at the Cu/weld metal interface compared to those at the SS/weld metal interface [15]. In the present case, the melting of Cu is suppressed during conduction mode, i.e., the reason for the presented interface. The mentioned article studied the thermodynamic situation at the interface to show the distinctions in composition. These distinctions or microstructural bands at the weld interfaces are attributed to the chemical potential and the peritectic reaction at the SS/weld and Cu/weld metal interfaces. The growth of grains starts at the interface, followed by the thermal effect at the Fe side. The microstructure's overall cellular growth with a smaller part of the intercellular region is experienced [15]. The welding mode suggests that the quantity of Cu particles decreases when moving from the Cu/weld metal interface to the SS/weld metal interface. It is the sign of microstructural banding that includes the compositional gradient as well. Furthermore, the fine cellular structure supports the claim of better intermixing and mechanical properties. However, mechanical properties are not reported.

B. R. Moharana et al. [26] studied the effect of limited interaction (see Table 4.1) between the laser beam and Cu material on joint behaviors. The interfaces of joints are in line with the published work elsewhere [15, 16, 25, 27, 29, 30, 35, 36]. However, the reason for the uneven curvature is not only thermal conductivities or melt patterns. There is more than this that governs the geometry of interfaces. In short, a thermodynamic study is needed based on the Biot and Nasselt numbers, which can shed some light on the formation of the shape of the interface. However, thermodynamic research has been done earlier [15] on the compositional gradient

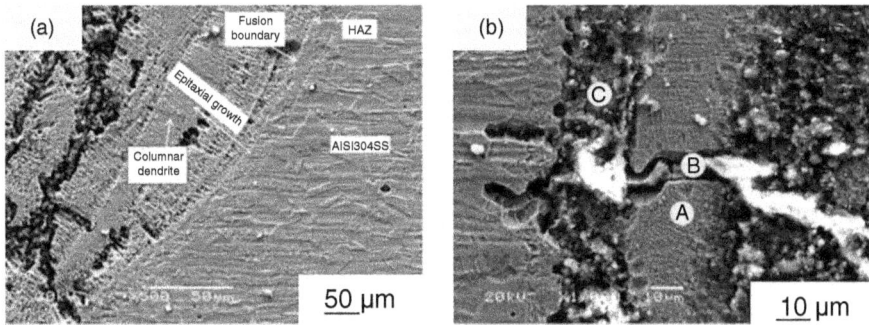

FIGURE 4.6 (a) Welding pool growth adjacent to the fusion boundary and (b) micro-channel [26].

near the interfaces. The epitaxial growth leading to a columnar dendrite is obtained near the SS/weld metal interface (Figure 4.6a). This microstructural feature indicates better joint strength (see Table 4.1). However, the microchannel (Figure 4.6b) is reported, and its effect on static strength is adverse. The reason for the defect can be confirmed through the EDS analysis. The EDS results show a higher quantity of Cu, followed by an interruption of Fe particles in it. The epitaxial growth of grains surrounds the channel. Upon close analysis, it is clear that this defect is formed due to the thermal stress mismatch, which is predictably due to the prolonged interaction of immiscible liquid and solid material. It further makes the surface tension criterion complex. Above all, the rejection of Cr from the dendritic structure results in carbide precipitation. Hence, the microchannel consists of SS at the outer boundary, followed by carbides and Cu at the core. This compositional gradient and degree of undercooling, along with surface tension, promote the miscibility gap. For the experiments to determine the surface tension criterion for Cu and SS, the reader can refer to the published literature [43, 44]. The exploited parametric conditions (see Table 4.1) result in good mechanical properties. However, the parametric situation for the presented results is not exact. On the other hand, the work on heat transfer and welding modes is still in its future scope.

The intermixing and microstructure of materials govern better joint integrity during LBW of Cu/SS bimetallic material. The weld quality is measured on the scale of metallurgical and mechanical properties. The metallurgical properties and mechanical properties are direct reflections of the process parameters. Hence, it is necessary to identify the desired joint parameters. The discussion above talks about microstructure formation. the following discussion highlights the effect of the parameters on the mechanical properties. The parameters such as laser power [27, 29], beam waist position [29], pulse duration [29], laser scanning speed [16, 25–27, 29], laser beam offset [25, 27, 29, 30, 35, 36], and laser oblique angle [27] have been varied to know their effect on the joint formation.

The laser power is the primary source of heat during LBW. Hence, the heat increases with an increase in laser power, keeping other parameters constant. The higher power promotes porosity and groove formation at the weld's face [29]. The mentioned problem can be overcome by increasing pulse duration while keeping laser power constant. The higher pulse duration suppresses the heat transfer of the laser into the weld, resulting in the material's restricted vaporization at the face. Subsequently, weld penetration is also suppressed [29]. Interestingly, a higher laser power weld groove is not observed [15, 16, 26, 27, 29, 30, 35, 36, 37, 45]. It is attributed to the higher welding speed than the work done earlier [29]. However, the fixed-parameter set with the variable laser power is not reported. By comparing the joints' macrofeature, it is predicted that the heat input governs the groove formation. At the same time, full penetration can be obtained by setting optimum laser power and weld speed. Weld penetration is a function of laser power [15, 16, 26, 27, 29, 30, 35, 36, 45, 46].

The heat input and weld penetration increase as the welding speed decreases, keeping other parameters constant [25, 27]. Hence, the weld penetration follows the same trend as the heat input. Interestingly, contradiction is shown in the trend of weld penetration and weld width in the published work elsewhere [25, 27]. The detailed experimental work is needed to know the real effect of heat input and welding speed on the weld width [29]. Weld penetration is the direct outcome of employing laser power; 3000 W in the present case [25, 27] compared to nearly 500 W in the article published earlier [29]. As mentioned earlier, the groove-like feature at the face is not observed in this study and is attributed to other set parameters, peimarily high laser scanning speed [15, 25, 27]. However, thermal shrinkages leading to undercutting at the weld's face are presented at 30 m/min welding speed [25]. The excessive speed promotes a high solidification rate, which solidifies the surface of the joint, followed by the subsurface. It creates a distinct surface tension at the boundary of liquid and solid [43, 44]. It apparently promotes the solidification of the subsurface from the liquid phase. This cooling pattern shrinks material at the surface, which is the reason for the undercut.

Similarly, porosities or solidification cracking may arise due to Cu diffusion at or near the weld surfaces or region (precisely weld interfaces) in higher quantities. Immiscibility between Cu and Fe materials plays a significant role in defect formation. Interestingly, at 15 m/min weld speed, the macro defects are not present. However, the microstructural feature is not discussed in the mentioned study. Simultaneously, microhardness results do not show signs of martensitic formation [25]. The tensile properties are not reported in the study. The combination of independent variable study parameters is not mentioned in the article [29]. At the same time, penetration is limited due to the lower heat input. Therefore, the double-side LBW has been carried out.

The detailed study on the different parameters and their effect on the joint properties and the reasons associated with them is not precise. However, the article studied the various parameters listed in Table 4.1 and obtained the highest joint properties more than the softer base material, i.e., Cu. The joint has passed the assessment of the vacuum test. However, the testing condition was unknown. Similarly, the laser beam's offsetting toward the Cu side is recommended to avoid solidification cracking due to both Cu and Fe's immiscibility, as mentioned earlier [29]. Equal weld penetration was reported when the laser beam shifted toward the Cu side without any concrete evidence [29]. The reported penetration feature may be applicable for a particular laser power or set of parameters. The same is not mentioned elsewhere;

shifting the laser beam toward the Cu side does not create any significance for weld penetration [25–27, 30, 35, 36]. It concludes that the other parameters, such as beam diameter, laser power, and welding speed, are vital. However, the comparison can be made based on the heat input calculations.

The amount of the Cu material within the joint can also be identified through the results of hardness [15, 16, 25–27, 29, 30]. In other words, laser beam offset and welding mode can be predicted through the hardness results. The significant fluctuation in the hardness value may be due to the laser beam focused at the joint line. On the other hand, hardness fluctuation is relatively low when Cu participation within the weld material is high, i.e., a laser beam is offset toward the Cu side. The intermediate hardness variation is observed in welding-brazing mode. Moreover, the lowest hardness value corresponds to the Cu HAZ, and this is the reason for the fracture initiation from this area.

The fracture locations reported so far are Cu BM, Cu HAZ, interface, and FZ [15, 16, 25–27, 29, 30, 35, 36]. The incomplete fusion resulting from the too-high laser power of the beam offset promotes the fracture at the interface. In this case, deformation at Cu interfaces while tensile loading is marginal. Furthermore, the extensive participation of Cu particles within the weld material promotes the stress concentration at FZ while tensile loading, leading to fracture in the area. The appropriate degree of mixing shows the fracture, either from Cu HAZ or from Cu base material.

The unique technique of LBW with MIG to weld Cu/SS bimetallic material is exploited [37]. The reported microstructure formation is in line with the work done earlier [35]. The Fe-rich area near the SS/weld metal interface and distributed Fe particles are proofs of liquid separation, as illustrated earlier (Figure 4.2). It concludes that fusion welding of dissimilar materials or alloys results in elemental separation or microsegregation at the fusion welding zone [35]. On the other hand, significant melting restrains the Fe detachment from the Fe blocks. It will have an adverse effect on the joints' elongation and, apparently, on their tensile properties. The report says [35] that eliminating the Fe blocks and promoting the Fe-rich small spherical particles within the Cu metal matrix at the weld area increases the tensile property marginally. It can be done by shifting the laser beam torch toward the Cu side or avoiding the laser beam's direct contact with SS. It will promote the melting of SS through heat conducted through Cu. In this case, the melting of SS was suppressed significantly. Simultaneously, the Fe particles detached from the interface and diffused into the weld area. However, an extension of such particle distribution is not clear from the study [35]. The uniform hardness distribution suggested that the particle distribution within a limited space compared to fusion welding [37]. There is no doubt that the amount of melted SS decreases with an increased laser beam offset toward the Cu side. Table 4.1 incorporates the details of the experimental study [37]. The highest joint property is obtained when the laser beam shifts toward the Cu side by 0.5 mm. However, the tensile property in the mentioned article still has scope for improvement. The work can also be done to optimize the MIG and LBW parameters to achieve better joint properties.

By reviewing the work summarized in Table 4.1, it is clear that good joints can be achieved by shifting the laser beam on the either side. Interestingly, positioning the beam at the joint line results in better joint properties. No doubt, the highest tensile strength is reported when the laser beam shifts toward the Cu side by 0.4 mm. However, the aforementioned double-sided welding has been done to obtain full penetration and a groove defect with porosities. From the obtained results by all the researchers, it is safe to predict that the quantity of Cu in the joint is not the measure for the defect-free sound joint. Still, the appropriate melting of both materials and their diffusion within the weld area is desired. The intermixing ratio of the material should be maintained by shifting the laser beam toward the SS side. The considerable exposure of the laser beam to SS base material (when shifted toward the SS base material side) melts SS more. Simultaneously, it saves the energy that is wasted due to the high reflectivity and thermal conductivity of Cu. The laser beam can also be focused at the joint line with an appropriate oblique angle toward the SS. It can serve the purpose of reducing laser contact with Cu, and it can also reduce the degree of heat conduction. That can help the diffusion of Cu efficiently compared to that of the welding-brazing mode. However, work should be done to identify the quantity of melting and intermixing of both materials related to the heat input and cooling rate.

4.3 ELECTRON BEAM WELDING OF STEEL–COPPER

Owing to advantages such as high heat density, appropriate beam positioning, short HAZ, and most importantly, contamination-free weld formation, EBW is the competitive joining technique for bimetallic combinations such as Cu/SS [46]. Sun and Karppi [47] presented the scope of EBW for dissimilar materials. The report [47] also indicates the advantages of EBW over conventional fusion welding technology. Unlike the LBW, the EBW eliminates reflectivity while welding highly reflective materials like Cu and Al. However, successful LBW for Cu/SS dissimilar material has been demonstrated by employing appropriate parameters (see Table 4.1).

On the other hand, the necessity of a vacuum environment restricts the joint geometry. Despite this fact, the EBW of Cu/SS was exploited previously and summarized in Table 4.1. It is clear that while EBW, positioning of the beam at the joint line or either side of the joint is attempted to optimize the degree of intermixing. On the other hand, thickness ranging from 1.2 to 71 mm has been welded. Obviously, with an increase in thickness, the parameter was exploited, indicating an increase in heat input. Undoubtedly, the researcher employed the single-pass and double-pass welding methods to achieve full penetration [38]. They observed that the heterogeneity increased with the increased thickness of the joint. It can be attributed to the diversified properties and thermal gradient throughout the joint's cross-section while welding. Hence, the thermal gradient promotes a faster cooling rate or uneven solidification. It restrains the filling of microfissures by Cu, resulting in microcracks and porosities. Additionally, the penetration of Cu into steel HAZ and the lack of full penetration on either side may affect the joints' dynamic properties. In contrast, very high current at a lower voltage can mitigate lower penetration for high section thickness during EBW [41].

The Cu/SS bimetallic joint's microstructure after employing the EBW process shows mainly two sections: Fe-Cr-Ni reach and Cu reach phases. It means there is a dominance of microsegregation. It is primarily due to turbulent flow, which promotes widespread diffusion of Cu and Fe adjacent to opposite interfaces, resulting in

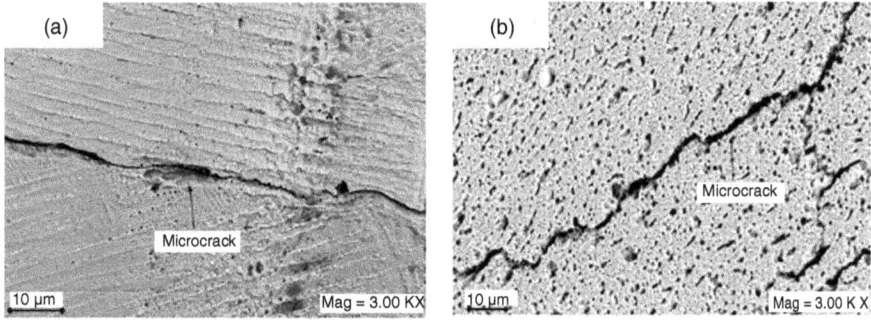

FIGURE 4.7   Microcracks while EBW of Cu/SS at (a) no beam oscillation (b) with 1 mm beam oscillation [20].

microsegregation. Hence, the beam oscillation while the EBW of Cu/SS is exploited [20] reduces or eliminates the degree of separation or microsegregation. The result shows that the oscillation effect does not affect much on the tensile strength at room temperature and elevated temperature. The room-temperature elongation properties under external tensile loading are also marginally affected. However, the percentage elongation at elevated temperatures and impact strength at room temperature were significantly affected. The increase in beam oscillation diameter from 1 to 2 mm shows decreases in impact strength at room temperature and percentage elongation at elevated temperatures. The higher oscillation diameter allows more exposure of the beam toward the Cu, i.e., the reason for extended Cu melting at the weld area. It ends with macro or micro discontinuities because of compositional separation, referring to Cu and Fe's limited mutual solubility at room temperature, which deteriorates the joint properties because of metallurgical heterogeneity in the weld area.

The EDS and the hardness result reported in the mentioned article [20] confirm the same. Microcracks are observed for all the welded joints (Figure 4.7). However, good mechanical properties are reported [20]. Similar results are reported in the article published elsewhere [19]. The sufficient amount of melted Cu in the weld region, i.e., 86% obtained during oscillating beam diameter, is set at 1 mm. It fills the cracks. It is seen that the reason for the higher percentage elongation at elevated temperatures is the high impact strength at room temperature. The lower and higher amounts of Cu within the weld area cause the segregated surfaces to further create cracks. It is the result of a complex thermal stress concentration phenomenon. This shows that the work has to be done in detail to identify the proper degree of intermixing of both materials, as mentioned earlier.

The exploitation of filler material was done to define the intermixing phenomenon [39]. The Cu filler material is exploited along with other parametric conditions (Table 4.1). The study reported a maximum tensile strength of 274 MPa among the available literature on the EBW of a Cu/SS bimetallic joint. The microstructure of the joint reveals a clear, distinct zone, which is the Cu and Fe reach. There is no doubt that Cu is a limited candidate in the weld zone. At the same time, UMZ is formed adjacent to the FZ's Cu/weld metal interface. It is made of Cu filler material and not Cu base material. The parametric condition, especially a beam diameter of

TABLE 4.1
Summary of Literature on Beam Welding of Cu/SS Bimetallic Joint

Substrate	Thickness (mm)	Joint Design	Parameters	Maximum Obtained Properties of the Joint	Reference
			Laser Beam Welding		
Fe/Cu	7	Butt	Laser power (kW): 4.5 & 5.5 Welding speed (mm/min): 1260 & 1500 Laser type: CO_2 Shielding gas: Ar + He Beam diameter (mm): 0.5 (focused) & 2 (defocused) Power calibration factor: 0.778 Laser beam position: At joint line	VHN: 700 HV	[15]
Pure Cu/AISI 304 SS	2	Butt	Laser power (kW): 2.3–2.9 Welding speed (mm/min): 2100–3000 Beam diameter: 400 μm Shielding gas: He Gas flow rate: 17 L/min Laser type: disk Laser beam position: At joint line	VHN: 240–250 HV	[16]
Commercially pure Cu/AISI 304 SS	3	Butt	Laser power (kW): 3.0, 3.5 Scan speed (mm/min): 2000, 3000 Laser beam position: 50 μm toward SS Beam diameter (mm): 0.18 Focal length (mm): 300 Laser beam angle: 90° Gas flow rate: 20 L/min Focal position at the surface	UTS: 190–201 MPa at 0.01 mm/min % EL: 19 VHN: >350 HV	[26]

(Continued)

TABLE 4.1 (Continued)
Summary of Literature on Beam Welding of Cu/SS Bimetallic Joint

Substrate	Thickness (mm)	Joint Design	Parameters	Maximum Obtained Properties of the Joint	Reference
T2 Cu/201 SS	2	Butt	**Fixed parameter** Laser type: CO_2 Beam diameter: 0.2 mm Focal distance: 200 mm **Study on welding speed** Laser power (kW): 3 Beam offset: 0 mm Beam oblique angle: 0° Welding speed (mm/min): 500, 1000, 1500 **Study on laser power** Beam offset: 0.1 mm toward SS Beam oblique angle: 0° Welding speed (mm/min): 1000 Laser power (kW): 1.5, 2, 2.5 **Study of laser oblique angle** Laser power (kW): 3 Welding speed (mm/min): 1500 Beam offset: 0.1 mm toward SS Beam oblique angle: 0°, 2°, 4°, 6°, 8° **Study on beam offset** Laser power (kW): 3 Beam oblique angle: 4° Welding speed (mm/min): 1500 Beam offset (mm): 0, 0.1, 0.2, 0.3, 0.4	Tensile strength: 260 MPa at 0.05 mm/s VHN: 230 HV **Optimum variable for each study** Beam oblique angle: 2° Beam offset: 0.1 mm toward SS side Welding speed: 1500 mm/min Laser power (KW): 2	[27]

(Continued)

TABLE 4.1 (Continued)
Summary of Literature on Beam Welding of Cu/SS Bimetallic Joint

Substrate	Thickness (mm)	Joint Design	Parameters	Maximum Obtained Properties of the Joint	Reference
Electrolytic Cu/SS	1.2	Butt	**Fixed parameter** Laser type: Nd: YAG Beam diameter: 400 μm Laser power (kW): 3 Pulse duration: 3 ms Operating distance: 100 mm Shielding gas: Argon Gas flow rate: 8 L/min Energy input per unit 12 J/mm **Study of beam offset** Beam offset: 100 μm toward steel and Cu, 0 μm **Study of welding speed** Welding speed (mm/min): 0, 15,000, 30,000	VHN: 300 HV **Optimum condition** Beam offset: 100 μm toward Cu side Welding speed: 15,000 mm/min	[25]
T1 Cu/E235A steel	7, 10	Butt	**Common parameters** Beam diameter: 0.7 mm Shielding gas: He Laser type: CO_2 **Sample 1** Laser power (kW): 8 Beam focusing distance: 3 mm below workpiece surface Beam offset: 0.5 mm toward steel side **Sample 2** Laser power (kW): 11 Beam focusing distance: 4 mm below workpiece surface Beam offset: 1 mm toward steel side	UTS: 233 MPa % EL: 29 **Optimum weld**: sample 2	[36]

(Continued)

TABLE 4.1 (*Continued*)
Summary of Literature on Beam Welding of Cu/SS Bimetallic Joint

Substrate	Thickness (mm)	Joint Design	Parameters	Maximum Obtained Properties of the Joint	Reference
T2 Cu/201 SS	2	Butt	Laser power (kW): 3 Welding speed (mm/min): 1500 Laser type: CO_2 Welding head focal distance: 200 mm Oblique angle: 2° toward SS Beam offset (mm): 0, 0.2 toward SS side	**Optimum weld** At 0.2 mm beam offset toward SS side	[35]
OFC Cu/ 316 SS	4.2	Butt	**Fixed parameters** Laser type: Nd:YAG Beam diameter: 550 μm Shielding gas: He Pulse duration (s): 10, 15 Laser beam energy: 3 J Laser power (kW): 0.1–0.6 Welding speed (mm/min): 12–120 Beam offset (mm): 0–0.4 toward Cu side Beam waist position (mm): 0.2 below to 0.6 above the work piece surface	UTS: 312 MPa VHN: 150–170 HV **Optimum condition** Beam offset (mm): 0.4 mm toward Cu side Welding speed (mm/s): 1 Laser power (kW): 0.4 Beam waist position (mm): 0.6 above the workpiece surface	[29]

(*Continued*)

TABLE 4.1 (*Continued*)
Summary of Literature on Beam Welding of Cu/SS Bimetallic Joint

Substrate	Thickness (mm)	Joint Design	Parameters	Maximum Obtained Properties of the Joint	Reference
M1 Cu/ 12Cr18Ni10Ti austenitic steel	3	Butt	Laser type: Fiber Focusing distance (mm): 0, 3,6,9,12,15,18,21 Focusing length (mm): 450 Laser power (kW): 2.5 Welding speed (mm/min): 600 Beam offset (mm): 0.5–1.5 toward steel side	UTS: 270 MPa JE: 93% VHN: 200–230 HV **Optimum condition** Beam offset (mm): 1 mm toward steel side	[30]
			Hybrid Laser MIG Welding Processes		
T2 Cu/ 304 SS	2	Butt	Filler wire type: HS201 Filler wire diameter (mm): 1.2 Shielding gas: Argon Gas flow rate (L/min): 20 Laser type: IPG YLR-6000 fiber laser Beam spot diameter (mm): 0.33 Laser beam wavelength (nm): 1070 Laser power (kW): 3 Welding speed (mm/min): 1000 Laser beam offset: 0, 0.5, 1, 1.5 toward Cu side Laser inclinational angle to the vertical axis: 5 **MIG parameter** Welding current (A): 60° Arc welding, torch inclination angle with vertical axis: 45° Filler wire extension (mm): 1.2 Laser – arc distance (mm): 3	UTS: 215 MPa at 2 mm/min strain rate JE (%): 90 VHN: 160 HV at 0 mm beam offset **Optimum Beam offset** 0.5 mm toward Cu side	[46]

(*Continued*)

TABLE 4.1 (Continued)
Summary of Literature on Beam Welding of Cu/SS Bimetallic Joint

Substrate	Thickness (mm)	Joint Design	Parameters	Maximum Obtained Properties of the Joint	Reference
			Electron Beam Welding		
Joint 1 ETP Cu/304L SS **Joint 2** ETP Cu/304 SS **Joint 3** ETP Cu/316L	**Joint 1** 33 Cu/29 SS **Joint 2** 50 **Joint 3** 71	Butt	**Fixed parameters** Cathode current (A): 2.35 Voltage (kV): 150 Gun vacuum (mbar):10–5 Chamber vacuum (mbar) 10–4 Beam focus point (mm): 0 Welding speed (mm/min): 360 **Joint 1** Power input (kW): 13.5 Beam current (mA): 90 Gun-specimen distance (mm): 320 Welding pass: 1 **Joint 2** Power input (kW): 18 Beam current (mA): 120 Gun-specimen distance (mm): 450 Welding pass: 2 **Joint 3** Power input (kW): 19.5 Beam current (mA): 130 Gun-specimen distance (mm): 575 Welding pass: 2	**Joint 1** VHN: 175 HV **Joint 2** VHN:200 HV **Joint 3** VHN: 300 HV **Optimum weld** Joint 1	[38]

(Continued)

TABLE 4.1 (Continued)

Summary of Literature on Beam Welding of Cu/SS Bimetallic Joint

Substrate	Thickness (mm)	Joint Design	Parameters	Maximum Obtained Properties of the Joint	Reference
QCr0.8 Cu/304 SS	2.7	Butt	Filler material: Cu Filler wire diameter(mm): 1.2 Beam diameter (mm): 0.2 Wire feeding angle: 60° Beam current (mA): 25, 30, 35 Welding speed (mm/min): 100, 200, 300 Wire feed rate (m/min): 1, 1.6, 2 Beam offset (mm): 0.3 toward SS side and toward Cu side, 0	UTS: 274 MPa VHN: 240 HV **Optimum condition** Beam current (mA):30 Beam offset (mm): 0.3 mm toward SS side Wire fees rate (m/min): 1 Welding speed (mm/min): 100	[39]
OFHC Cu/AISI 304L	25	Butt	Beam power (kW): 15 Beam current (mA): 395 Welding speed (cm/s): 0.6 Beam offset (mm): 0	UTS: 210 MPa JE: 98% VHN: 180 HV	[41]

(*Continued*)

TABLE 4.1 (Continued)
Summary of Literature on Beam Welding of Cu/SS Bimetallic Joint

Substrate	Thickness (mm)	Joint Design	Parameters	Maximum Obtained Properties of the Joint	Reference
C10300/AISI 304 SS	3	Butt	Beam voltage (kV): 60 Welding speed (mm/min): 1000 Gun-specimen distance (mm): 460 Welding chamber vacuum (mbar): $<5 \times 10^{-5}$ Welding pass: 1 Beam offset (mm): 0 **Joint 1** Beam oscillation: No Beam current (mA): 65 **Joint 2** Beam oscillation frequency (Hz): 600 Beam current (mA): 73 Beam oscillation diameter (mm): 1 **Joint 3** Beam oscillation frequency (Hz): 600 Beam oscillation diameter (mm): 2 Beam current (mA): 80	Hot tensile strength (MPa): 183 at 0.5 mm/min strain rate Joint efficiency: 97 % EL: 14 Normal tensile strength (MPa): 252 JE: 97 % EL: 19 Bend test: no defect at 180° bending Charpy impact strength: 118 J/cm^2 VHN: 190 HV	[17, 20]
Pure Cu/AISI 304 SS	4	Pipe joint	Beam welding current (mA): 33.75 Beam focusing current (mA): 515 Welding speed (mm/s): 25 Beam offset (mm): 0.1 toward SS	VHN: 180 HV	[18]
Cu/ 304 SS	6	Butt	Beam current (mA): 50 Beam voltage (kV): 60 Welding speed (mm/s): 40 Work Distance (mm): 260 Beam offset (mm): 0	UTS: 243 MPa	[19]

0.2 mm and a beam offset of 0.3 mm toward the SS side, was confirmed. A higher beam offset than the beam diameter toward the Cu side suggests the solid-state of the Cu base material. Moreover, the fine Cu and Fe particles formed due to the high cooling rate contributed to the higher tensile strength.

The article [39] employed the beam offset on either side of the joint line. The beam offset toward the SS side by 0.3 mm presented the best joint quality. There is no doubt that the amount of melted Cu base material is limited, i.e., a higher beam offset toward the Cu side than the beam diameter. However, the EDS results presented elsewhere [16] show the efficient melting of Cu and its participation in the weld area even after the beam is offset toward the SS side. The reason for the phenomenon is predicted to be the interaction of the beam with Cu base material. It cannot be evaluated, i.e., due to a lack of information on beam diameter. On the other hand, Cu filler material stabilizes the situation of intermixing [39]. It can be confirmed through the UMZ of Cu near the Cu/weld interface. The UMZ is also the evidence of the absence of Fe diffusion in the area, which may lead to separation and defects. At the same time, it has been studied that the effect of UMZ on the mechanical and metallurgical properties is adverse [48, 49]. The joint's top surface showed a rich phase of Cu material, attributed to the efficient melting of the filler material. The hardness results are in line with the observation. The cooling rate of the weld area has also increased significantly, and that can be confirmed by the fine grain structure at the reported FZ.

4.4 CONCLUSION

The LBW parameters except oblique beam angle were extensively exploited to join Cu/SS dissimilar material. Similarly, the effect of the reflective nature of Cu on heat loss is not addressed.

The unique upgrade of hybrid LBW-MIG welding claims better intermixing with a decrease in heat input compared to conventional or hybrid arc welding processes.

The LBW of Cu/SS focuses on suppressing the melting of either Cu or SS by offsetting the beam toward the SS and Cu sides, respectively. Similarly, beam focusing parameters are studied to obtain better intermixing by placing a beam at the joint line.

The vacuum environment, while EBW reduces the chance of weld contamination, i.e., joint oxidation.

The electron beam oscillation at 1 mm diameter shows the better intermixing and melting of both Cu and SS materials.

The Cu- and Fe-rich regions within the FZ and spherical Cu and Fe particles signify desirable liquid separation during beam welding.

REFERENCES

1. Martinsen, K.; Hu, S. J.; Carlson, B. E. Manufacturing Technology Joining of Dissimilar Materials. *CIRP Annals*, 2015. https://doi.org/10.1016/j.cirp.2015.05.006.
2. Kumar, N.; Yuan, W.; Mishra, R. S. Introduction. In *Friction Stir Welding of Dissimilar Alloys and Materials*; Elsevier, 2015; pp 1–13. https://doi.org/10.1016/B978-0-12-802418-8.00001-1.

3. Mazul, I. V.; Belyakov, V. A.; Giniatulin, R. N.; Gervash, A. A.; Kuznetsov, V. E.; Makhankov, A. N.; Sizenev, V. S. Preparation to Manufacturing of ITER Plasma Facing Components in Russia. *Fusion Eng. Des.*, 2011, *86* (6–8), 576–579. https://doi.org/10.1016/j.fusengdes.2011.02.022.

4. Li, Y.; Liu, X. *Vacuum Science and Technology for Accelerator Vacuum Systems*; New York: Cornell University Ithaca, 2013.

5. Bertinelli, F.; Favre, G.; Ferreira, L. M. A.; Mathot, S.; Rossi, L.; Savary, F.; Boter, E. Design and Fabrication of Superfluid Helium Heat Exchnager Tubes for the LHC Superconducter Magnets. In *Proceedings of EPAC*; Lucerne, 2004; pp 1837–1839.

6. Bhanumurthy, K. Diffusion Bonding of Nuclear Materials. *BARC Newsl.*, 2013, *331*, 19–25.

7. Merola, M.; Escourbiac, F.; Raffray, R.; Chappuis, P.; Hirai, T.; Martin, A. Overview and Status of ITER Internal Components. *Fusion Eng. Des.*, 2014, *89* (7–8), 890–895. https://doi.org/10.1016/j.fusengdes.2014.01.055.

8. Ghodke, S. R.; Barnwal, R.; Mondal, J.; Dhavle, A. S.; Parashar, S.; Kumar, M.; Nayak, S.; Jayaprakash, D.; Sharma, V.; Acharya, S.; et al. Machining and Brazing of Accelerating RF Cavity. *Proceedings - International Symposium on Discharges and Electrical Insulation in Vacuum, ISDEIV*, 2014, 101–104. https://doi.org/10.1109/DEIV.2014.6961629.

9. Singh, R.; Pant, K. K.; Lal, S.; Yadav, D. P.; Garg, S. R.; Raghuvanshi, V. K.; Mundra, G. Vacuum Brazing of Accelerator Components. *J. Phys. Conf. Ser.*, 2012, *390*, 012025. https://doi.org/10.1088/1742-6596/390/1/012025.

10. Sabetghadam, H.; Zarei Hanzaki, A.; Araee, A.; Hadian, A. Microstructural Evaluation of 410 SS/Cu Diffusion-Bonded Joint. *J. Mater. Sci. Technol.*, 2010, *26* (2), 163–169. https://doi.org/10.1016/S1005-0302(10)60027-8.

11. Velikanova, T.; Turchanin, M. Chromium - Copper - Iron. In *Iron Systems, Part 3*; Springer Berlin Heidelberg: Berlin, Heidelberg; 2008, pp 88–128. https://doi.org/10.1007/978-3-540-74199-2_6.

12. Shirinyan, A.; Wautelet, M.; Belogorodsky, Y. Solubility Diagram of the Cu-Ni Nanosystem. *J. Phys. Condens. Matter*, 2006, *18* (8), 2537–2551. https://doi.org/10.1088/0953-8984/18/8/016.

13. Yang, B.; Guo, C.; Li, C.; Du, Z. Experimental Investigation and Thermodynamic Modelling of the Fe-Ni-Ta System. *J. Phase Equilibria Diffus.*, 2020, *41* (4), 500–521. https://doi.org/10.1007/s11669-020-00807-3.

14. Jucken, S.; Martin, M. H.; Irissou, E.; Davis, B.; Guay, D.; Roué, L. Cold-Sprayed Cu-Ni-Fe Anodes for CO_2-Free Aluminum Production. *J. Therm. Spray Technol.*, 2020, *29* (4), 670–683. https://doi.org/10.1007/s11666-020-01002-z.

15. Phanikumar, G.; Manjini, S.; Dutta, P.; Chattopadhyay, K.; Mazumder, J. Characterization of a Continuous CO2 Laser-Welded Fe-Cu Dissimilar Couple. *Metall. Mater. Trans. A*, 2005, *36* (8), 2137–2147. https://doi.org/10.1007/s11661-005-0334-6.

16. Sahul, M.; Sahul, M.; Turňa, M.; Zacková, P. Disk Laser Welding of Copper to Stainless Steel. *Adv. Mater. Res.*, 2014, *1077*, 76–81. https://doi.org/10.4028/www.scientific.net/AMR.1077.76.

17. Kar, J.; Dinda, S. K.; Roy, G. G.; Roy, S. K.; Srirangam, P. X-Ray Tomography Study on Porosity in Electron Beam Welded Dissimilar Copper-304SS Joints. *Vacuum*, 2018, *149*, 200–206. https://doi.org/10.1016/j.vacuum.2017.12.038.

18. Turna, M.; Sahul, M.; Ondruska, J.; Lokaj, J. Electron Beam Welding of Copper to Stainless Steel. In *Annals of DAAAM and Proceedings of the International DAAAM Symposium*; Katalinic, B., Ed.; DAAAM International: Vienna, 2011; Vol. 22, pp 833–834.

19. Raj, R. A. Microstructural Evaluation and Tensile Characterization of Electron Beam Welded Dissimilar Metal Joints of Copper and Stainless 304. *Technology*, 2018, *9*(3), 519–528.

20. Kar, J.; Roy, S. K.; Roy, G. G. Effect of Beam Oscillation on Electron Beam Welding of Copper with AISI-304 Stainless Steel. *J. Mater. Process. Technol.*, 2016, *233*, 174–185. https://doi.org/10.1016/j.jmatprotec.2016.03.001.

21. Metzger, G.; Lison, R. Electron Beam Welding of Dissimilar Metals. *Weld. Res. Counc. Bull.*, 1974, *215* (196), 230–240.

22. Shu, X.; Chen, G.; Liu, J.; Zhang, B.; Feng, J. Microstructure Evolution of Copper/Steel Gradient Deposition Prepared Using Electron Beam Freeform Fabrication. *Mater. Lett.*, 2018, *213*, 374–377. https://doi.org/10.1016/j.matlet.2017.11.016.

23. Dinda, S. K.; Kar, J.; Roy, G. G.; Kockelmann, W.; Srirangam, P. Texture Mapping in Electron Beam Welded Dissimilar Copper-Stainless Steel Joints by Neutron Diffraction. *Vacuum*, 2020, 109668. https://doi.org/10.1016/j.vacuum.2020.109668.

24. Osipovich, K. S.; Astafurova, E. G.; Chumaevskii, A. V.; Kalashnikov, K. N.; Astafurov, S. V.; Maier, G. G.; Melnikov, E. V.; Moskvina, V. A.; Panchenko, M. Y.; Tarasov, S. Y.; et al. Gradient Transition Zone Structure in "Steel-Copper" Sample Produced by Double Wire-Feed Electron Beam Additive Manufacturing. *J. Mater. Sci.*, 2020, *55* (22), 9258–9272. https://doi.org/10.1007/s10853-020-04549-y.

25. Weigl, M.; Schmidt, M. Influence of the Feed Rate and the Lateral Beam Displacement on the Joining Quality of Laser-Welded Copper-Stainless Steel Connections. *Phys. Procedia*, 2010, *5* (2), 53–59. https://doi.org/10.1016/j.phpro.2010.08.029.

26. Moharana, B. R.; Sahu, S. K.; Sahoo, S. K.; Bathe, R. Experimental Investigation on Mechanical and Microstructural Properties of AISI 304 to Cu Joints by CO2 Laser. *Eng. Sci. Technol. an Int. J.*, 2016, *19* (2), 684–690. https://doi.org/10.1016/j.jestch.2015.10.004.

27. Chen, S.; Huang, J.; Xia, J.; Zhao, X.; Lin, S. Influence of Processing Parameters on the Characteristics of Stainless Steel/Copper Laser Welding. *J. Mater. Process. Technol.*, 2015, *222*, 43–51. https://doi.org/10.1016/j.jmatprotec.2015.03.003.

28. Moharana, B. R.; Sahu, S. K.; Maiti, A.; Sahoo, S. K.; Moharana, T. K. An Experimental Study on Joining of AISI 304 SS to Cu by Nd-YAG Laser Welding Process. *Mater. Today Proc.*, 2020, *33*, 5262–5268. https://doi.org/10.1016/j.matpr.2020.02.953.

29. Shen, H.; Gupta, M. C. Nd:Yttritium-Aluminum-Garnet Laser Welding of Copper to Stainless Steel. *J. Laser Appl.*, 2004, *16* (1), 2–8. https://doi.org/10.2351/1.1642635.

30. Kuryntsev, S. V.; Shiganov, I. N. Welding Austenitic Steel to Copper with the Defocused Radiation of a Fibre Laser. *Weld. Int.*, 2018, *32* (1), 50–53. https://doi.org/10.1080/0950 7116.2017.1382074.

31. Li, J.; Cai, Y.; Yan, F.; Wang, C.; Zhu, Z.; Hu, C. Porosity and Liquation Cracking of Dissimilar Nd:YAG Laser Welding of SUS304 Stainless Steel to T2 Copper. *Opt. Laser Technol.*, 2020, *122* (October 2019), 105881. https://doi.org/10.1016/j.optlastec.2019.105881.

32. Nguyen, Q.; Azadkhou, A.; Akbari, M.; Panjehpour, A.; Karimipour, A. Experimental Investigation of Temperature Field and Fusion Zone Microstructure in Dissimilar Pulsed Laser Welding of Austenitic Stainless Steel and Copper. *J. Manuf. Process.*, 2020, *56* (February), 206–215. https://doi.org/10.1016/j.jmapro.2020.03.037.

33. Ramachandran, S.; Lakshminarayanan, A. K. An Insight into Microstructural Heterogeneities Formation between Weld Subregions of Laser Welded Copper to Stainless Steel Joints. *Trans. Nonferrous Met. Soc. China*, 2020, *30* (3), 727–745. https://doi.org/10.1016/S1003-6326(20)65249-9.

34. Joshi, G. R.; Badheka, V. J. Processing of Bimetallic Steel-Copper Joint by Laser Beam Welding. *Mater. Manuf. Process.*, 2019, *34* (11), 1232–1242. https://doi.org/10.1080/104 26914.2019.1628262.

35. Chen, S.; Huang, J.; Xia, J.; Zhang, H.; Zhao, X. Microstructural Characteristics of a Stainless Steel/Copper Dissimilar Joint Made by Laser Welding. *Metall. Mater. Trans. A Phys. Metall. Mater. Sci.*, 2013, *44* (8), 3690–3696. https://doi.org/10.1007/s11661-013-1693-z.

36. Yao, C.; Xu, B.; Zhang, X.; Huang, J.; Fu, J.; Wu, Y. Interface Microstructure and Mechanical Properties of Laser Welding Copper-Steel Dissimilar Joint. *Opt. Lasers Eng.*, 2009, *47* (7–8), 807–814. https://doi.org/10.1016/j.optlaseng.2009.02.004.

37. Meng, Y.; Li, X.; Gao, M.; Zeng, X. Microstructures and Mechanical Properties of Laser-Arc Hybrid Welded Dissimilar Pure Copper to Stainless Steel. *Opt. Laser Technol.*, 2019, *111* (October), 140–145. https://doi.org/10.1016/j.optlastec.2018.09.050.

38. Magnabosco, I.; Ferro, P.; Bonollo, F.; Arnberg, L. An Investigation of Fusion Zone Microstructures in Electron Beam Welding of Copper-Stainless Steel. *Mater. Sci. Eng. A*, 2006, *424* (1–2), 163–173. https://doi.org/10.1016/j.msea.2006.03.096.

39. Zhang, B.-G.; Zhao, J.; Li, X.-P.; Chen, G.-Q. Effects of Filler Wire on Residual Stress in Electron Beam Welded QCr0.8 Copper Alloy to 304 Stainless Steel Joints. *Appl. Therm. Eng.*, 2015, *80*, 261–268. https://doi.org/10.1016/j.applthermaleng.2015.01.052.

40. Zhang, B. G.; Zhao, J.; Li, X. P.; Feng, J. C. Electron Beam Welding of 304 Stainless Steel to QCr0.8 Copper Alloy with Copper Filler Wire. *Trans. Nonferrous Met. Soc. China (English Ed.*, 2014, *24* (12), 4059–4066. https://doi.org/10.1016/S1003-6326(14)63569-X.

41. Tosto, S.; Nenci, F.; Jiandong, H.; Corniani, G.; Pierdominici, F. Microstructure of Copper-AISI Type 304L Electron Beam Welded Alloy. *Mater. Sci. Technol.*, 2003, *19* (4), 519–522. https://doi.org/10.1179/026708303225010722.

42. Chen, Y. Z.; Liu, F.; Yang, G. C.; Xu, X. Q.; Zhou, Y. H. Rapid Solidification of Bulk Undercooled Hypoperitectic Fe-Cu Alloy. *J. Alloys Compd.*, 2007, *427* (1–2), L1–L5. https://doi.org/10.1016/j.jallcom.2006.03.012.

43. Matsumoto, T.; Misono, T.; Fujii, H.; Nogi, K. Surface Tension of Molten Stainless Steels Under. *J. Mater. Sci.*, 2005, *40*, 2197–2200.

44. Brillo, J.; Egry, I. Surface Tension of Nickel, Copper, Iron. *Calphad Comput. Coupling Phase Diagrams Thermochem.*, 2005, *0*, 2213–2216.

45. Mvola, B.; Kah, P.; Martikainen, J.; Suoranta, R. State-of-the-Art of Advanced Gas Metal Arc Welding Processes: Dissimilar Metal Welding. *Proc. Inst. Mech. Eng. Part B J. Eng. Manuf.*, 2015, *229* (10), 1694–1710. https://doi.org/10.1177/0954405414538630.

46. Cottrell, C. L. M. Electron Beam Welding - a Critical Review. *Mater. Des.*, 1985, *6* (6), 285–291. https://doi.org/10.1016/0261-3069(85)90009-3.

47. Sun, Z.; Karppi, R. The Application of Electron Beam Welding for the Joining of Dissimilar Metals: An Overview. *J. Mater. Process. Technol.*, 1996, *59* (3 SPEC. ISS.), 257–267. https://doi.org/10.1016/0924-0136(95)02150-7.

48. Cui, Y.; Xu, C.; Han, Q. Effect of Ultrasonic Vibration on Unmixed Zone Formation. *Scr. Mater.*, 2006, *55* (11), 975–978. https://doi.org/10.1016/j.scriptamat.2006.08.035.

49. Baeslack, B. Y. W. A. Unmixed Zone Formation in Austenitic Stainless Steel Weldments. *Weld. Res. Suppl.*, 1979, *58*, 168–176.

5 Studies on Cold Metal Transfer Welding of Aluminum 5083 Alloy to Pure Titanium

*Pankaj Kaushik, Ranjan Kumar,
Manjaiah M., and Ajith G. Joshi*

5.1 INTRODUCTION

The welding of titanium (Ti) alloys with aluminum (Al) could have a significant impact on the aerospace and automotive industries, owing to increasing demand for light-weight components. Manufacturing industries ever-increasing demands have sparked business interest in systems capable of welding metal combinations that were previously thought to not be weldable. The basic prerequisite for fusion welding of dissimilar metals is that they are soluble in each other and that their melting points are close. The Ti and Al have melting points of 1667°C and 660°C, respectively. Because of the large disparity in melting temperatures between the two metals, traditional fusion welding methods have a difficult time fusing them.

Several researchers have tried a variety of additional ways for combining Al and Ti materials, including pressure welding, diffusion bonding, and friction welding. Besides the cold metal transfer (CMT) process, modified metal inert gas welding is one of the most emerging welding technologies. It is suited for welding aluminum over other dissimilar metals due to its no-spatter characteristics. Feng et al. (2009) successfully demonstrated the CMT process based on the short-circuiting transfer technique to weld thin weld aluminum sheets. The minimum heat generated during the process promotes thin cross-section weldability and gap-bridging ability. Huang et al. (2022) joined the Ti6Al4V alloy with T2 Cu joints by the CMT process. The β-Ti phase and other secondary intermetallic phases were formed during the process. However, $AlCu_2Ti$ was the major intermetallic phase that governed the manipulation of the mechanical properties of joints. The occurrence of $AlCu_2Ti$ with a stable thickness and its dispersive distribution have significantly improved the mechanical properties. Babu et al. (2019) show that the CMT process

DOI: 10.1201/9781003327769-5

is a promising approach to joining dissimilar metals, such as Al, with steel surfaces. The interface formed in the process is also intermetallic-free, which yields better microstructural characteristics.

In addition, the CMT process yields better microstructural characteristics and mechanical properties, such as ultimate tensile strength (UTS), elongation, and hardness of the joined parts, compared to the gas metal arc welding process (Nagasai et al. 2022). The process parameters of CMT, such as wire feed and torch position, influence significantly the strength, fracture behavior, and mode of welded joints. Mou et al. (2019) illustrated that increased wire feed speed has enhanced the tensile strength and increased the interface thickness. Further, fracture mode classifications are proposed to identify and understand the crack propagation in the joints. The CMT-processed joint microstructure is classified into coarser and finer grain areas. Studies have shown that microhardness at the interface varies based on the type and composition of microstructure elements. Thus, they directly influence the mechanical properties and fracture behavior of materials (Zhou et al. 2019). Ortega et al. (2019) obtained dendritic microstructure aluminum deposits from the CMT process. However, initially fine grain structure and later coarser grain structure were obtained during the processing of subsequent layers of superposition. Rajeev et al. (2019) stated that dynamic and arc length correction parameters have a significant influence on the deposition characteristics of the CMT process. Also, the relatively small arc length correction parameter has a greater influence than the dynamic parameter on deposition characteristics. Jing et al. (2013) studied the CMT welding process of AZ31B magnesium alloy and 6061 aluminum alloy using ER4043 filler metal. They stated that the microhardness of the weld decreases as the presence of intermetallic compounds is reduced from Mg to Al.

During CMT of Ti6Al4V and AlSi5 alloys, the interface layer is comprised of different types of layers containing compounds of Ti-Al-Si with varying compositions. Cracks were initiated at the interface and propagated to the Al layer owing to a gradient in shrinkage temperature (Tian et al. 2019). Yang et al. (2019) concluded that interfacial energy at interphase boundaries has a significant effect on bonding strength. Low heat input, yielding low interfacial energy, has resulted in good bonding strength. Sambasiva Rao et al. (2011) employed gas tungsten arc welding of Al and Ti-6Al-4V alloys with Al-Si filler metal. The authors concluded that the transverse strength of the welded zone is greater than that of Al alloy. Also, intermetallic compounds were absent in the interface region. Zhang et al. (2020) investigated the influence of electrode polarity ratio on microstructure and mechanical properties. The ratio was found to be less significant in the study; however, larger positivity duty cycles have yielded greater hardness, and increased negativity duty cycles have deteriorated the mechanical properties due to insufficient melting of material. The CMT process operation with optimal process parameter levels aids in achieving enhanced mechanical properties and corrosion-resistant joints of Al/steel components processed by CMT. Also, optimized process parameters helps regulate the intermetallic layer thickness at the welding interface zone (Sravanthi et al. 2019). Selvamani (2021) demonstrated that changing phases for Al_2SiO_3 precipitates in the fusion zone primarily contributed to enhancing the strength of the welded region.

Titanium has huge applications in the aerospace, shipbuilding, and spacecraft industries due to its high strength and corrosion resistance properties. More research and exploration can be done on it. Aluminum is light weight, low cost, etc.; it can be a good replacement for other metals in cost-cutting and weight-reduction processes. Aluminum alloy of this series is known to be the marine industries main alloy. The joining of these dissimilar metals has been in great scope due to cost-cutting in industries and the increased strength of the base metal due to the mixing of the other metal in fusion welding processes. Further, limited work is carried out on the joining of aluminum and titanium series of dissimilar metals in cold metal transfer welding. Thus, in the current work, joining of aluminum 5083 alloy and pure titanium using the CMT welding process was done. The hardness, tensile strength, and microstructure of the welded joints at different process parameters were evaluated. The phase and composition analysis of Al5083-Ti welded materials was investigated using Energy-Dispersive Spectroscopy (EDS) and X-Ray Diffraction analysis. Fractography analysis of the fractured Al5083-Ti weldments was studied to understand the fracture behavior.

5.2   EXPERIMENTAL METHODOLOGY

5.2.1   MATERIALS AND MECHANICAL CHARACTERIZATION

In the present research, aluminum alloy 5083 and pure titanium were welded using the Fronius 400i CMT welding process with a plate dimension of $75 \times 75 \times 3$ mm, as shown in Figure 5.1, using ER4043 of diameter 1.6 mm as a filler material. A groove angle of 45° has been made on both Al-Ti plates using milling operations. The experiments were conducted using Argon as a shielding gas and a torch angle of 70° manually. Trials were carried out to achieve suitable welding parameters by adjustment of the weld parameters such as welding current (A), arc correction length (mm), and gas flow rate (L/min) to provide defect-free welds, as illustrated in Table 5.1. The composition of the alloys welded is illustrated in Tables 5.2 and 5.3. After successful completion of welding trials, nine samples were chosen for mechanical strength and microstructural characterization. The three tensile test samples were cut from each welded sample according to the ASTM E8 standard through a wire EDM machine. Samples of the tensile test and the loaded sample are shown in Figure 5.2.

The Vickers microhardness test was employed to examine the microhardness of joints and various zones generated after welding an Aluminum 5083-Titanium sample. The testing of the sample was done on the polished sample using a Via-S Matsuzawa Japan machine with an optical microscope attached to it. While diagonal lengths were measured to calculate the hardness value of the sample, Vickers hardness was calculated using the formulae (Eq. 5.1)

$$HV = 1.854 \left(F / D^2 \right) \tag{5.1}$$

where $F =$ applied load (Kgf) and $d =$ average length of the diagonals (mm).

The testing was done using a load of 200 g and a dwell time of 10 seconds.

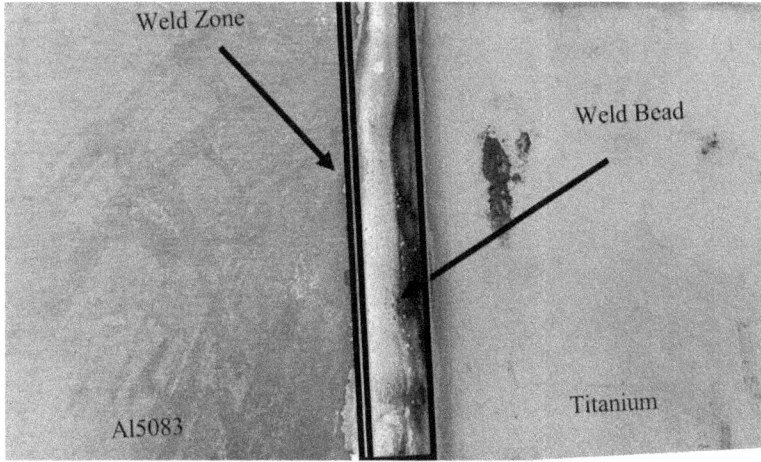

FIGURE 5.1 Welded sample image of Aluminum 5083-Titanium.

TABLE 5.1
Process Parameters for CMT Welding Process and Resultant Strength

Sample No.	Current (A)	Gas Flow Rate (L/min)	Arc Correction Length (mm)	Ultimate Tensile Strength (MPa)
1	50	15	0	36.7
2	50	20	0.2	57.1
3	50	25	0.5	83
4	55	15	0.2	95.7
5	55	20	0.5	85.4
6	55	25	0	81
7	60	15	0.5	79.1
8	60	20	0	74.9
9	60	25	0.2	62

TABLE 5.2
Chemical Composition of Aluminum 5083 Alloy

Element	Al	Mg	Si	Fe	Cu	Mn	Zn	Cr	Ti	Pb	Bi
Wt. %	94.747	4.300	0.080	0.190	0.026	0.560	0.004	0.077	0.015	0.001	0.0

TABLE 5.3
Chemical Composition of Filler Metal ER4043 Alloy

Element	Si	Fe	Cu	Mn	Cr	Ni	Zn	Ti	Mg	Al
Wt. %	4.5–6.0	0.8	0.30	0.05	-	-	0.10	0.20	0.05	Balance

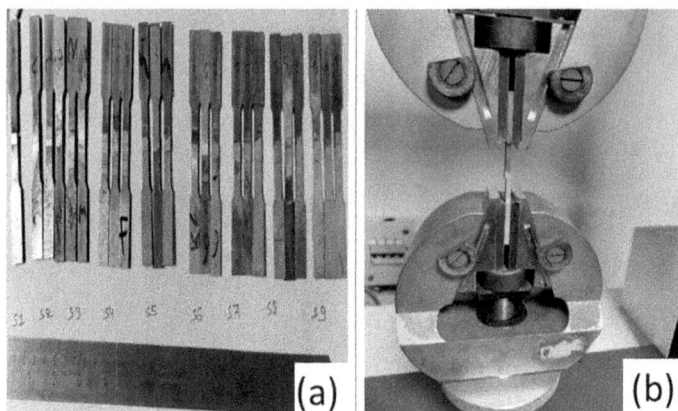

FIGURE 5.2   (a) Tensile test samples of CMT-welded joints; (b) Loaded sample on the UTM machine.

5.2.2 MICROSTRUCTURE ANALYSIS

Welded joints are treated to normal metallographic sample preparation before being examined under an optical microscope for their microstructure. The sample for the microcharacterization section was cut from a welded sample with dimensions of 20 × 5 mm and then mounted to hold the sample for grinding and polishing. After this, etching was done on the mounted sample using Keller's reagent (190 mL of deionized water, 2 mL of hydrofluoric acid, 3 mL of hydrochloric acid, and 5 mL of nitric acid) for about 45 seconds. The sample prepared is scanned under an optical microscope. Analysis was conducted to establish the scope of extension in the elemental diffusion in as well as around the interface region, as well as to identify the probable intermetallic compound combinations created as a result of the diffusion in the elements of the samples over the joint. The optical microscopic images (Figure 5.3) were taken at three different locations, viz., Al alloy and Ti alloy surfaces captured with 100× magnification and welded zones with 200× magnification. The grain boundaries are more prominent on the Al surface, while dark phase dispersion was observed on the Ti surface. The interface zone is clearly distinguishable with more fusion of both metals.

5.2.3 SEM AND EDS ANALYSIS

Further, the samples were studied under a scanning electron microscope to examine the degree of elemental diffusion in and around the interface region, as well as the probable intermetallic compound combinations created as a result of elemental diffusion over the joint. An automatic X-ray diffraction analysis was carried out on the sample, with the working parameters for X-Ray Diffraction analyses being a voltage of 45 kV and a current of 30 mA. The start angle and end angle have been taken as 20° and 120°, respectively. For the microhardness analyses, a Vickers microhardness setup was used, with readings taken at a distance of 0.3 mm up to 2 mm on each side of the weld zone, and analyses were done based on the values obtained. The tensile test was

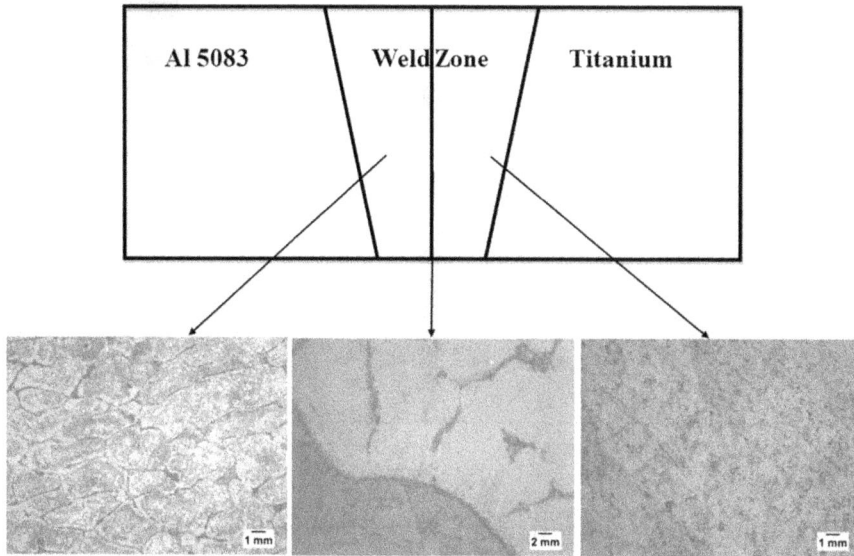

FIGURE 5.3   Showing optical microscope images taken at 100× magnification of Al and Ti alloys and 200× magnification at the welding zone.

carried out using the ESTM-E8 standard for the nine good welded samples at room temperature, and the values were recorded to determine the welding strength. EDS analysis was conducted on the sample to determine the presence of various elements and elemental mapping throughout the welding zones. To study the description, analysis, and interpretation of fracture surface morphologies (fracture topographies), as well as the linkages between them and the causal forces, mechanisms, and subsequent fracture progression, the fractured UTS samples were examined under a scanning electron microscope.

5.3   RESULTS AND DISCUSSION

5.3.1   ULTIMATE TENSILE STRENGTH

The UTS of nine samples studied with different combinations is illustrated in the last column of Table 5.1. Further variation of tensile strength against current is depicted in Figure 5.4. It was noticed that UTS increased with applied current from 50 to 60 A by approximately two times. The heat provided to the specimen increases due to increased current, causing greater fusion of the specimen material. Perhaps it yields larger intermetallic compounds in the alloy microstructure owing to the higher strength of the specimen. Whereas, further increase in current has led to a marginal drop in the strength of welded joints. It may be due to possible remelting with sufficient heat energy, subsequently dispersion, and adequate Ti elemental diffusion into the Al alloy instead of the occurrence of intermetallic compounds. In addition, refinement of grain structure has resulted in a decline in strength. Similar results were reported elsewhere by Gopal et al. (2022).

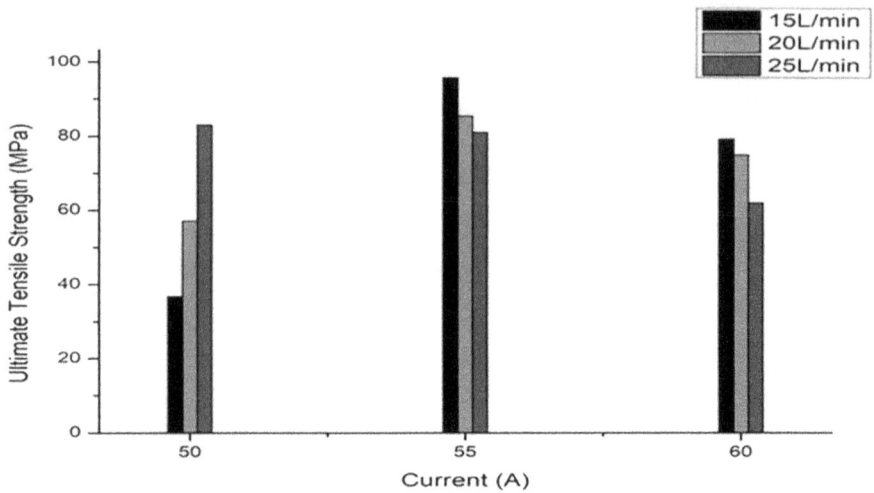

FIGURE 5.4   Column chart showing the effect of current on the ultimate tensile strength of the Al5083-Ti sample.

5.3.2   MICROHARDNESS

The indentation done on the parent metal side and on the weld zone is illustrated in Figure 5.5. Microhardness values plotted at different distances from the welded zone are depicted in Figure 5.6. The higher hardness values were observed on the Ti side, close to the weld zone. It is possibly due to the occurrence of brittle intermetallic compounds. Further, solid solution strengthening is evidently greater with the greater existence of precipitates formed due to fusion. The hardness value of the sample was found to be in the range of 61.6–169.7 HV. The decrease in the microhardness value on the aluminum 5083 side is probably due to the relatively lower hardness of the Al alloy. In addition, it is due to the grain growth in the heat-affected zone of Al metal, the lack of precipitate presence, the minimum existence of alloy elements, and the coarser grain size (Elrefaey, 2015).

5.3.3   SEM AND EDS ANALYSIS OF THE WELDED SAMPLE

Figure 5.7 shows the SEM images of a welded sample. The grain boundary and grains on the parent metal surface on either side of the welded zone are also illustrated. SEM analysis depicts that grain boundaries are coarser on Al side with distinguishable grain boundaries. The β-phase presence on the Al-microstructure and porosity are more pronouncing. The interdendritic microstructure of the ternary phase is also evident on the surface of the Al alloy. However, the lack of porosity was noticed on the Ti side with clear grain boundaries. Further, the Ti layer is relatively more continuous compared to the Al alloy side. Similar observations were reported by Tian et al. (2019). Possibly the continuous layer is due to better stirring and fusion phenomena that occurred during welding.

The EDS analyses of the CMT-welded sample on the aluminum 5083 side (Figure 5.8a) showed the presence of elements like aluminum (the base metal),

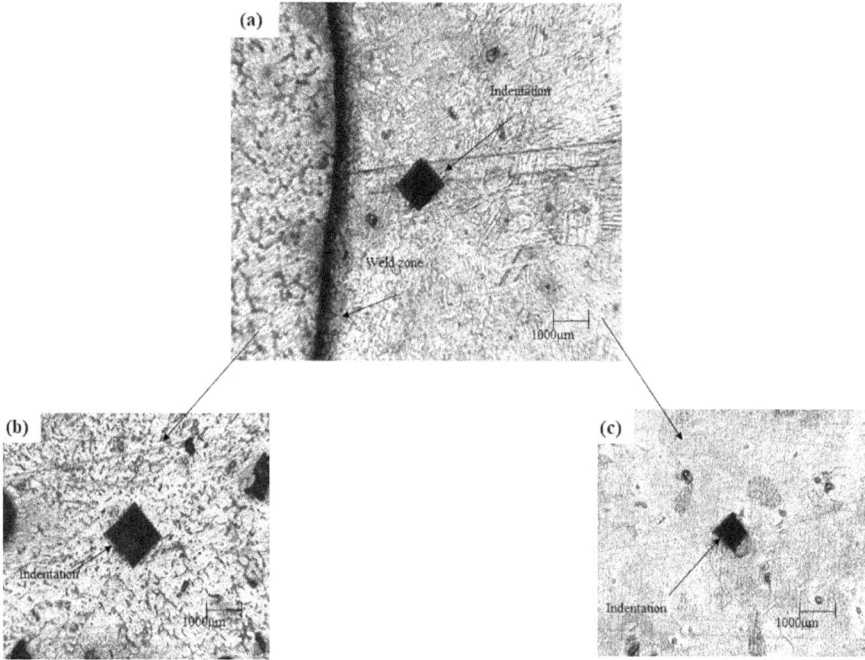

FIGURE 5.5   Micro-Vickers hardness (a) indentation on the Al5083 side, (b) indentation around the weld zone, and (c) indentation on the Ti side.

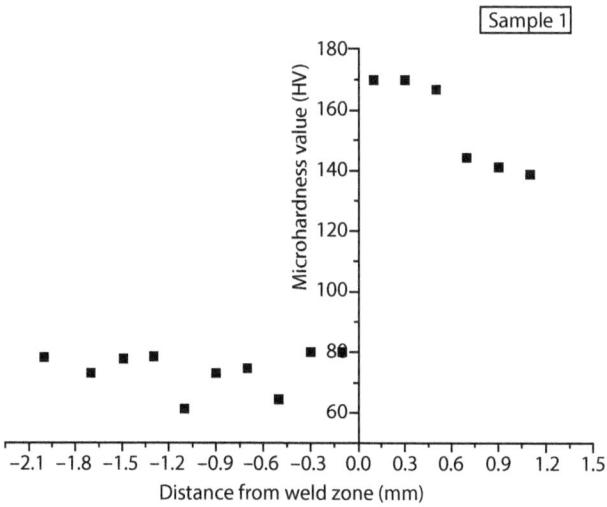

FIGURE 5.6   Microhardness values of the Aluminum 5083-Titanium welded joint.

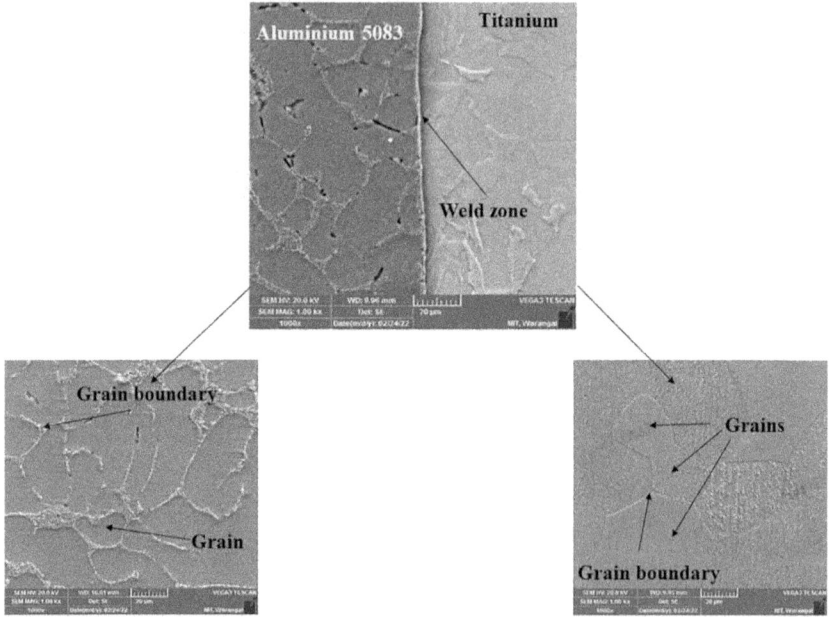

FIGURE 5.7 SEM micrographs of Al5083-Ti of cold metal transfer-welded samples.

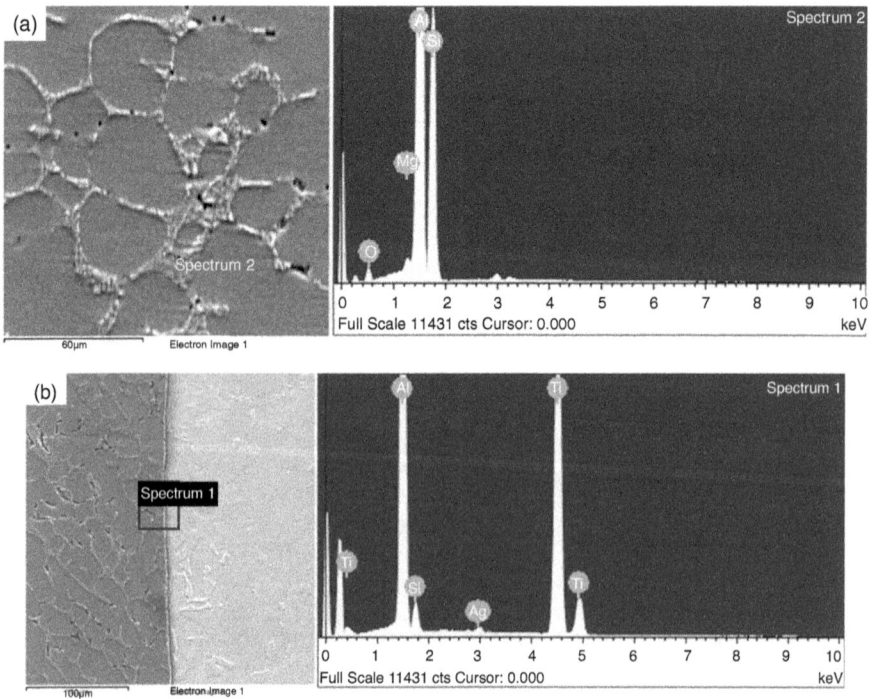

FIGURE 5.8 EDS analysis showing the different material compositions of the welding specimen: (a) Al5083 side and (b) weld zone of Al5083-Ti.

silicon, magnesium and a molecule of oxygen. On the other hand, the analyses on the weld zone side (Figure 5.8b) showed the presence of elements like aluminum (base metal), titanium (base metal), silicon, and traces of silver. The presence of silicon metal was noticed due to the filler metal used. On the weld zone side, the absence of oxygen shows no formation of oxides in the welding zone.

5.3.4 FRACTOGRAPHY

Failure analysis (FA) is a multidisciplinary, multidimensional study of scientific reports. It connects various disciplines of engineering and attempts to inter-relate numerical modeling to surface science and, finally, tribology. The main aim of fractography is the measurement of the fracture in surface topographic features. Thus, it reveals the prominent characteristics of the fractured surface. Although optical methods (stereomicroscopy) are commonly used in traditional macro-fractography, research analyses are encouraged increasingly to get better and more quantitative results in SEM microfractography. The following is the pattern observed after the fractography analyses of the fractured CMT samples (Figure 5.9).

During the tensile test of Al5083-Ti, the fracture occurred on the aluminum side of the welded sample. It was observed under the scanning electron microscope that the fractured surface of the CMT-welded samples exhibits a river-like structure. Perhaps, due to the lower thickness of the intermetallic compounds, most of the sample's failure was ductile in nature. Also, limited ductility fractures were observed with intergranular fractures. Transgranular (cleavage) fracture is characterized by river-like patterns, which are topographically depicted as plateaus joined by shear ledges, indicating the direction of the crack propagation. River patterns are formed when the mode of fracture mainly changes from 1st mode to 3rd mode with the increasing component of mode III.

5.4 CONCLUSION

Based on the current studies on the CMT welding process of Al-Ti, the following conclusions were drawn:

- High tensile strength was observed in the welding sample of Al5083-Ti at a welding current of 55 A and gas flow rate of 15 L/min, and an arc correction length of 0.2 mm. Further increases in current resulted in a decline in strength, possibly due to adequate alloying of elements during fusion instead of intermetallic compound formation.
- An increase in hardness in the weld zone of Al5083-Ti is due to the formation of intermetallic compounds of $TiAl_3$.
- The interdendritic structure was observed on the Al – side of the welded zone with coarser grains, contributing to its lower hardness. While continuous grains with less porosity on the Ti side of the weld zone owe their higher hardness.

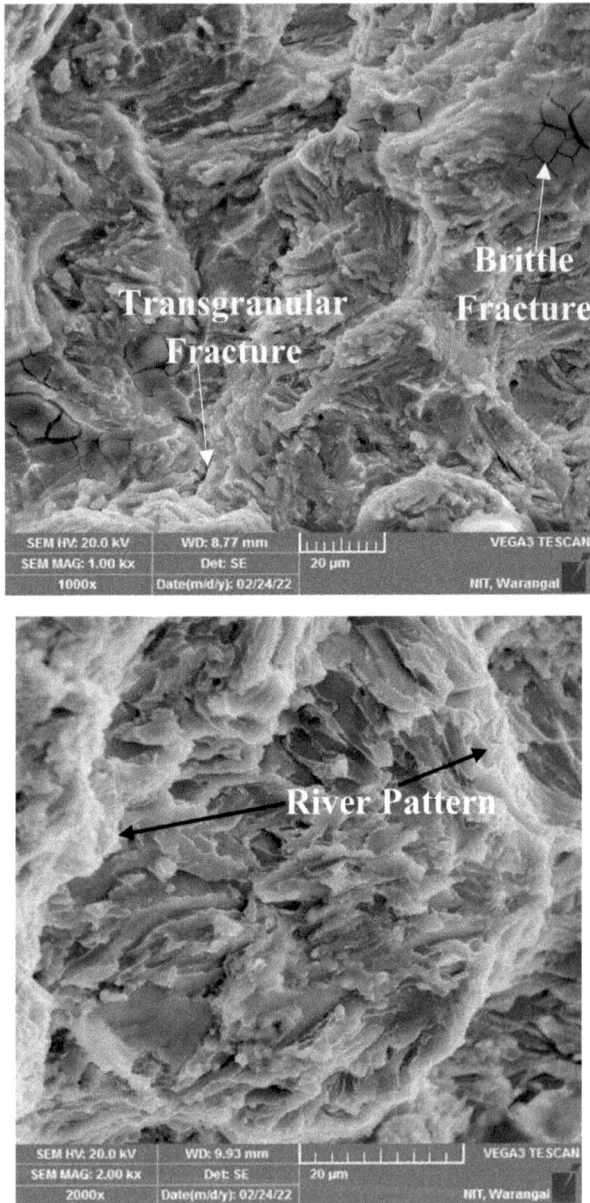

FIGURE 5.9 Fractography images of the fractured Al5083-Ti samples after a tensile test with (a) 1000× and (b) 2000× magnification.

- SEM and EDS analyses of the welded zone confirm the presence of inter-metallic compounds. Further, the presence of porosity is relatively greater on Al side of the welded zone.
- The fractography analysis revealed a river-like pattern on the fractured surface with transgranular fracture occurrence.

REFERENCES

Babu, S. et al. 2019. "Cold Metal Transfer Welding of Aluminium Alloy AA 2219 to Austenitic Stainless Steel AISI 321." *Journal of Materials Processing Technology* 266(May 2018): 155–64.

Elrefaey, A. 2015. "Effectiveness of Cold Metal Transfer Process for Welding 7075 Aluminium Alloys." *Science and Technology of Welding and Joining* 20(4): 280–85.

Feng, J., H. Zhang, and P. He. 2009. "The CMT Short-Circuiting Metal Transfer Process and Its Use in Thin Aluminium Sheets Welding." *Materials & Design* 30(5): 1850–52.

Gopal, D. et al. 2022. "Optimization of Processing Parameters of Cold Metal Transfer Joined 316L and Weld Bead Profile Influenced by Temperature Distribution Based on Genetic Algorithm." *Proceedings of the Institution of Mechanical Engineers, Part C: Journal of Mechanical Engineering Science* 236(19): 10271–80.

Huang, L. et al. 2022. "Interfacial Layer Regulation and Its Effect on Mechanical Properties of Ti6Al4V Titanium Alloy and T2 Copper Dissimilar Joints by Cold Metal Transfer Welding." *Journal of Manufacturing Processes* 75(January): 1100–1110.

Jing, S. et al. 2013. "Microstructure Characteristics and Properties of Mg/Al Dissimilar Metals Made by Cold Metal Transfer Welding with ER4043 Filler Metal." *Rare Metal Materials and Engineering* 42(7): 1337–41.

Mou, G. et al. 2019. "Microstructure and Mechanical Properties of Cold Metal Transfer Welding-Brazing of Titanium Alloy (TC4) to Stainless Steel (304L) Using V-Shaped Groove Joints." *Journal of Materials Processing Technology* 266: 696–706.

Nagasai, B. P., S. Malarvizhi, and V. Balasubramanian. 2022. "Mechanical Properties and Microstructural Characteristics of Wire Arc Additive Manufactured 308 L Stainless Steel Cylindrical Components Made by Gas Metal Arc and Cold Metal Transfer Arc Welding Processes." *Journal of Materials Processing Technology* 307(May): 117655.

Ortega, A. G. et al. 2019. "Characterisation of 4043 Aluminium Alloy Deposits Obtained by Wire and Arc Additive Manufacturing Using a Cold Metal Transfer Process." *Science and Technology of Welding and Joining* 24(6): 538–47.

Rajeev, G. P., M. Kamaraj, and S. R. Bakshi. 2019. "Effect of Correction Parameters on Deposition Characteristics in Cold Metal Transfer Welding." *Materials and Manufacturing Processes* 34(11): 1205–16.

Sambasiva Rao, A., G. M. Reddy, and K. S. Prasad. 2011. "Microstructure and Tensile Properties of Dissimilar Metal Gas Tungsten Arc Welding of Aluminium to Titanium Alloy." *Materials Science and Technology* 27(1): 65–70.

Selvamani, S. T. 2021. "Microstructure and Stress Corrosion Behaviour of CMT Welded AA6061 T-6 Aluminium Alloy Joints." *Journal of Materials Research and Technology* 15: 315–26.

Sravanthi, S. S., S. G. Acharyya, K. V. Phani Prabhakar, and J. Joardar. 2019. "Effect of Varying Weld Speed on Corrosion Resistance and Mechanical Behavior of Aluminium - Steel Welds Fabricated by Cold Metal Transfer Technique." *Materials and Manufacturing Processes* 34(14): 1627–37.

Tian, Y. et al. 2019. "Microstructure and Mechanical Properties of Wire and Arc Additive Manufactured Ti-6Al-4V and AlSi5 Dissimilar Alloys Using Cold Metal Transfer Welding." *Journal of Manufacturing Processes* 46(May): 337–44.

Yang, J. et al. 2019. "Heat Input, Intermetallic Compounds and Mechanical Properties of Al/Steel Cold Metal Transfer Joints." *Journal of Materials Processing Technology* 272(May): 40–46.

Zhang, P., G. Li, H. Yan, and Y. Tian. 2020. "Effect of Positive/Negative Electrode Ratio on Cold Metal Transfer Welding of 6061 Aluminum Alloy." *International Journal of Advanced Manufacturing Technology* 106(3–4): 1453–64.

Zhou, J. et al. 2019. "Microstructure and Mechanical Properties of AISI 430 Ferritic Stainless Steel Joints Fabricated by Cold Metal Transfer Welding." *Materials Research Express* 6(11): 0–15.

6 Diffusion Bonding for Dissimilar Metals and Alloys

Gaurav Sharma

6.1 INTRODUCTION

Dissimilar metal joining is a rapidly expanding field since many sectors demand joins made of different materials as part of machines, equipment, and many other items to produce a hybrid structure. Welding is the most typical method for constructing such a hybrid structure. To build such buildings, numerous factors must be addressed. The primary issue of combining different metals is an incompatibility in mechanical, physical, and chemical characteristics, which results in an unacceptable joint in terms of joint strength, structural integrity, and efficiency. Because fusion welding normally melts portions of both bases, the melting temperatures of both metals influence the type of welding performed and the filler material employed. The welder must employ a temperature high enough to melt both metals. Using approaches that do not melt the metals might potentially help to address problems caused by different melting points. Another issue with dissimilar metal joining is the difference in the coefficients of thermal expansion of different metals. When the thermal expansion coefficients of the metals differ significantly, variations in temperature surrounding the welded connection might create undue strain on the weld. The stresses will be concentrated in the intermetallic zone, where the two metals combine with the filler material. In this intermetallic zone, the welded connection is more prone to thermal fatigue, especially in applications with a large number of temperature cycles. The most significant aspect to consider in dissimilar metal joining is solubility. Ideally, the metals should be interpolable with one another. However, some metals are not compatible in this fashion. In some cases, employing a third metal that is soluble with both can aid in the formation of the weld.

Diffusion bonding, a solid-state joining procedure, has acquired popularity in the aerospace, chemical, and nuclear sectors for the connecting of diverse comparable and dissimilar materials. It is a sophisticated solid-state joining technology in which the joint is formed by the interdiffusion of atoms across the interface caused by heating in a controlled environment under a pre-defined pressing stress. Diffusion bonding may create a link between similar and different metals and alloys, as well as nonmetallic materials, which would be difficult or impossible to join using traditional welding procedures (Kazakov, 1985). During bonding, the faying surfaces are

 DOI: 10.1201/9781003327769-6

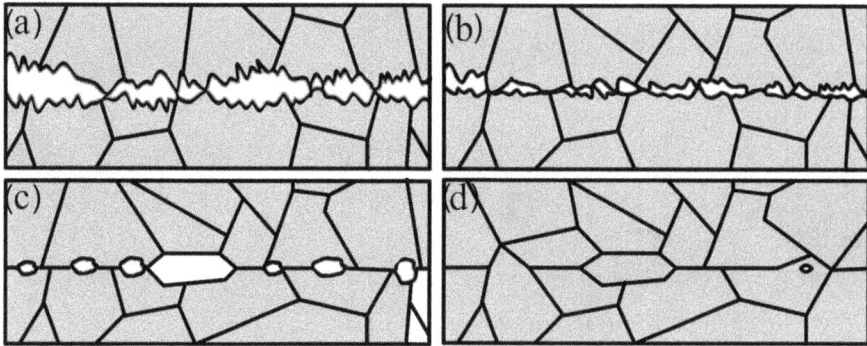

FIGURE 6.1 (a–d) Evolution of voids at the diffusion bond interface (Sharma & Dwivedi, 2018).

brought into contact so that atomic diffusion can take place through the interface. The surface of the mating parts should be flat and free from contaminants like oil or dust. However, the real surface contains microasperities and contaminated layers. Contaminated surface layer inhibits diffusion. This layer has to be removed before bonding. If anyone on the mating surface has contamination or an oxide layer, then bonding will not take place.

When the two bonding surfaces come into contact with each other, they have initial contact at some points of surface asperities (Figure 6.1a). In the first stage, deformation of the asperities takes place by yielding at these points, and actual metal comes into contact over a large area, and a joint forms in these locations with some voids between them (Figure 6.1b). In the second stage, diffusion of the atoms becomes faster due to the increase in actual contact area, and the interfacial grain boundary migrates by eliminating more voids (Figure 6.1c). In the third stage, the voids that remained at the interface are eliminated by volume diffusion of atoms by the vacancy diffusion mechanism (Figure 6.1d).

6.2 DIFFUSION MECHANISMS

There are different mechanisms proposed for the diffusion of atoms in lattice structures:

- Vacancy mechanism
- Ring mechanism
- Interstitial mechanism
- Exchange mechanism

6.2.1 Vacancy Mechanism

According to this mechanism, as the metals to be joined are heated, the atoms present at the surface of one material interchange their position with the adjacent vacant position of another material. The motion of vacancies is in the opposite direction to that of the motion of the atoms. In this mechanism, both self and interdiffusion take place through the movement of atoms as well as vacancies (Figure 6.2).

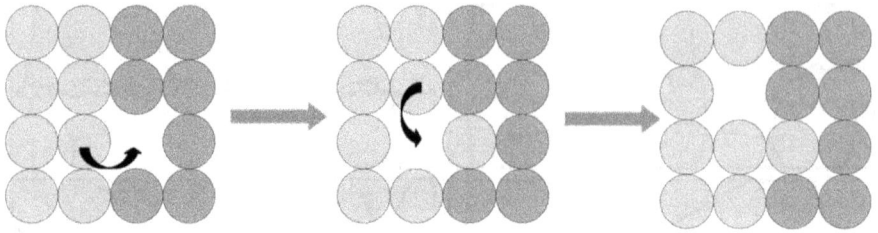

FIGURE 6.2   Schematic of the vacancy mechanism.

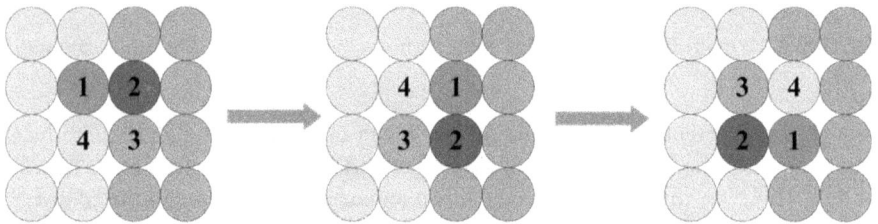

FIGURE 6.3   Schematic of the ring mechanism.

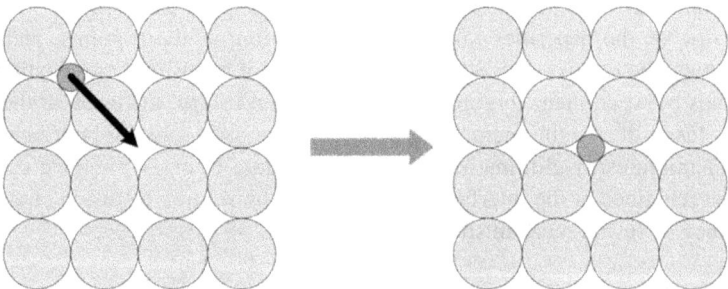

FIGURE 6.4   Schematic of the interstitial mechanism.

6.2.2   Ring Mechanism

This mechanism involves a circular exchange of four atoms. Four neighboring atoms move around a ring at a time. This mechanism exists in metals with a closely packed lattice (Figure 6.3).

6.2.3   Interstitial Mechanism

This mechanism involves the movement of an atom from an interstitial position to another neighboring vacant position. It can further move from one position to another. For this mechanism, an appreciable amount of energy must be imparted for the movement of an atom. So this mechanism generally exists in the case of interdiffusion in solid solutions having an atom smaller than the parent atoms (Figure 6.4).

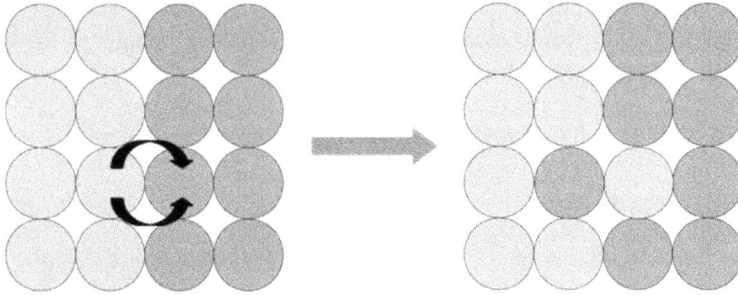

FIGURE 6.5   Schematic of the exchange mechanism.

6.2.4   Exchange Mechanism

In the exchange mechanism, interchanges of position between two adjacent atoms take place. This type of movement requires more energy because each atom has to move a distance of two atomic diameters, and energy is spent on local distortion of the lattice (Figure 6.5).

6.3   PROCESS VARIABLES

The success of the diffusion bonding process depends mainly on three adjustable variables: bonding temperature, bonding pressure, and bonding time. The quality of the achieved bonding joint depends also on the bonding environment (for example, vacuum or reducing atmosphere), the material properties, and the preparation of the contact surfaces. With appropriate process variables and setup, the diffusion bond will have the same tensile strength and ductility as the parent material.

6.3.1   Bonding Temperature

Bonding temperatures are typically between 0.5 and 0.8 times the absolute melting point of the bonded composition's most fusible constituent. Temperature increases the plastic deformation of surface asperities and speeds up atomic diffusion across the bonding contact. Bonding temperatures can be significantly higher in specific situations. Thus, to establish a sound joint, just a modest bonding pressure is required. This section discusses the influence of bonding temperature on the metallurgical and mechanical characteristics of dissimilar diffusion bonds. According to numerous studies, increasing the bonding temperature improves joint performance (Kundu et al., 2005; Kurt et al., 2007a; Kurt et al., 2007b; Velmurugan et al., 2016). Kurt and Çalik (2009) investigated the effect of bonding temperature on diffusion bonding of 304 stainless steel and 4140 steel in the temperature range of 750°C–900°C and obtained a maximum shear strength of 472 MPa at the 900°C bonding temperature. Some Cr-carbide precipitates were also present at the interface due to the diffusion of C and Cr from 304 stainless steel to 4140 steel. Noh et al. (2012) investigated the diffusion bonding of ODS steel and F82H steel at 1100°C (solid-state diffusion bonding) and 1240°C (liquid-phase diffusion bonding) under a 15 MPa uniaxial pressure for

1 hour of bonding time in vacuum and found no macroscopic deformation at the bond interface at 15 MPa pressure. It was reported that the tensile strength of the liquid-phase diffusion bond was 735 MPa with 13.5% ductility and a fracture from the F82H side. In the case of solid-state diffusion bond, the maximum strength of 650 MPa was obtained with 12.5% of the ductility having fractured from the interface.

6.3.2 BONDING PRESSURE

To secure the tight fit between the mating surfaces, bonding pressure is applied. The specimens are compressed together with a force able to disperse the oxide films on the contact surfaces. Even though applied stress is generally below the yield strength of the parent material, the yield point can be exceeded at the peaks of surface asperities. Plastic deformation of the machining ridges and surface asperities enables the surface crystals to reach contact within an atomic distance, which is an important pre-condition for the diffusion mechanisms to take place. Sufficient bonding pressure also ensures that all gaps and voids are filled at the interface during bonding. The effect of bonding pressure on the metallurgical and mechanical properties of diffusion bonds are discussed in this section. It is reported by the various researchers that an increase in bonding temperature increases joint performance, but too much increase in bonding pressure ceases the vacancies at the faying surfaces. To avoid any unwanted macroscopic deformations, bonding pressure is usually kept as low as possible while still achieving a sufficient strength of the diffusion bond. An increase in bonding pressure reduces the size of microvoids and unbounded regions by enhancing local plastic deformation. In the initial stage of bonding, an increase in bonding pressure enhances the deformation of surface asperities at the point of initial contact and increases the diffusion rate (Du et al., 2018; Ren et al., 2018). It is reported that an increase in bonding pressure enhances the strength of the diffusion bond. However, initially the rate of improvement is high, but it decreases with further increases in bonding pressure (Sheng, 2005). Gao et al. (2018) studied the effect of bonding pressure on diffusion bonding of TC-6 and copper alloys using nickel as an interlayer. It was reported that shear strength first increased with an increase in bonding pressure, but after a certain limit, it dropped sharply due to the deformation of the copper base metal (Figure 6.6).

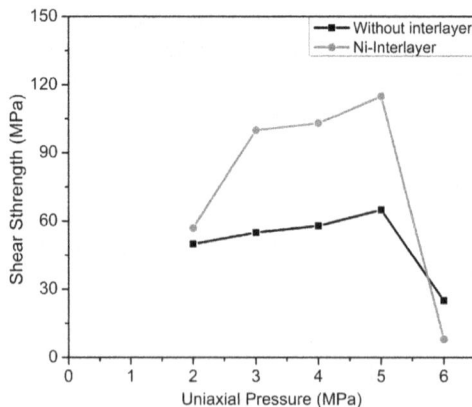

FIGURE 6.6 Variation in tensile shear strength with bonding pressure (Gao et al., 2018).

6.3.3 BONDING TIME

The bonding time at the effective bonding temperature and pressure is also reasonable to keep to a minimum value, with the limitation that it still has to be long enough for the diffusion processes to take place. Shorter holding time is more sustainable in terms of energy consumption, which makes it economically and environmentally profitable. Controlling and optimizing the holding time is very important, especially in mass production applications of diffusion bonding. The various researchers stated that the performance of the diffusion bond increases with an increase in holding time (Liu et al., 2017; Yılmaz & Yılmaz, 2016). The effect of bonding time on the shear strength of the diffusion bond of Ti-6Al-4V and AISI 304 was studied by Özdemir and Bilgin (2009). They reported that as the diffusion bonding time increased, the strength of the diffusion bond also increased.

6.3.4 BONDING ENVIRONMENT

The bonding environment affects the joint quality. Vacuum cleans the surfaces and sublimes the oxide films, improving the material properties at the diffusion zone. A similar reaction can be achieved by reducing the atmosphere with hydrogen. Sometimes an inert gas is used as a medium in the bonding process chamber.

6.4 CHALLENGES IN JOINING DISSIMILAR METALS BY DIFFUSION BONDING

Diffusion bonding of dissimilar metals and alloys is also a difficult operation since most couples have different physical, mechanical, and chemical characteristics. The melting temperature, coefficient of thermal expansion, and other properties of most diffusion bonding materials vary significantly. Refectory and chemically active metals like titanium and niobium, etc., when heated, react with the gases present and affect joint integrity. For metal combinations that have no or limited solubility in each other, extra care is taken for joining by diffusion bonding. For example, in the diffusion bonding of steel and copper, the diffusion of iron to copper or vice versa produces a solid solution with limited solubility, so to join such combinations, an intermediate layer can be used that has good solubility with both base metals.

6.4.1 USE OF INTERLAYERS IN DIFFUSION BONDING

Interlayers are commonly employed in diffusion bonding to unite incompatible materials or materials that are difficult to bind directly. It aids in the prevention/reduction of the production of brittle intermetallic compounds at the contact. Interlayers containing melting point depressants (e.g., B, Si, or P) or forming a eutectic with the parent metal being bonded can be used.

One of the most valuable criteria in the selection of an interlayer is that it should be metallurgically compatible with the materials to be bonded. Use of interlayers in diffusion bonding minimizes or eliminates the following problems of conventional diffusion bonding in dissimilar material joining:

- Restricts the development of intermetallic compounds
- Residual stresses in the diffusion bond
- Formation of voids at the joint interface

Using interlayer in diffusion bonding typically entails inserting a thin interlayer between the substrates; however, the interlayer material is occasionally positioned outside the joint to flow in through capillarity, as in several brazing procedures. The interlayer material can come in a variety of forms:

- Thin foil (Elrefaey & Tillmann, 2009; Yıldız et al., 2016; Zakipour et al., 2015a)
- Amorphous foil (Noto et al., 2013; Yuan et al., 2009; Zhou et al., 2015)
- Fine powders (Hdz-García et al., 2014; Hosseinabadi et al., 2014; Wang et al., 2015)
- Powder compact (Fillabi et al., 2008)
- Paste (Amirnasiri et al., 2017; Du & Shiue, 2009; Jiang et al., 2010; Soltani Tashi et al., 2014)
- Electroplating (Cooke et al., 2012; Gawde et al., 2010)

This section discusses the impact of various interlayers on the metallurgical and mechanical characteristics of dissimilar diffusion bonds. Noto et al. (2013) investigated the diffusion bonding of 9-Cr ODS steel with a Fe-3B-2Si-0.5C filler at 1180°C for 0.5–4 hours. It was claimed that the thickness of the original foil remained constant during the holding period. Grain coarsening was observed in the diffusion-affected zone of melting and isothermal solidification. Silicon diffusion into the base metal was found to be slower than that of boron and chromium because the foil was swiftly resolidified owing to a rise in melting temperature as the boron concentration reduced due to rapid boron diffusion and silicon remained in the bonding zone due to its slower diffusivity. Di Luozzo et al. (2014) studied the diffusion bonding of EN-235 steel and employed Fe-B-Si foil as an interlayer and reported similar tensile properties as base metals having failure far away from the interface.

The selection of material and the thickness of the interlayer are also important parameters in dissimilar metal joining. It is well accepted that the residual stresses at the diffusion bond interface decrease with an increase in interlayer thickness (Vaidya et al., 1998). But at the same time, increase in interlayer thickness impairs the mechanical properties. (Han et al., 2003; Tra et al., 2002; Zakipour et al., 2015b). Guo et al. (2016) investigated the influence of interlayer thickness in the bonding of cemented carbide to steel and discovered that an increase in thickness leads to poor strength (Figure 6.7).

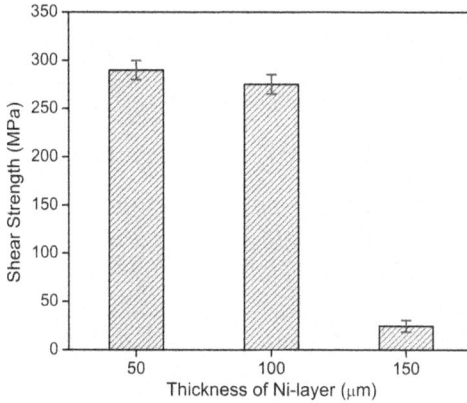

FIGURE 6.7   Bar chart showing the effect of interlayers on tensile strength (Guo et al., 2016).

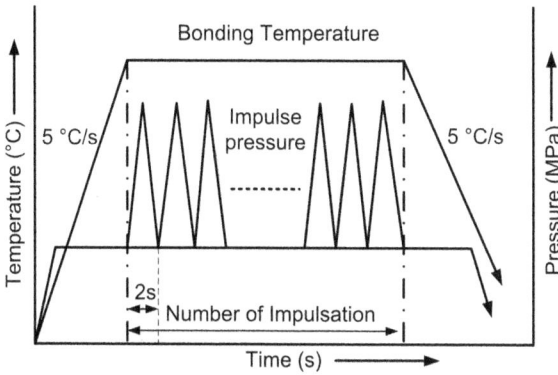

FIGURE 6.8   Schematic of the impulse pressuring diffusion bonding process (Yuan et al., 2013).

6.5   APPROACHES USED TO IMPROVE THE EFFECTIVENESS OF DISSIMILAR METAL BONDING

6.5.1   DIFFUSION BONDING USING PRESSURE PULSES

Diffusion bonding generally requires a highly polished surface to develop a sound diffusion bond. To overcome these preparation requirements and reduce bonding time, impulse pressure diffusion bonding can be used. This process is a modification of the conventional diffusion bonding process. The very first experiment that was conducted using pressure pulses was performed by the Ukraine Barton Welding Institute, and now researchers are using the same approach to obtain a sound joint in comparatively less bonding time using machined surfaces (Dai et al., 2013; Sharma et al., 2018; Sharma & Dwivedi, 2017, 2019; Wang et al., 2016). In this process, impulse pressure is applied instead of continuously applied constant pressure, (Figure 6.8). Due to the application of impulse pressure the oxide film breaks easily and the voids at the bonding interface reduce significantly, which in turn results in

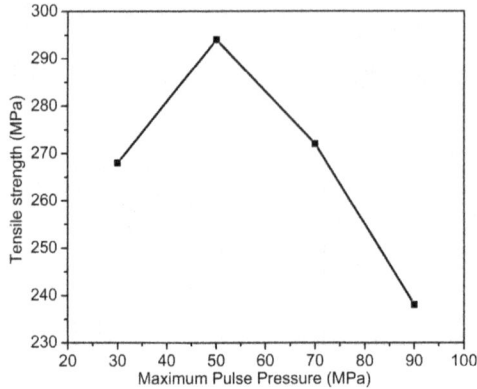

FIGURE 6.9   Effect of maximum impulse pressure on joint tensile strength (Yuan et al., 2008).

an accelerated bonding process. The duration of bonding decreases due to the accelerated bonding. Some studies are available for impulse pressure diffusion bonding.

Yuan et al. (2013) used impulse pressure and a pure nickel interlayer to conduct diffusion bonding of copper to stainless steel. Because of the impulse pressure, the time required for diffusion bonding was greatly shortened. The bond's strength was enhanced from 270 to 321 MPa. The interlayer impact was also seen on joint strength. At 15 μm interlayer thickness, joint strength improved by 42%. However, when interlayer thickness increases, joint strength decreases due to the production of intermetallic compounds.

In another study, Yuan et al. (2008) used pressure pulsation to perform diffusion bonding of titanium alloy to stainless steel and discovered that bonding time was greatly shortened due to impulse pressure. Figure 6.9 shows that the joint strength initially rises with maximal impulse pressure, but at a certain threshold, it begins to decrease owing to the production of microcracks.

6.5.2 FRICTION-ASSISTED DIFFUSION BONDING

In friction-assisted diffusion bonding, friction is used to break the oxide layer by the relative motion at the initial point of contact under pressure at a high temperature. Due to the breakdown of the oxide layer, diffusion bonding is accelerated. Chen et al. (2012, 2013) bonded $Zr_{55}Cu_{30}Ni_5Al_{10}$ BMG plates with aluminum alloy via pre-friction-aided diffusion bonding. To break the oxide deposit, a pre-friction treatment was conducted at 300 °C. When the pressure was applied, the plastic flow of the material at a high temperature reduced the volume of voids at the original interface and caused rapid growth in the contact area until sufficient area was achieved to support the pressure. During diffusion, voids are eliminated. It was found that in the conventional method, at 80 MPa pressure for a period of 18 minutes, bonding was incomplete, whereas for FADB, a complete joint was obtained at a pressure of 10 MPa for a period of 2 minutes (Table 6.1). They concluded that the friction treatment removes the oxide layer and improves the performance of diffusion bonding.

TABLE 6.1

Comparison of Process Parameter and Joint Strength of the Processes

S.N.	Process	Temperature (°C)	Time (min)	Pressure (MPa)	Strength (MPa)
1	CDB	440	18	80	2.10 ± 0.83
2	FADB	440	2	10	58.42 ± 5.29

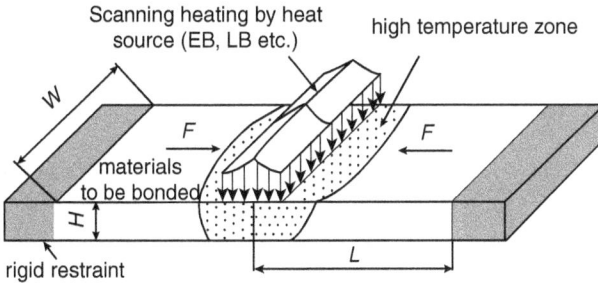

FIGURE 6.10   Setup of rigid restraint thermal self-compressing diffusion bonding (Deng et al., 2014).

6.5.3   SELF-COMPRESSING DIFFUSION BONDING

In this method, the polished plates are fixed rigidly in a fixture, and the interface is heated with the help of a scanning heat source. During the heating of the material near the bond interface, thermal expansion took place, and compressive stresses were generated because the fixed restrained plates resisted the expansion. Due to the development of compressive pressure in the thermoplastic material near the interface, diffusion across the interface took place, resulting in a sound joint (Figure 6.10). Deng et al. (2014) used an electron beam heat source to achieve stiff self-compressing diffusion bonding in Ti-alloy. In a vacuum, the interface was heated to 850°C. According to reports, the strength of the formed diffusion bond was comparable to that of the base metal.

6.5.4   FRICTION STIR WELDING-ASSISTED DIFFUSION BONDING

It is a hybrid joining technology that employs the heat generated by friction stir welding as a heat source to create a lap joint via diffusion bonding. To produce the heat, a friction stir welding tool is penetrated in the upper plate. The pin of the rotating tool is penetrated in such a way that it is always 0.1–0.05 mm above the interface to avoid intermetallic formation at the interface. Haghshenas et al. (2014) used friction stir-driven diffusion bonding to bond an aluminum alloy to a Zn-coated high-strength steel (Figure 6.11). Lap joints were developed using the generated heat of FSW as a heat source for diffusion bonding. This heat encourages interdiffusion at the interface. The diffusion bond produced at the 16 mm/min traverse speed of friction stir welding has a maximum shear strength of 61.6 MPa.

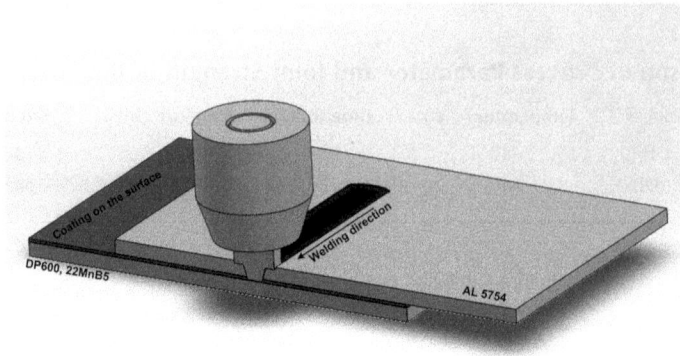

FIGURE 6.11 Schematic of setup used for friction stir welding-assisted diffusion bonding (Haghshenas et al., 2014).

6.6 CONCLUSION

This chapter reported dissimilar metal joining by diffusion bonding. Considering the importance of the area and problems associated with the fusion welding of dissimilar metals, diffusion bonding was found to be a promising candidate as a solid-state joining technology. The process variables: bonding temperature, pressure, and time, as well as the bonding environment, material properties, and the preparation method of the contact surfaces, were found to be important. Some innovative approaches, like friction-assisted and self-compressing diffusion bonding, are promising techniques to fabricate sound joints with improved quality characteristics. Future attempts are recommended to explore many other novel techniques in diffusion bonding for joining various types of materials, to optimize the process, and to achieve sustainability.

REFERENCES

Amirnasiri, A., Parvin, N., & haghshenas, M. S. (2017). Dissimilar diffusion brazing of WC-Co to AISI 4145 steel using RBCuZn-D interlayer. *Journal of Manufacturing Processes*, *28*, 82–93. https://doi.org/10.1016/j.jmapro.2017.06.001.

Chen, H. Y., Cao, J., Song, X. G., Shi, Y. L., & Feng, J. C. (2012). Effect of pre-friction surface treatment on the pre-friction assisted diffusion bonding of Zr55Cu30Ni5Al10 bulk metallic glass to aluminum alloy. *Materials Letters*, *74*, 125–127. https://doi.org/10.1016/j.matlet.2012.01.106.

Chen, H. Y., Cao, J., Song, X. G., Si, G. D., & Feng, J. C. (2013). Pre-friction diffusion hybrid bonding of Zr55Cu30Ni5Al10 bulk metallic glass. *Intermetallics*, *32*, 30–34. https://doi.org/10.1016/j.intermet.2012.07.030.

Cooke, K. O., Khan, T. I., & Oliver, G. D. (2012). Transient liquid phase diffusion bonding Al-6061 using nano-dispersed Ni coatings. *Materials and Design*, *33*(1), 469–475. https://doi.org/10.1016/j.matdes.2011.04.051.

Dai, G. Q., WQ, Q., & Zhuang, H. (2013). Structure performance and diffusion mechanism of aluminum alloy heat pipe low-temperature diffusion brazing joints. *Chinese Journal of Rare Metals*, *37*(6), 851.

Deng, Y., Guan, Q., Wu, B., Wang, X., & Tao, J. (2014). Study on rigid restraint thermal self-compressing bonding - A new solid state bonding method. *Materials Letters, 129*, 43–45. https://doi.org/10.1016/j.matlet.2014.05.029.

Di Luozzo, N., Doisneau, B., Boudard, M., Fontana, M., & Arcondo, B. (2014). Microstructural and mechanical characterizations of steel tubes joined by transient liquid phase bonding using an amorphous Fe-B-Si interlayer. *Journal of Alloys and Compounds, 615*(S1), S18–S22. https://doi.org/10.1016/j.jallcom.2013.11.161.

Du, Y. C., & Shiue, R. K. (2009). Infrared brazing of Ti - 6Al - 4V using two silver-based braze alloys. *Journal of Materials Processing Technology, 209*, 5161–5166. https://doi.org/10.1016/j.jmatprotec.2009.03.001.

Du, Z., Zhang, K., Lu, Z., & Jiang, S. (2018). Microstructure and mechanical properties of vacuum diffusion bonding joints for γ-TiAl based alloy. *Vacuum, 150*, 96–104. https://doi.org/10.1016/j.vacuum.2018.01.035.

Elrefaey, A., & Tillmann, W. (2009). Solid state diffusion bonding of titanium to steel using a copper base alloy as interlayer. *Journal of Materials Processing Technology, 209*(5), 2746–2752. https://doi.org/10.1016/j.jmatprotec.2008.06.014.

Fillabi, M. G., Simchi, A., & Kokabi, A. H. (2008). Effect of iron particle size on the diffusion bonding of Fe-5%Cu powder compact to wrought carbon steels. *Materials and Design, 29*(2), 411–417. https://doi.org/10.1016/j.matdes.2007.01.004.

Gao, L., Li, X., Hua, P., Wang, M., & Zhou, W. (2018). Nickel interlayer on the microstructure and property of TC6 to copper alloy diffusion bonding. *Journal of Adhesion Science and Technology, 32*(14), 1548–1559. https://doi.org/10.1080/01694243.2018.1429860.

Gawde, P. S., Kishore, R., Pappachan, A. L., Kale, G. B., & Dey, G. K. (2010). Low temperature diffusion bonding of stainless steel. *Transactions of the Indian Institute of Metals, 63*(6), 853–857. https://doi.org/10.1007/s12666-010-0130-x.

Guo, Y., Wang, Y., Gao, B., Shi, Z., & Yuan, Z. (2016). Rapid diffusion bonding of WC-Co cemented carbide to 40Cr steel with Ni interlayer: Effect of surface roughness and interlayer thickness. *Ceramics International, 42*(15), 16729–16737. https://doi.org/10.1016/j.ceramint.2016.07.145.

Haghshenas, M., Abdel-Gwad, A., Omran, A. M., Gökçe, B., Sahraeinejad, S., & Gerlich, A. P. (2014). Friction stir weld assisted diffusion bonding of 5754 aluminum alloy to coated high strength steels. *Materials and Design, 55*, 442–449. https://doi.org/10.1016/j.matdes.2013.10.013.

Han, J. U. N. H., Ahn, J. A. E. P., & Shin, M. C. (2003). Effect of interlayer thickness on shear deformation behavior of AA5083 aluminum alloy / SS41 steel plates manufactured by explosive welding. *Journals of Materials Science, 8*, 13–18.

Hdz-García, H. M., Martinez, A. I., Muñoz-Arroyo, R., Acevedo-Dávila, J. L., García-Vázquez, F., & Reyes-Valdes, F. A. (2014). Effects of silicon nanoparticles on the transient liquid phase bonding of 304 stainless steel. *Journal of Materials Science and Technology, 30*(3), 259–262. https://doi.org/10.1016/j.jmst.2013.11.006.

Hosseinabadi, N., Sarraf-Mamoory, R., & Mohammad Hadian, A. (2014). Diffusion bonding of alumina using interlayer of mixed hydride nano powders. *Ceramics International, 40*(2), 3011–3021. https://doi.org/10.1016/j.ceramint.2013.10.006.

Jiang, W., Gong, J. M., & Tu, S. T. (2010). Effect of holding time on vacuum brazing for a stainless steel plate-fin structure. *Materials and Design, 31*(4), 2157–2162. https://doi.org/10.1016/j.matdes.2009.11.001.

Kazakov, N. F. (1985). *Diffusion Bonding of Materials*. Moscow: Mir Publications.

Kundu, S., Ghosh, M., Laik, A., Bhanumurthy, K., Kale, G. B., & Chatterjee, S. (2005). Diffusion bonding of commercially pure titanium to 304 stainless steel using copper interlayer. *Materials Science and Engineering A, 407*(1–2), 154–160. https://doi.org/10.1016/j.msea.2005.07.010.

Kurt, B., & Çalik, A. (2009). Interface structure of diffusion bonded duplex stainless steel and medium carbon steel couple. *Materials Characterization*, *60*(9), 1035–1040. https://doi.org/10.1016/j.matchar.2009.04.011.

Kurt, B., Orhan, N., & Hasçalik, A. (2007a). Effect of high heating and cooling rate on interface of diffusion bonded gray cast iron to medium carbon steel. *Materials and Design*, *28*(7), 2229–2233. https://doi.org/10.1016/j.matdes.2006.06.002.

Kurt, B., Orhan, N., Evin, E., & Çalik, A. (2007b). Diffusion bonding between Ti - 6Al - 4V alloy and ferritic stainless steel. *Materials Letters*, *61*, 1747–1750. https://doi.org/10.1016/j.matlet.2006.07.123.

Liu, K., Li, Y., Xia, C., & Wang, J. (2017). Effect of bonding time on interfacial microstructure and shear strength of vacuum diffusion bonding super-Ni / NiCr laminated composite to Ti-6Al-4V joint without interlayer. *Vacuum*, *143*, 195–198. https://doi.org/10.1016/j.vacuum.2017.06.025.

Noh, S., Kim, B., Kasada, R., & Kimura, A. (2012). Diffusion bonding between ODS ferritic steel and F82H steel for fusion applications. *Journal of Nuclear Materials*, *426*(1–3), 208–213. https://doi.org/10.1016/j.jnucmat.2012.02.024.

Noto, H., Kasada, R., Kimura, A., & Ukai, S. (2013). Grain refinement of transient liquid phase bonding zone using ODS insert foil. *Journal of Nuclear Materials*, *442*(1–3 SUPPL.1), S567–S571. https://doi.org/10.1016/j.jnucmat.2013.04.054.

Özdemir, N., & Bilgin, B. (2009). Interfacial properties of diffusion bonded Ti-6Al-4V to AISI 304 stainless steel by inserting a Cu interlayer. *International Journal of Advanced Manufacturing Technology*, *41*(5–6), 519–526. https://doi.org/10.1007/s00170-008-1493-6.

Ren, X., Li, S., & Xiong, Z. (2018). Isostatic diff usion bonding and post-solution treatment between Cr22Ni5Mo3MnSi and Cr30Ni7Mo3MnSi duplex stainless steels. *Journal of Manufacturing Processes*, *34*, 215–224. https://doi.org/10.1016/j.jmapro.2018.06.010. Isostatic.

Sharma, G., & Dwivedi, D. K. (2017). Effect of pressure pulsation on bond interface characteristics of 409 ferritic stainless steel diffusion bonds. *Vacuum*, *146*, 152–158. https://doi.org/10.1016/j.vacuum.2017.09.049.

Sharma, G., & Dwivedi, D. K. (2018). Impulse pressure-assisted diffusion bonding of ferritic stainless steel. *International Journal of Advanced Manufacturing Technology*, *95*(9–12), 4293–4305. https://doi.org/10.1007/s00170-017-1490-8.

Sharma, G., & Dwivedi, D. K. (2019). Study of metallurgical and mechanical properties of CSEF P92 steel diffusion bonds developed using pressure pulsation. *Journal of Manufacturing Processes*, *38*(December 2018), 196–203. https://doi.org/10.1016/j.jmapro.2019.01.017.

Sharma, G., Tiwari, L., & Dwivedi, D. K. (2018). Impulse pressure assisted diffusion bonding of low carbon steel using silver interlayer. *Transactions of the Indian Institute of Metals*, *71*(1), 11–21. https://doi.org/10.1007/s12666-017-1136-4.

Sheng, G. M. (2005). An experimental investigation of phase transformation superplastic diffusion bonding of titanium alloy to stainless steel. *Journal of Materials Science*, *0*, 6385–6390. https://doi.org/10.1007/s10853-005-1629-0.

Soltani Tashi, R., Akbari Mousavi, S. A. A., & Mazar Atabaki, M. (2014). Diffusion brazing of Ti-6Al-4V and austenitic stainless steel using silver-based interlayer. *Materials and Design*, *54*, 161–167. https://doi.org/10.1016/j.matdes.2013.07.103.

Tra, D., Ferrante, M., & Ouden, G. Den. (2002). Diffusion bonding of aluminium oxide to stainless steel using stress relief interlayers. *Materials Science & Engineering A*, *337*, 287–296.

Vaidya, R. U., Rangaswamy, P., Bourke, M. A. M., & Butt, D. P. (1998). Measurement of bulk residual stresses in molybdenum disilicide/stainless steel loints uning neutron scattering. *Acta Materialia*, *46*(6), 2047–2061.

Velmurugan, C., Senthilkumar, V., Sarala, S., & Arivarasan, J. (2016). Low temperature diffusion bonding of Ti-6Al-4V and duplex stainless steel. *Journal of Materials Processing Technology*, *234*, 272–279. https://doi.org/10.1016/j.jmatprotec.2016.03.013.

Wang, W., Fan, D., Huang, J., Cui, B., Chen, S., & Zhao, X. (2015). A new partial transient liquid-phase bonding process with powder-mixture interlayer for bonding Cf/SiC composite and Ti-6Al-4V alloy. *Materials Letters*, *143*, 237–240. https://doi.org/10.1016/j.matlet.2014.12.110.

Wang, F. L., Sheng, G. M., & Deng, Y. Q. (2016). Impulse pressuring diffusion bonding of titanium to 304 stainless steel using pure Ni interlayer. *Rare Metals*, *35*(4), 331–336. https://doi.org/10.1007/s12598-014-0368-2.

Yıldız, A., Kaya, Y., & Kahraman, N. (2016). Joint properties and microstructure of diffusion-bonded grade 2 titanium to AISI 430 ferritic stainless steel using pure Ni interlayer. *International Journal of Advanced Manufacturing Technology*, *86*(5–8), 1287–1298. https://doi.org/10.1007/s00170-015-8244-2.

Yılmaz, O., & Yılmaz, O. (2016). Effect of welding parameters on diffusion bonding of type 304 stainless steel - copper bimetal. *Journal of Materials Science & Technology*, *0836*(March). https://doi.org/10.1179/026708301101510834.

Yuan, X., Kim, M. B., & Kang, C. Y. (2009). Characterization of transient-liquid-phase-bonded joints in a duplex stainless steel with a Ni-Cr-B insert alloy. *Materials Characterization*, *60*(11), 1289–1297. https://doi.org/10.1016/j.matchar.2009.05.012.

Yuan, X. J., Sheng, G. M., Qin, B., Huang, W. Z., & Zhou, B. (2008). Impulse pressuring diffusion bonding of titanium alloy to stainless steel. *Materials Characterization*, *59*(7), 930–936. https://doi.org/10.1016/j.matchar.2007.08.003.

Yuan, X., Tang, K., Deng, Y., Luo, J., & Sheng, G. (2013). Impulse pressuring diffusion bonding of a copper alloy to a stainless steel with/without a pure nickel interlayer. *Materials and Design*, *52*, 359–366. https://doi.org/10.1016/j.matdes.2013.05.057.

Zakipour, S., Halvaee, A., Amadeh, A. A., Samavatian, M., & Khodabandeh, A. (2015a). An investigation on microstructure evolution and mechanical properties during transient liquid phase bonding of stainless steel 316L to Ti-6Al-4V. *Journal of Alloys and Compounds*, *626*, 269–276. https://doi.org/10.1016/j.jallcom.2014.11.160.

Zakipour, S., Samavatian, M., Halvaee, A., Amadeh, A., & Khodabandeh, A. (2015b). The effect of interlayer thickness on liquid state diffusion bonding behavior of dissimilar stainless steel 316/Ti-6Al-4V system. *Materials Letters*, *142*, 168–171. https://doi.org/10.1016/j.matlet.2014.11.158.

Zhou, X., Dong, Y., Liu, C., Liu, Y., Yu, L., Chen, J., Li, H., & Yang, J. (2015). Transient liquid phase bonding of CLAM/CLAM steels with Ni-based amorphous foil as the interlayer. *Materials and Design*, *88*, 1321–1325. https://doi.org/10.1016/j.matdes.2015.09.104.

7 Friction Stir Welding
A Solution for Dissimilar Material Joining

Sunil Sinhmar and Kallol Mondal

7.1 INTRODUCTION

Friction stir welding (FSW) comes under the category of solid-state joining techniques. The term 'solid state' indicates that the joining takes place below the melting point of the base materials. Figure 7.1 shows the schematic of the FSW process. It uses a non-consumable tool with a negligible consumption rate that has a pin, shoulder, and shank (tool body) as its integral parts (Mehta et al., 2019). The plates to be welded are clamped in the butt position, which is the most common joint configuration. However, other configurations, such as lap, can also be used. The tool under rotation on its axis is inserted at the abutting line and moves in the welding direction (Thapliyal & Dwivedi, 2020). A weld joint generally comprises various zones, such as the nugget zone (NZ; pin-stirred region), the thermo-mechanically affected zone (TMAZ), the heat-affected zone (HAZ), and the unaffected base

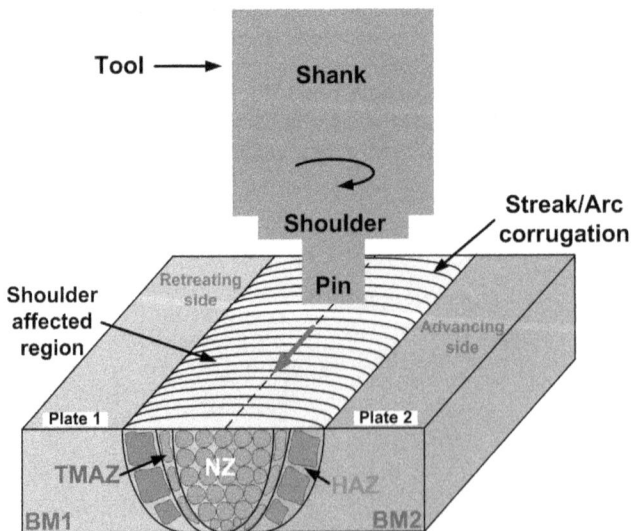

FIGURE 7.1 Schematic of friction stir welding of two dissimilar materials.

DOI: 10.1201/9781003327769-7

material (Sinhmar & Dwivedi, 2020a). The NZ experiences dynamic plastic deformation (PD) and undergoes recrystallization. This is the region where the actual joining takes place. The region next to the NZ is the TMAZ, a very narrow region experiencing PD without recrystallization. The next region is the HAZ, which only experiences thermal effects during welding without PD and recrystallization. The region at the top surface that experiences stirring due to shoulder rotation is called the shoulder-affected zone. The curvature-type impression in this region is called streak or arc corrugation. Advancing and retreating sides have been discussed in the later portion of the chapter.

Present industrial manufacturing requires lightweight structures with a good strength-to-weight ratio, corrosion resistance, and wear resistance (Fattah-alhosseini et al., 2019). In the interest of low-weight designs and increasing fuel productivity, particularly within the transportation fields, different joining strategies have been adapted to enhance the design performance and working of distinctive joint structures. Similarly, numerous applications in defense sectors as well as power industries need joints of different materials fabricated by several conventional and solid-state methods. The dissimilar joints are important due to their various economic and technical advantages, such as weight reduction, diminishing the emission of greenhouse gases by lowering fuel consumption, and hence, improving the overall performance of the product. Therefore, the demand for an eco-friendly joining technique is increasing among researchers (Zhang et al., 2019).

Welding is the best alternative to replace bolting and riveting in the aerospace industry (Jonckheere et al., 2013). Presently, there is no other better option than FSW to join dissimilar materials (Kwon et al., 2008). As the difference between the metallurgical as well as mechanical properties of the two materials increases, the joining of two becomes difficult. Differences in the chemical, thermal, and physical properties also add to the difficulty of joining dissimilar materials (Mehta & Badheka, 2016a). The joining of dissimilar alloys, such as Al alloys, Cu alloys, and Mg alloys, etc., gives an advantage in obtaining the combined properties of both alloys in the joint. Such joints are more desirable in complex loading conditions, and their demand is increasing with the discoveries of newer materials and the requirements of new designs (Pabandi et al., 2018).

The joint of dissimilar alloys indicates the joining of either different grades of the same alloy (e.g. 2024 aluminum alloy to 7075 aluminum alloy) or different alloy systems, e.g. Al alloy to Mg alloy. Figure 7.2 shows a FSW joint of dissimilar materials (aluminum and magnesium) at different rotational speeds (Shi et al., 2017). The variation in the stir zone with changes in rotational speeds can be noticed. The mixing is not significant at 600 and 700 rpm, as indicated in Figure 7.2a and b, respectively, while mixing between Al and Mg becomes significant with an increase in the rotational speed, as shown in Figure 7.2c and d. Banded structures and severely deformed zones can be observed at high rpm. Banded structure comprises dark and gray bands of both materials at the NZ/TMAZ interface. The severely deformed zone is located at the bottom part of the NZ and formed due to the severe intermixing at the lower part of the pin. The points a, b, c, d, e, f, and g show the different locations at the NZ/TMAZ interfaces on the advancing side at different speeds, while point 'h' shows the shoulder/material interface.

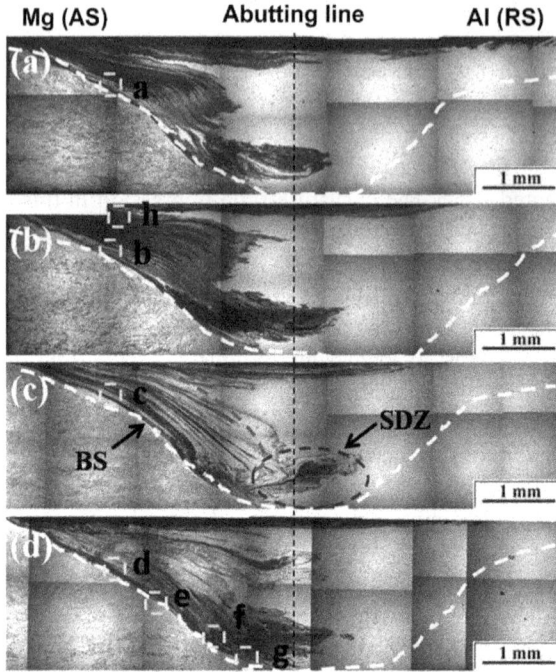

FIGURE 7.2 Macrostructure showing the cross section of FSW joint of Al and Mg alloys welded at a constant traverse speed of 100 mm/min and a rotational speed of (a) 600 rpm, (b) 700 rpm, (c) 800 rpm, and (d) 900 rpm.

The combination of many dissimilar materials typically joined by fusion welding leads to the formation of intermetallic compounds in the joint and results in poor mechanical performance, especially in terms of tensile strength and ductility. Formation of brittle phases in the fusion zone occurs depending on the type of base material, and this significantly reduces the mechanical properties, such as strength and ductility, along with fatigue resistance (Shanmuga Sundaram & Murugan, 2010). Formation of the intermetallics can be easily controlled by controlling the time and temperature of the joint at optimized process parameters in a solid-state joining process.

Researchers have been trying numerous solid-state processes, such as friction welding, ultrasonic welding, explosive welding, and diffusion bonding, to achieve a dissimilar joint with fewer intermetallics. However, some factors, like high power consumption, size limits, high cost of equipment and maintenance, etc., constrain their use. Among various solid-state techniques, FSW is a novel and more effective technique. High repeatability, low cost of equipment, cheaper running and maintenance, as well as high weld quality, make the FSW a profound welding technique for dissimilar materials. The characteristics, like lower heat input, homogenization, refined microstructure, and densification in the case of FSW, improve the strength of the joint as compared to other joining techniques.

The properties and soundness of the FSW joint of dissimilar materials are highly dependent on the positions of the base materials, i.e., advancing and retreating sides. It has been observed that generally, the material with lower strength positioned on the advancing side results in better mechanical performance than the one on the retreating side (Jamshidi Aval et al., 2011a). The decision to keep material on either side is complicated because of the different thermal and conductive nature of the materials in cases of dissimilar joints. The position of the material affects the material flow mechanism due to the unsymmetrical nature of deformation by the rotating tool on the advancing and rotating sides. Therefore, the quality of the mixing of the materials in the stir zone is highly dependent on the relative position. The position is also important to obtain better microstructure and mechanical strength with a defect-free joint. According to literature, the stir zone comprises the material positioned on the retreating side as a major portion, and thus the performance of the weld joint mainly depends on the material kept on the retreating side.

It has also been found that the placement of high-resistive material (in terms of strength and corrosion susceptibility) on the retreating side ultimately gives a high-resistive joint (Jonckheere et al., 2013). Similarly, the placement of softer material on the advancing side results in a fracture on the same side after the joining (Fattahalhosseini et al., 2019). The softer material on the advancing side also results in high yield, tensile strength, and fatigue strength with better mixing of the material. Strength of the joint is dictated by the weaker material. However, the region, which experiences a drastic reduction in strength owing to the annealing phenomenon, undergoes failure (Amancio-Filho et al., 2008).

On the other hand, opposite recommendations are also available in the literature (Khajeh et al., 2021). Figure 7.3 shows a dissimilar FSW joint of Al and Cu. The harder copper sheet is kept on the advancing side, whereas the softer material, i.e., aluminum is positioned on the retreating side. This combination results in a sound joint. In this case, the shoulder is plunged up to a depth of 0.05 mm in the surface of the base material, and the tilt angle is kept at 2°.

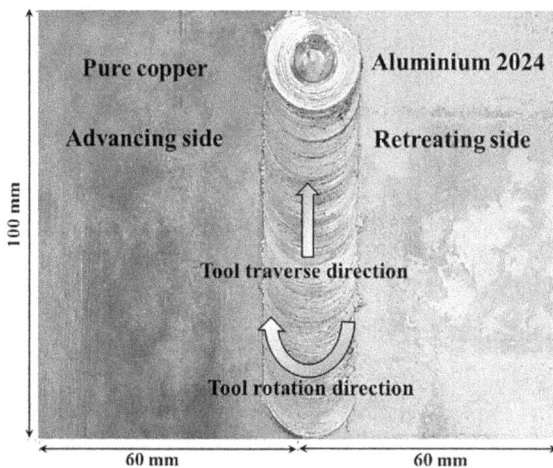

FIGURE 7.3 Friction stir-welded plates of Al alloy and copper.

Tool offset becomes more important to produce simultaneous PD in both materials despite having dissimilar thermal properties and melting points. However, optimization of the process parameters, such as rotational and traverse speeds, can help to skip the tool offsetting.

In an FSW process, the material mixing by the pin due to the PD takes place under the shoulder of a rotating tool, which is known as an FSW tool. Diffusion of the materials at interfaces occurs at high temperatures during FSW, and this leads to intermetallic formation. High material movement during stirring in FSW of dissimilar materials results in more interfaces, and hence diffusion increases, further accelerating the intermetallic formation.

The dissimilar joints are severely prone to corrosion attack. Corrosion occurs due to the formation of stronger galvanic cells because of the presence of dissimilar materials as compared to the similar materials used in welding. The schematic shows the preferential corrosion attack in the anodic or active region than the cathodic region to the galvanic couple (Figure 7.4) (Sinhmar & Dwivedi, 2018). In this case, the Al matrix behaves as the anode, whereas the Al_2Cu precipitate behaves as the cathode. The material around the precipitate gets dissolved preferentially because of the potential difference, and the precipitate comes out of the surface, resulting in a pit. Large difference in the potential of two joining materials would increase the severity of the corrosion attack. The formation of different regions and microstructural and metallurgical heterogeneities also leads to the development of a micro-galvanic cell in a FSW joint. Regional heterogeneities cannot be avoided in a dissimilar FSW joint.

7.2   MAJOR PARAMETERS

There are many parameters that need to be taken care of during FSW. The main parameters that need special attention are rotational speed, traverse speed, dwell time, tilt angle, position of the plate (advancing side or retreating side), tool offset, tool shoulder dimension and feature, and shape and profile of the pin. Out of these parameters, position of the plate to be welded and tool offset become extremely important in the case of dissimilar FSW. These parameters are briefly discussed below.

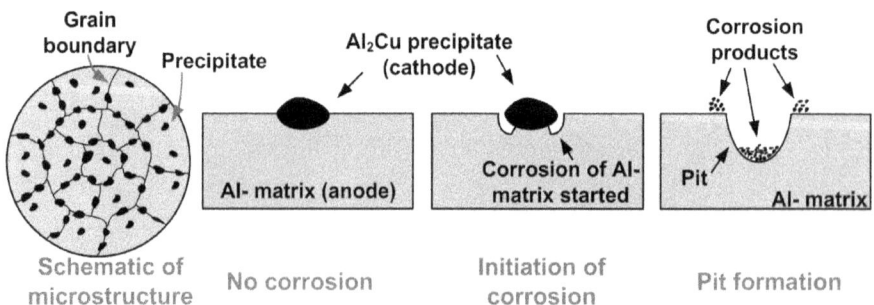

FIGURE 7.4   Schematic showing the propagation of corrosion attack due to formation of anode and cathode.

7.2.1 ROTATIONAL SPEED

Rotation of the tool in terms of rotation per minute (rpm) is needed to perform the stirring action by the pin in the weld region. This causes PD and frictional heating during the welding (Sinhmar & Dwivedi, 2019). Tool is rotated at a particular rotational speed by considering the type of material as a workpiece, the thickness of the plates to be welded, the cooling condition, etc.

7.2.2 TRAVERSE SPEED

Traverse speed is the relative linear movement between the tool and workpiece. This is the speed with which welding is performed, and this is also known as 'welding speed'. Most of the time, bed of the machine is moved at traverse speed. In other words, the workpiece moves, whereas the tool remains stationary (just rotating). Generally, traverse speed is expressed in terms of millimeters per minute (mm/min) in the FSW. This parameter plays an important role in deciding the heat input into the weld.

7.2.3 DWELL TIME

Before starting the welding, workpieces at the region of joining should be heated properly so that they become soft and result in a sound weld joint. Therefore, to achieve this softness, the tool is kept in a rotating condition without traversing after plunging into the workpiece. The time for which it is kept in this condition before traversing is called dwell time. It is expressed in terms of a second(s). Dwell time depends on various factors, such as type of workpiece materials, thickness of workpiece, rotation speed, etc.

7.2.4 TILT ANGLE

This is also called 'tool tilt angle'. It is an angle between the central axis of the tool and a vertical line perpendicular to the workpiece to be welded (Figure 7.5a). This is provided by giving an inclination to the machine spindle. Generally, this is provided on the trailing side (back side) of the tool. It is not always necessary to give a tilt angle. Sometimes, FSW is performed without providing the tilt angle (zero tilt angle). Tool tilt angle provides the forging action by the shoulder on the workpiece surface, which, in turn, reduces the possibility of defect formation, enabling the development of a sound weld joint. It is expressed in degrees and generally kept in the range of 0°–3°. However, it depends on various factors, such as workpiece materials, tool shoulder profile, etc.

7.2.5 ADVANCING SIDE

The material flow behavior in and around the rotating pin in the case of FSW/friction stir processing is not similar (Thapliyal & Dwivedi, 2018). This causes a significant difference in the microstructure as well as properties (mechanical and corrosion)

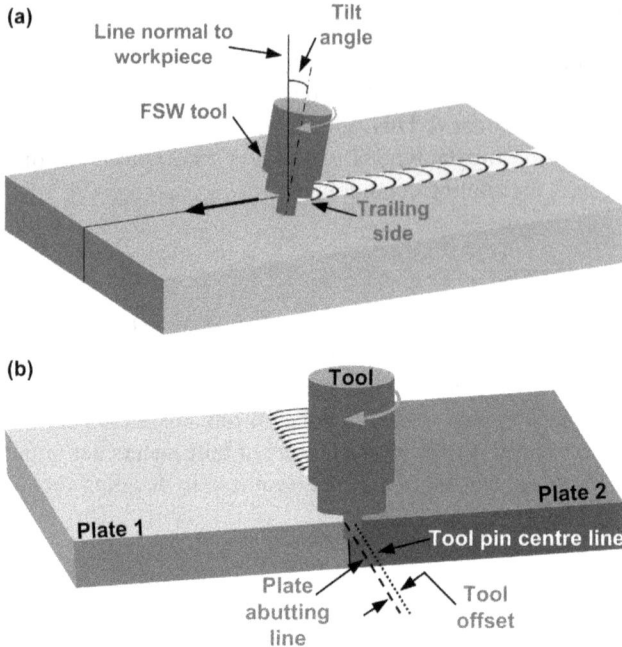

FIGURE 7.5 Schematic of the FSW process, showing (a) tilt angle and (b) tool offset.

of both sides. The side on which rotation of the tool and traversing are in the same direction is called the advancing side (Figure 7.1).

7.2.6 RETREATING SIDE

The side on which rotation of the tool and movement of the tool are in opposite directions is called the retreating side. This can be easily understood by visualizing the tool's rotation and movement in Figure 7.1. Consideration of both sides becomes more important in the case of dissimilar FSW than the FSW of similar materials. Material is positioned on a particular side (either retreating or advancing) by considering their various properties, such as hardness, etc.

7.2.7 TOOL OFFSET

Generally, in the conventional FSW of similar materials, center of the tool pin is kept at the abutting line of the plates to be welded. Here, the abutting line means the line at which two plates are kept in butt weld position, and the welding is carried out along this line. It indicates that no offset of the tool has been used in this condition. Therefore, tool offset is the amount of distance by which the center of the tool pin is intentionally shifted on either side, i.e., advancing and retreating sides from the abutting line. Figure 7.5b shows the tool offset used during the FSW. In this case, the welding is not carried out along the abutting line when tool offset is used. Generally, this parameter is used in the dissimilar FSW and is expressed in 'mm'. It significantly affects the intermetallic formation, generation of heat during the welding, material flow, and defect formation in the weld joint, such as root defects, lack of penetration, etc. Tool offsetting reduces the extent of mixing of dissimilar materials because the tool pin is concentrated toward one material, and hence, reduces the degree of intermetallic formation.

7.2.8 SHOULDER DIAMETER

A large amount of heat is generated due to friction between the tool shoulder and the surface of the workpiece during FSW. Therefore, the shoulder is the major source of heat generation during the FSW (Sinhmar & Dwivedi, 2020b). Generally, its dimension is kept three times to the plate thickness. For example, for a plate of 6 mm thickness, the preferred shoulder diameter is 18 mm. However, it is not necessary to follow this convention. A large diameter generates more heat than a small one. Various types of features of the shoulder surface are used in the literature, for example, curved, concave, flute, etc. (Verma & Saha, 2023). However, studies on some other kinds of tools, such as shoulderless tools and double-shoulder tools, are also available in the literature (Eslami et al., 2015; Mirzaei et al., 2020).

7.2.9 PIN PROFILE

Shape of the pin of the FSW tool is also an important parameter to obtain a sound weld joint. Pin can be cylindrical, conical, or trapezoidal in shape. Different shapes possess different effects on the material flow. The pin can be threaded or non-threaded. Generally, the intensity of PD is greater in the case of the threaded pin than the non-threaded pin. The effect of pin shape and profile varies according to the type of material. The pin profile becomes more significant in the case of the dissimilar FSW due to different properties of the base materials, the requirement of tool offsetting, and the different material flow behavior on the advancing and retreating sides.

7.2.10 PITCH RATIO

This is the ratio of the tool traverse speed (v) to the tool rotational speed (w) and is generally represented as v/w. The pitch ratio would increase with an increase in traverse speed and a decrease in tool rotational speed, and vice versa.

7.3 FSW OF DISSIMILAR MATERIALS

7.3.1 DISSIMILAR ALUMINUM ALLOYS

The joint of wrought and cast aluminum alloys is commonly used in various industries. Such types of dissimilar joints have the potential to increase the usage of economical castings. Aluminum alloys are the alloys on which FSW has been used more frequently than other material systems. Most of the time, FSW joints exhibit lower strength than the base material in the case of aluminum alloys. As the gap between flow and material characteristics of two materials increases, the optimization of the FSW process for those combinations becomes more difficult.

Vast literature is available on the FSW joints of dissimilar Al alloys (Shah et al., 2018). It has been claimed that the position of the stronger Al alloy on the advancing side increases the joint efficiency of the dissimilar FSW joint, and the combination of the materials affects the tensile strength of the joint significantly. The effect of various pin profiles on the performance of FSW joints made of Al alloys has also been studied (Palanivel et al., 2012). In an FSW joint between AA7050 and AA6061, the majority of the joints have fractured from the AA6061 side and HAZ. However, poor mixing of the materials at low rotational speed results in stir zone fracture of the joint.

In literature, the heat-treatable Al alloy is friction stir-welded with non-heat-treatable Al alloy, and the effect of natural aging as well as residual stresses on the performance of the joint has been investigated (Jamshidi Aval et al., 2011b). The side of the joint, which comprises heat-treatable Al alloy, experiences softening due to the heat input during welding. In this case, 100% joint efficiency has been achieved in a FSW joint between boron carbide containing AA1100 and heat-treatable Al alloys. However, in this case, the heat-treatable Al alloy (AA6063) is positioned on the advancing side (Guo et al., 2012). Similarly, fatigue strength is also dependent on the types of alloys placed on the advancing side of the joint. Furthermore, an increase in tool speed results in a high fatigue strength of the joint.

A complex vortex pattern has been noticed in the friction stir seam welding of dissimilar Al alloys due to dissimilar intermixing (Gerlich et al., 2008). The Al alloy with low flow strength has been positioned on the advancing side. Dissimilar intermixing is more profound at high rotational speed and low traverse speed, and this is attributed to the significant stirring due to the threaded pin, traversing motion, and in-situ extrusion during the friction stir seam welding.

The mixing process between AA7075 (material used for the stringer, which is a thin strip fastened to the skin material in the aircraft) and AA2024 (skin material, that covers the outer body of the aircraft) is considerably improved by enhancing the rotational speed of the FSW tool (Ahmed et al., 2021). Low rotation speed affects the stringer material within the stir zone only. However, an increase in the rotational speed allows the AA7075 to get distributed and mixed with the AA2024. The ratio of shoulder diameter to pin diameter at a value of 3 results in a higher mechanical strength than the weld produced at other values, such as 2, 2.5, 3.5, and 4 (Saravanan et al., 2016). The morphology and size of the grains of both types of materials used in the joints are also important factors. Rotation speed affects material mixing as well as torque to a greater extent than traverse speed.

In a study of dissimilar FSW of 2024 and 7075 Al alloys, severe corrosion attack has been observed on the 7075 side, owing to the more anodic behavior (D'Urso et al., 2017). The FSW joint exhibited different corrosion behavior in different regions (NZ, TMAZ, HAZ, and BM) due to the microstructural variation. Generally, the region with low hardness experiences more pitting attacks than the region with high hardness.

7.3.2 Aluminum to Non-aluminum Alloys

7.3.2.1 Aluminum-Copper

The combination of Cu and Al alloys is used in automobiles, electrical, and aerospace industries. Currently, wide research is being dedicated to joining the dissimilar alloys such as Al and Cu to attain their combined beneficial properties. Conventional fusion welding of Al to Cu could result in some intermetallics, which can deteriorate the mechanical performance of the joint due to the noteworthy dissimilarity in the physical and chemical properties of both materials.

The welding of Al to Cu is always a difficult task due to the differences in physical properties, melting temperatures, and chemical compositions. The quality of Al-Cu FSW joints is highly dependent on the position of the materials and tool pin offset (Mehta & Badheka, 2016b). In a study, it has been reported that the offsetting of the tool toward the Al side (placed on the advancing side) results in a defect-free joint at lower heat input. In such cases, Cu was kept on the retreating side (Galvão et al., 2010). Offsetting of the pin toward the Al alloy causes further stirring on the advancing side, and some of the particles are detached from the Cu side. This helps in good metallurgical bonding among Al matrix and Cu particles to give defect-free joints (Galvão et al., 2012). The high heat input due to the high tool rotational speed and large shoulder diameter increase the electrical resistivity of the Al-Cu FSW joint by more than 9% (Akinlabi & Akinlabi, 2012). Therefore, it has been concluded that the increase in heat input into the joint results in a high electrical resistivity of the joint.

Owing to their differences, the friction stir weld of copper and aluminum is vulnerable to bimetallic or galvanic corrosion. The presence of Al_2Cu in the weld region promotes corrosion attack due to its noble behavior than the surrounding Al matrix. This causes the local potential difference and accelerates corrosion. The increase in anodic activity in the NZ has been reported at low rpm, whereas HAZ is severely affected at high rpm (Akinlabi et al., 2014).

7.3.2.2 Aluminum-Magnesium

The generation of intermetallics, such as $Al_{30}Mg_{23}$, $Al_{12}Mg_{17}$, and Al_2Mg_3, inhibits the use of conventional welding processes for joining Al-Mg dissimilar joints. Therefore, it has been concluded that discontinuous and uniform disseminative intermetallics could be helpful to enhance joint strength (Ji et al., 2017). In most of the Al-Mg FSW joints, Al kept on the advancing side, resulting in a sound joint. Dissimilar FSW of Mg and Al alloys would help to achieve weight reduction in the structure. The weld shows a great volume of $Al_{12}Mg_{17}$ and higher hardness in the center of the weld than in other regions. PD during dissimilar FSW usually leads to alternating bands (layers) of both base materials in the center of the weld. Hence, in the case of the Al-Mg joint, the presence of alternating bands of Mg and Al alloys at the weld center was reported in the literature (Yan et al., 2010). The peak temperature during the FSW also helps Al and Mg atoms diffuse at the interfaces of the bands. Generally, the gap between two consecutive bands/rings is equal to the traverse of the tool per unit rotation. This zone (banded) becomes the weakest part of the Al-Mg joint due to the formation of intermetallics (Shi et al., 2017). Sometimes, the FSW of low-melting-temperature material to a comparatively high-melting-temperature material causes local melting of low-melting-point material in the stir zone, which is termed liquation. The difference in the melting temperature of aluminum and magnesium is not much; therefore, the direct liquation of either material does not take place during the FSW. However, according to the Al-Mg phase diagram, the eutectic temperature of Al-Mg constituents is low, which leads to melting of the eutectic, especially at high rpm (Azizieh et al., 2016). Hence, the parameters resulting in low heat input, such as the low rpm of the tool, can help reduce/prevent the liquation in the FSW joint.

In the case of corrosion, the severity of galvanic attack is associated with the ratio of the anode-to-cathode surface area. For a given cathode surface area (noble material), a smaller anode (active material) would corrode heavily as compared to the situation where the anode area is large with respect to the cathodic area. This is termed the 'area effect' in the corrosion. The prime locations of the severity of electrochemical attack are observed in the narrow zones of AZ31 alloy (Mg alloy) next to Al2024 (Al alloy) areas, where a low ratio of anode-to-cathode surface area is observed. The attack is fundamentally caused by the formation of a strong galvanic couple amid Al and Mg alloys in the dissimilar FSW joints (Liu et al., 2009). Residual stresses are found in the welded workpieces owing to mechanical restraints, severe PD, and thermal loading - unloading experienced during FSW. The residual stresses (transverse and longitudinal) are usually found more on the advancing side than the retreating side, especially when the tool is offset toward the advancing side. Tool offset applied to the Mg side results in higher maximum residual stresses on the retreating side as compared to the advancing side.

In literature, avoiding high heat input by using some cooling systems has shown great improvement in joint reliability owing to the suppression of intermetallic formation. Numerous approaches for cooling-aided FSW need to be encouraged for future work (Singh et al., 2020). The most common cooling methods used for the FSW joints are forced cooling using water, air, liquid nitrogen, liquid CO_2, etc. Figure 7.6 shows the use of a cooling source during the FSW. A nozzle (cooling source) moves behind the tool at a fixed distance to supply the cooling media. It continuously supplies the cooling media at a constant flow rate to increase the cooling rate of the FSW joint. The use of a high-conductive material for the back plate also helps to increase the cooling rate of the weld joints.

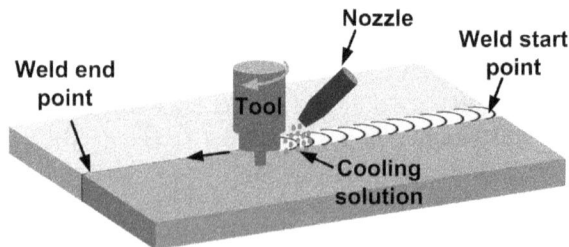

FIGURE 7.6 Schematic showing the use of a cooling source during friction stir welding.

7.3.2.3 Aluminum-Steel

The dissimilar combination of Al and steel is one of the outstanding structural combinations considered by industries such as marine, automotive, railway, and aerospace (Kaushik & Dwivedi, 2021). Some intermediate materials, such as zinc foil, are sometimes used to produce a sandwich structure of steel and Al to produce a dissimilar FSW joint. The rotating tool is offset toward the Al side, and the pin is plunged only into the Al workpiece (Ogawa et al., 2019). Figure 7.7 shows the variation in tensile strength of the Al-steel joint with respect to tool offset (Watanabe et al., 2006). The schematic shown in Figure 7.7 indicates the position of the pin in the base metal plates. The number '0' indicates the abutting line, whereas the symbols '−' and '+' show the aluminum and mild steel sides, respectively. An offset of 0.2 mm toward the Fe plate results in a higher tensile strength than the other offset values (−0.2, 0, 0.4, 0.6, 1, and 2 mm) in a mild steel-Al joint. The ultimate tensile strength is drastically reduced as the pin shifts toward the positive/Fe side after reaching the tool offset value of '0.4'. The problem in welding steel with Al alloys is primarily due to the enormously low dilution among the major constituents, i.e., Al and Fe. Therefore, the strong affinity of aluminum and iron to form brittle intermetallics reduces the dynamic as well as static strength of the joint.

On the contrary, it is true that the limited presence of intermetallics at the Al-steel interface is good for obtaining the joint strength. However, a large increase in intermetallic thickness reduces the same. Figure 7.8 shows the respective change in joint strength (ultimate tensile strength) with respect to intermetallic thickness for Al-steel joints (Tanaka et al., 2009). The solid symbols show the failure of the joint at the weld interface, while the open symbols show the failure from the aluminum parent material side. The joint strength improves exponentially with the reduction in the intermetallic thickness, as shown by the curved line in Figure 7.8. In the case of the steel-7075 Al alloy joint, the maximum joint strength is less than the ultimate tensile strength of the 7075 base metal.

Intermetallics grow quickly through interaction among the FSW tool and steel workpiece, and therefore, it is essential to control such interactions to prevent their growth.

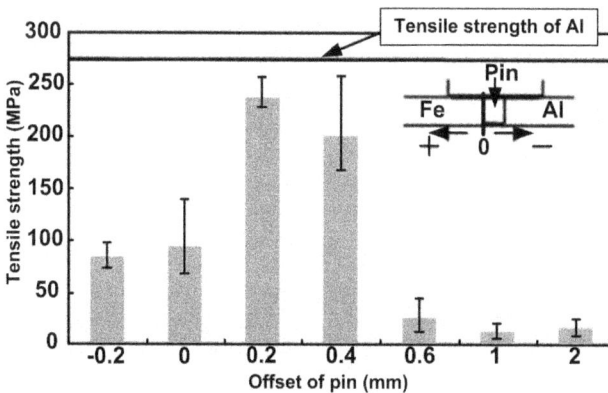

FIGURE 7.7   Variation of tensile strength of the dissimilar FSW joints with a change in tool offset.

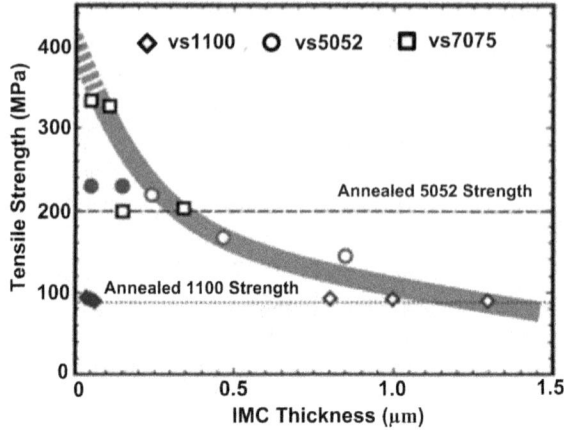

FIGURE 7.8 Curve showing the variation in tensile strength of a dissimilar FSW joint with respect to IMC thickness.

A strategy such as forced cooling to control the intermetallic growth rate could help increase the strength of joints during the production of lightweight structures in the automobile and aerospace industries. Variants of FSW, such as the friction stir scribe technique, can also be used to decrease the heat input to render the intermetallic formation (T. Wang et al., 2019). The presence of intermetallics of Al-Fe increases the corrosion susceptibility of the Al-steel FSW joint. Only a small portion of the pin (due to offsetting) rotates on the steel side, which results in small fragments of steel and intermetallics in the Al matrix at the top surface of the weld in the presence of high heat input in this region. The rotation of the shoulder at the top surface of the weld results in high heat input in this region. Therefore, the top surface experiences a higher corrosion rate than the remaining portion of the weld joint.

7.3.2.4 Aluminum-Titanium

Initially, corner joints using FSW are tried for Ti alloys. Thereafter, dissimilar FSW joints of Ti with other light alloys have been produced and investigated, especially for automobile and aerospace applications (Dressler et al., 2009). The weld joint of Ti with other metal systems is also susceptible to intermetallic compound formation. In the case of FSW of Ti with Al, the major phases like $TiAl_3$ and TiAl are formed as a result of the weld's heat input, which is affected by the rotational speed variation (Shehabeldeen et al., 2021). The huge difference between the material properties, like thermal conductivity, crystal structure, coefficient of linear expansion, melting point, etc., between Ti and Al makes their welding difficult. Titanium particles are found dispersed in the stir zone of the FSW joint of Al alloy and Ti alloy (Kar et al., 2020). The dissimilar welding of Ti alloy to Al alloy could have a foremost application in the automobile and aerospace industries, where cost savings, weight reduction, and high strength are desirable. For example, the joining of skin-stringer is gradually substituting the riveted fuselage assemblies in airplane manufacturing.

The thermal conductivity of the material to be welded is an important property, and the same is true in the case of Ti and its alloys, which have a lower value than steels. Therefore, there are greater difficulties in attaining a sound quality dissimilar FSW joint of Al-Ti as compared to Al-steel (Li et al., 2014). The thickness of the intermetallic compound layer at the Al-Ti interface plays a very important role in the performance of Al-Ti dissimilar joints. Some external element additions, like Nb, reduce the brittleness of the joint by restricting the intermetallic formation, such as $TiAl_3$ in the weld joint. It has also been reported that the tool offset toward the Ti alloy workpiece results in a sound joint (Bang et al., 2011).

7.3.3 ALUMINUM TO NON-METALS

7.3.3.1 Aluminum-Polymer

The use of polymers offers easier manufacturability, enhanced thermal insulation properties, and further flexibility in design, which are the main factors in reducing the overall cost of manufacturing (Moshwan et al., 2015). In an Al-polymer joint, the load-bearing capacity of the product gets affected by the development of bubbles and holes along with polymer degradation during FSW due to the raised local temperature (Derazkola & Simchi, 2019). The key joining mechanisms are recommended to be the generation of nano-scale pores in a thin zone inside the metal surface oxide near the interface, which accommodates intrusion of polymers, along with secondary van der Waals' bonding. The above-mentioned mechanism can be attributed to the chemical reactions at the interface, resulting in the substantial improvements in the performance of the FSW joint, especially in terms of mechanical properties compared to the polymer substrate.

7.3.4 MISCELLANEOUS

7.3.4.1 Steel-Titanium

The lap type of joints of steel-Ti have been produced using FSW. Generally, the Ti at the top of the lap results in better performance than the steel at the top, owing to the higher softness of the Ti than the steel. An optimized welding process parameter is needed to achieve a minimum level of heat combination to attain a sound weld joint (Campo et al., 2014). The Ti is a highly reactive material, especially at elevated temperatures, and therefore, oxygen pick-up can be observed at high rotational speeds owing to the high temperature in the NZ of the FSW joint. In a lap FSW joint of Ti-304SS, the presence of a solid solution of oxygen in Ti and the development of oxide films can reduce the Ti flow and its plasticity. This, in turn, weakens the NZ and thwarts direct contact of Ti with 304SS (Fazel-Najafabadi et al., 2010).

In a study, the presence of interlock type features at the interface was found to be responsible for the improved shear strength of the FSW joint of Ti to stainless steel (Fazel-Najafabadi et al., 2011). The corrosion behavior of the FSW joints of steel-Ti alloys has not been studied much in the literature. A FSW joint of stainless steel-NiTi alloy shows a marginal effect on the corrosion resistance of the FSW joint compared to both base metals (West et al., 2021). The SS-NiTi alloy joint promotes the formation of passive film and reduces the galvanic effect in the weld joint.

7.3.4.2 Dissimilar Copper Alloys

The fusion welding of the Cu alloys becomes difficult due to their high thermal conductivity along with the high melting temperature. These alloys require preheating to carry out the fusion welding. Generally, the FSW operates below the melting temperature of the base material, and this feature makes it more preferable than the fusion welding process for the Cu alloys. Mostly continuous as well as discontinuous types of dynamic recrystallization have been observed in the case of friction stir processing and FSW of dissimilar Cu alloy (Cu-brass) joints, and these mechanisms can be changed by varying the welding parameters and material composition of the materials (Heidarzadeh et al., 2018). However, the formation of new recrystallized grains in the case of pure Cu is due to continuous dynamic recrystallization. Both high heat input (high rpm/low traverse speed) and low heat input (low rpm/high traverse speed) lead to a defective weld joint with low tensile strength. A joint prepared at optimized parameters (900 rpm/40 mm/min) leads to balanced heat input, which in turn gives a high tensile strength to the Cu-Brass joint (Erdem, 2015).

7.3.4.3 Dissimilar Polymer Materials

Polymer is preferred in various structures due to its light weight. The use of reinforced composite polymers is growing in automobile and aerospace manufacturing applications. The FSW, among other joining methods such as adhesive joining, laser joining, ultrasonic joining, etc., is the most favorable method to join different polymers due to its ease of operation, low maintenance, low cost, and high joint performance (Kumar et al., 2018). The lap joint configuration is the most preferred configuration for the FSW joint of polymers based on their requirements, such as in structural applications. Therefore, most of the literature available is on the lap joint configuration (Eslami et al., 2018). In the case of the polymer-to-polymer FSW joint, a tool with a large diameter of pin, a low feed rate, and long-time pressure application was recommended to achieve the uniform microstructure of the polymer (Rezaee Hajideh et al., 2017). In a FSW joint of low-density polyethylene, the reinforcement of Fe particles during welding increased the tensile strength of the joint as compared to the parent materials (Singh et al., 2016). According to a study, the use of a stationary shoulder tool as compared to the conventional tool for the FSW of dissimilar polymer materials results in high lap shear strength (Eslami et al., 2018). Most of the samples were fractured on the retreating side during the lap shear test.

7.3.4.4 Dissimilar Steels

Some problems of conventional welding like hydrogen cracking and solidification cracking can be avoided by using FSW due to its low-melting and solid-state nature (Santos et al., 2018). A vortex-like shape is observed in a FSW joint due to the flow of material under the influence of tool movement (traverse and rotation). Similar features can also be observed in the dissimilar FSW joints of other alloys (Li et al., 2012). The presence of Cr or other particles responsible for chromium carbide formation or other undesirable products is not observed in the FSW joint. Hence, intermetallics or chromium carbide are not observed in the FSW joint (Jafarzadegan et al., 2012). It should be noted that no HAZ is observed on the stainless-steel side. In a FSW joint between stainless steel and S275 steel, mechanical as well as metallurgical bonding have been found as key bonding mechanisms at the interface

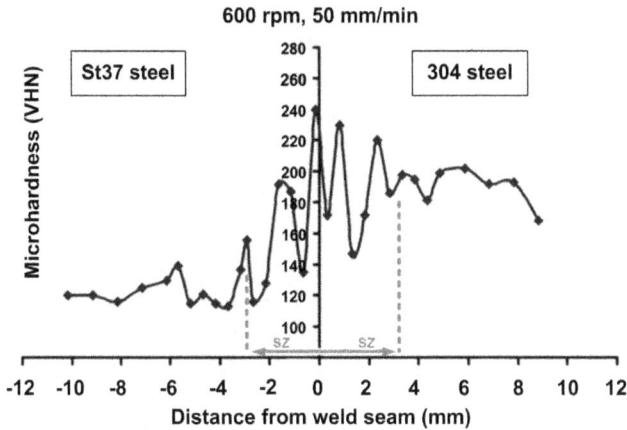

FIGURE 7.9 Microhardness variation in a dissimilar FSW joint of st37 and 304 steel.

(H. Wang et al., 2019). The grain refinement in the NZ of a dissimilar steel FSW joint results in high hardness among all the regions of the joints (Figure 7.9) (Jafarzadegan et al., 2013). The width of the NZ is almost close to the pin diameter, and the maximum hardness observed in this region is higher than the maximum hardness of both base metals. The presence of ferrite and pearlite in the NZ results in a variation in the hardness of this zone. The influence of pitch ratio on the variation of grain size has been found to be more significant than the tool offset. Pitch ratio directly affects the strain rate as well as material flow, and this leads to grain refinement (Pankaj et al., 2020).

7.3.4.5 Titanium-Carbon Fiber-Reinforced Polymer

The common methods available in the literature to carry out joining of metallic systems and carbon fiber-reinforced polymer (CFRP) are ultrasonic joining, riveting, friction spot joining, and laser joining (Balle et al., 2009). The melting during laser joining causes defects such as porosity, while ultrasonic and friction spot joining methods have limitations on the size of the joint. FSW can be a preferable alternative to producing a sound weld joint of CFRP in metallic systems due to its solid-state nature and more control over the interface temperature of the joint. Very limited work has been reported on the FSW joint of Ti-CFRP dissimilar materials. The dissimilar FSW joint between Ti and CFRP exhibited significant shear strength owing to the appropriate reactions between Ti and CFRP with modifications to their surfaces. The weld was produced at an intermediate temperature between the thermal decomposition temperature and the melting temperature of the CFRP. The defects toward the CFRP near the FSW interface were suppressed at this temperature.

7.4 MAJOR ISSUES

The way to respond at elevated temperatures while experiencing the deformation during FSW by the different materials is not identical. This unlike behavior of the different materials during FSW makes the joining difficult. It is very hard to obtain such joining parameters that are equally suitable for both materials. Even an inappropriate tool position can cause defects such as line remnants. Tool offsetting is required in cases where the dissimilar materials have a large gap in their metallurgical and physical properties. The formation of galvanic cells and electrochemical interactions between dissimilar materials in FSW joints are of great concern, and these need further investigation. The presence of various regions, such as the NZ, TMAZ, HAZ, and parent material, results in galvanic cell formation, which accelerates the corrosion. The interfaces where proper mixing is not observed attribute poor corrosion resistance to the improperly mixed region of the stir zone. As usual, in the case of the heat-treatable Al alloys, the NZ experiences a higher corrosion rate than the base metals. The existence of Al-Mg-Zn precipitates and Al-Cu precipitates in the joints having 2xxx and 7xxx series Al alloys develops the sites with potential differences and hence generates the galvanic cell. The development of brittle intermetallics leading to weakening of joints and cracking of welds in dissimilar joints is a major concern, and it is hard to control their formation at the interface. The performance of the weld can be enhanced by controlling the heat generation using some kinds of cooling systems, such as submerged FSW. Further, a wrong choice of FSW parameters in the fabrication of dissimilar materials results in problems like dispersion of the hard material particles (the ones that are hard among both materials) in NZ toward the softer material. This dispersion is highly dependent on the tool offset and rotation speed. The lack of repeatability and standardization of the FSW process is still a major issue and needs to be addressed for the industrialization of FSW.

7.5 FUTURE SCOPE

Over the course of 30 years, the use of FSW has expanded from aluminum to steel and even plastic. It progressed from a stage where the tool shoulder had a variety of attributes to a stage where it is shoulderless. Despite having many benefits, some of the problematic aspects of this technique have not been properly addressed in the literature. One of the major issues that needs to be tackled in the future is the presence of an exit hole at the end of the weld joint. This is the bigger hurdle to accepting the FSW for industrial applications. In the future, during the FSW of high-strength dissimilar materials, the use of a clamping device (such as a heating shoe) with the function of independent heat supply to both sides of the workpieces can be considered. The use of stationary shoulder FSW for dissimilar materials can be explored in more depth in the future. The stationary shoulder tool supplies concentrated heat to the base materials as compared to the conventional tool. The tool life in high-strength material welding is limited and highly susceptible to wear during the FSW. Therefore, a special coated FSW tool can be developed for such applications. A thorough study of intermetallics needs to be carried out in future work to identify the optimum thickness of the intermetallics. Moreover, the parameters responsible for producing such intermetallics, which would help to improve the performance of the weld joint, need to be explored in the future. FSW can become a pioneer in joining dissimilar materials with different thicknesses in the future. Work can be done to develop sophisticated tooling systems to produce weld joints from the sheets with varying thicknesses along their length. The literature lacks a study of the various forces involved in the FSW process. Therefore, a complete investigation can be carried out to study the effects of axial force, thrust force, torque, etc. Machine learning will be an important tool in the future to enhance the productivity of FSWs made of dissimilar materials. Various models and algorithms can be used to identify the possibility of defects in dissimilar weld joints before their occurrence. Machine learning can improve the efficiency and precision of the process. This technique can be helpful for real-time control of the FSW of dissimilar materials with varying properties in the future.

7.6   SUMMARY

The current manufacturing sectors have a critical demand for joining materials that are different from one another. The differences in various properties, such as thermal, physical, chemical, etc., make the joining of dissimilar materials difficult. In this scenario, FSW becomes a pioneer in producing defect-free and economical joints without complexity. FSW develops the joint below the melting point of the workpiece material. Thus, the chance of intermetallic formation in the case of the dissimilar FSW joints is greatly reduced. Among the various sectors, the scope for the automobile industry to use dissimilar materials' FSW is utmost. It helps in the weight reduction of the vehicle, which in turn reduces fuel consumption. Lightweight vehicles help achieve the goal of reducing carbon emissions in the future. Therefore, dissimilar weld joints are important from an economical as well as sustainable point of view. FSW is extensively used for the joining of aluminum alloys. The automobile industry commonly uses the dissimilar joint of aluminum with other metal systems like steel due to its higher strength-to-weight ratio. In view of this, the present chapter discusses various features of FSW joints of dissimilar aluminum alloys, aluminum to other metal systems, and dissimilar metal systems except aluminum, along with practical examples and major issues. The various studies of the metallurgical, mechanical, and electrochemical behavior of the FSW joints confirm the capability of the FSW technique to produce dissimilar material fabrication. Various approaches, such as parameter optimization and forced cooling, seem helpful to reduce the intermetallic formation, which in turn improves the performance of the FSW joint. This chapter also discusses the role of various parameters such as location of dissimilar workpieces, tilt angle, tool offset, etc. to give an insight to the students as well as research fellows for future research and development in this field.

ACKNOWLEDGEMENTS

Authors want to acknowledge Science and Engineering Research Board, India to provide the fund under project no. PDF/2022/001915.

REFERENCES

Ahmed, M. M. Z., El-Sayed Seleman, M. M., Zidan, Z. A., Ramadan, R. M., Ataya, S., & Alsaleh, N. A. (2021). Microstructure and mechanical properties of dissimilar friction stir welded AA2024-T4/AA7075-T6 T-butt joints. *Metals*, *11*(1), 1–19. https://doi.org/10.3390/met11010128.

Akinlabi, E. T., & Akinlabi, S. A. (2012). Effect of heat input on the properties of dissimilar friction stir welds of aluminium and copper. *American Journal of Materials Science*, 2(5), 147–152. https://doi.org/10.5923/j.materials.20120205.03.

Akinlabi, E. T., Andrews, A., & Akinlabi, S. A. (2014). Effects of processing parameters on corrosion properties of dissimilar friction stir welds of aluminium and copper. *Transactions of Nonferrous Metals Society of China (English Edition)*, 24(5), 1323–1330. https://doi.org/10.1016/S1003-6326(14)63195-2.

Amancio-Filho, S. T., Sheikhi, S., dos Santos, J. F., & Bolfarini, C. (2008). Preliminary study on the microstructure and mechanical properties of dissimilar friction stir welds in aircraft aluminium alloys 2024-T351 and 6056-T4. *Journal of Materials Processing Technology*, 206(1–3), 132–142. https://doi.org/10.1016/j.jmatprotec.2007.12.008.

Azizieh, M., Sadeghi Alavijeh, A., Abbasi, M., Balak, Z., & Kim, H. S. (2016). Mechanical properties and microstructural evaluation of AA1100 to AZ31 dissimilar friction stir welds. *Materials Chemistry and Physics*, 170, 251–260. https://doi.org/10.1016/j.matchemphys.2015.12.046.

Balle, F., Wagner, G., & Eifler, D. (2009). Ultrasonic metal welding of aluminium sheets to carbon fibre reinforced thermoplastic composites. *Advanced Engineering Materials*, 11(1–2), 35–39. https://doi.org/10.1002/adem.200800271.

Bang, K. S., Lee, K. J., Bang, H. S., & Bang, H. S. (2011). Interfacial microstructure and mechanical properties of dissimilar friction stir welds between 6061-T6 aluminum and Ti-6%Al-4% V alloys. *Materials Transactions*, 52(5), 974–978. https://doi.org/10.2320/matertrans.L-MZ201114.

Campo, K. N., Campanelli, L. C., Bergmann, L., Santos, J. F. dos, & Bolfarini, C. (2014). Microstructure and interface characterization of dissimilar friction stir welded lap joints between Ti-6Al-4V and AISI 304. *Materials and Design*, 56, 139–145. https://doi.org/10.1016/j.matdes.2013.11.002.

Derazkola, H. A., & Simchi, A. (2019). An investigation on the dissimilar friction stir welding of T-joints between AA5754 aluminum alloy and poly(methyl methacrylate). *Thin-Walled Structures*, 135, 376–384. https://doi.org/10.1016/j.tws.2018.11.027.

Dressler, U., Biallas, G., & Alfaro Mercado, U. (2009). Friction stir welding of titanium alloy TiAl6V4 to aluminium alloy AA2024-T3. *Materials Science and Engineering A*, 526(1–2), 113–117. https://doi.org/10.1016/j.msea.2009.07.006.

D'Urso, G., Giardini, C., Lorenzi, S., Cabrini, M., & Pastore, T. (2017). The effects of process parameters on mechanical properties and corrosion behavior in friction stir welding of aluminum alloys. *Procedia Engineering*, 183, 270–276. https://doi.org/10.1016/j.proeng.2017.04.038.

Erdem, M. (2015). Investigation of structure and mechanical properties of copper-brass plates joined by friction stir welding. *International Journal of Advanced Manufacturing Technology*, 76(9–12), 1583–1592. https://doi.org/10.1007/s00170-014-6387-1.

Eslami, S., de Figueiredo, M. A. V., Tavares, P. J., & Moreira, P. M. G. P. (2018). Parameter optimisation of friction stir welded dissimilar polymers joints. *International Journal of Advanced Manufacturing Technology*, 94(5–8), 1759–1770. https://doi.org/10.1007/s00170-017-0043-5.

Eslami, S., Ramos, T., Tavares, P. J., & Moreira, P. M. G. P. (2015). Shoulder design developments for FSW lap joints of dissimilar polymers. *Journal of Manufacturing Processes*, 20, 15–23. https://doi.org/10.1016/j.jmapro.2015.09.013.

Fattah-alhosseini, A., Naseri, M., Gholami, D., Imantalab, O., Attarzadeh, F. R., & Keshavarz, M. K. (2019). Microstructure and corrosion characterization of the nugget region in dissimilar friction-stir-welded AA5083 and AA1050. *Journal of Materials Science*, 54(1), 777–790. https://doi.org/10.1007/s10853-018-2820-4.

Fazel-Najafabadi, M., Kashani-Bozorg, S. F., & Zarei-Hanzaki, A. (2010). Joining of CP-Ti to 304 stainless steel using friction stir welding technique. *Materials and Design*, 31(10), 4800–4807. https://doi.org/10.1016/j.matdes.2010.05.003.

Fazel-Najafabadi, M., Kashani-Bozorg, S. F., & Zarei-Hanzaki, A. (2011). Dissimilar lap joining of 304 stainless steel to CP-Ti employing friction stir welding. *Materials and Design*, *32*(4), 1824–1832. https://doi.org/10.1016/j.matdes.2010.12.026.

Galvão, I., Leal, R. M., Loureiro, A., & Rodrigues, D. M. (2010). Material flow in heterogeneous friction stir welding of aluminium and copper thin sheets. *Science and Technology of Welding and Joining*, *15*(8), 654–660. https://doi.org/10.1179/1362171 10X12785889550109.

Galvão, I., Loureiro, A., Verdera, D., Gesto, D., & Rodrigues, D. M. (2012). Influence of tool offsetting on the structure and morphology of dissimilar aluminum to copper friction-stir welds. *Metallurgical and Materials Transactions A: Physical Metallurgy and Materials Science*, *43*(13), 5096–5105. https://doi.org/10.1007/s11661-012-1351-x.

Gerlich, A., Su, P., Yamamoto, M., & North, T. H. (2008). Material flow and intermixing during dissimilar friction stir welding. *Science and Technology of Welding and Joining*, *13*(3), 254–264. https://doi.org/10.1179/174329308X283910.

Guo, J., Gougeon, P., & Chen, X. G. (2012). Microstructure evolution and mechanical properties of dissimilar friction stir welded joints between AA1100-B 4C MMC and AA6063 alloy. *Materials Science and Engineering A*, *553*, 149–156. https://doi.org/10.1016/j. msea.2012.06.004.

Heidarzadeh, A., Laleh, H. M., Gerami, H., Hosseinpour, P., Shabestari, M. J., & Bahari, R. (2018). The origin of different microstructural and strengthening mechanisms of copper and brass in their dissimilar friction stir welded joint. *Materials Science and Engineering A*, *735*, 336–342. https://doi.org/10.1016/j.msea.2018.08.068.

Jafarzadegan, M., Abdollah-zadeh, A., Feng, A. H., Saeid, T., Shen, J., & Assadi, H. (2013). Microstructure and mechanical properties of a dissimilar friction stir weld between austenitic stainless steel and low carbon steel. *Journal of Materials Science and Technology*, *29*(4), 367–372. https://doi.org/10.1016/j.jmst.2013.02.008.

Jafarzadegan, M., Feng, A. H., Abdollah-Zadeh, A., Saeid, T., Shen, J., & Assadi, H. (2012). Microstructural characterization in dissimilar friction stir welding between 304 stainless steel and st37 steel. *Materials Characterization*, *74*, 28–41. https://doi.org/10.1016/j. matchar.2012.09.004.

Jamshidi Aval, H., Serajzadeh, S., & Kokabi, A. H. (2011a). Thermo-mechanical and microstructural issues in dissimilar friction stir welding of AA5086-AA6061. *Journal of Materials Science*, *46*(10), 3258–3268. https://doi.org/10.1007/s10853-010-5213-x.

Jamshidi Aval, H., Serajzadeh, S., & Kokabi, A. H. (2011b). Evolution of microstructures and mechanical properties in similar and dissimilar friction stir welding of AA5086 and AA6061. *Materials Science and Engineering A*, *528*(28), 8071–8083. https://doi. org/10.1016/j.msea.2011.07.056.

Ji, S., Meng, X., Liu, Z., Huang, R., & Li, Z. (2017). Dissimilar friction stir welding of 6061 aluminum alloy and AZ31 magnesium alloy assisted with ultrasonic. *Materials Letters*, *201*, 173–176. https://doi.org/10.1016/j.matlet.2017.05.011.

Jonckheere, C., de Meester, B., Denquin, A., & Simar, A. (2013). Torque, temperature and hardening precipitation evolution in dissimilar friction stir welds between 6061-T6 and 2014-T6 aluminum alloys. *Journal of Materials Processing Technology*, *213*(6), 826–837. https://doi.org/10.1016/j.jmatprotec.2013.01.001.

Kar, A., Yadav, D., Suwas, S., & Kailas, S. V. (2020). Role of plastic deformation mechanisms during the microstructural evolution and intermetallics formation in dissimilar friction stir weld. *Materials Characterization*, *164*, 110371. https://doi.org/10.1016/j. matchar.2020.110371.

Kaushik, P., & Dwivedi, D. K. (2021). Effect of tool geometry in dissimilar Al-Steel Friction Stir Welding. *Journal of Manufacturing Processes*, *68*, 198–208. https://doi.org/10.1016/j. jmapro.2020.08.007.

Khajeh, R., Jafarian, H. R., Seyedein, S. H., Jabraeili, R., Eivani, A. R., Park, N., Kim, Y., & Heidarzadeh, A. (2021). Microstructure, mechanical and electrical properties of dissimilar friction stir welded 2024 aluminum alloy and copper joints. *Journal of Materials Research and Technology*, *14*, 1945–1957. https://doi.org/10.1016/j.jmrt.2021.07.058.

Kumar, R., Singh, R., Ahuja, I. P. S., Penna, R., & Feo, L. (2018). Weldability of thermoplastic materials for friction stir welding- A state of art review and future applications. *Composites Part B: Engineering*, *137*, 1–15. https://doi.org/10.1016/j.compositesb.2017.10.039.

Kwon, Y. J., Shigematsu, I., & Saito, N. (2008). Dissimilar friction stir welding between magnesium and aluminum alloys. *Materials Letters*, *62*(23), 3827–3829. https://doi.org/10.1016/j.matlet.2008.04.080.

Li, B., Zhang, Z., Shen, Y., Hu, W., & Luo, L. (2014). Dissimilar friction stir welding of Ti-6Al-4V alloy and aluminum alloy employing a modified butt joint configuration: Influences of process variables on the weld interfaces and tensile properties. *Materials and Design*, *53*, 838–848. https://doi.org/10.1016/j.matdes.2013.07.019.

Li, X. W., Zhang, D. T., Qiu, C., & Zhang, W. (2012). Microstructure and mechanical properties of dissimilar pure copper/1350 aluminum alloy butt joints by friction stir welding. *Transactions of Nonferrous Metals Society of China (English Edition)*, *22*(6), 1298–1306. https://doi.org/10.1016/S1003-6326(11)61318-6.

Liu, C., Chen, D. L., Bhole, S., Cao, X., & Jahazi, M. (2009). Polishing-assisted galvanic corrosion in the dissimilar friction stir welded joint of AZ31 magnesium alloy to 2024 aluminum alloy. *Materials Characterization*, *60*(5), 370–376. https://doi.org/10.1016/j.matchar.2008.10.009.

Mehta, K. P., & Badheka, V. J. (2016a). Effects of tool pin design on formation of defects in dissimilar friction stir welding. *Procedia Technology*, *23*, 513–518. https://doi.org/10.1016/j.protcy.2016.03.057.

Mehta, K. P., & Badheka, V. J. (2016b). Effects of tilt angle on the properties of dissimilar friction stir welding copper to aluminum. *Materials and Manufacturing Processes*, *31*(3), 255–263. https://doi.org/10.1080/10426914.2014.994754.

Mehta, K. P., Carlone, P., Astarita, A., Scherillo, F., Rubino, F., & Vora, P. (2019). Conventional and cooling assisted friction stir welding of AA6061 and AZ31B alloys. *Materials Science and Engineering A*, *759*(January), 252–261. https://doi.org/10.1016/j.msea.2019.04.120.

Mirzaei, M. H., Asadi, P., & Fazli, A. (2020). Effect of tool pin profile on material flow in double shoulder friction stir welding of AZ91 magnesium alloy. *International Journal of Mechanical Sciences*, *183*. https://doi.org/10.1016/j.ijmecsci.2020.105775.

Moshwan, R., Rahmat, S. M., Yusof, F., Hassan, M. A., Hamdi, M., & Fadzil, M. (2015). Dissimilar friction stir welding between polycarbonate and AA 7075 aluminum alloy. *International Journal of Materials Research*, *106*(3), 258–266. www.hanser-elibrary.com.

Ogawa, D., Kakiuchi, T., Hashiba, K., & Uematsu, Y. (2019). Residual stress measurement of Al/steel dissimilar friction stir weld. *Science and Technology of Welding and Joining*, *24*(8), 685–694. https://doi.org/10.1080/13621718.2019.1588521.

Pabandi, H. K., Jashnani, H. R., & Paidar, M. (2018). Effect of precipitation hardening heat treatment on mechanical and microstructure features of dissimilar friction stir welded AA2024-T6 and AA6061-T6 alloys. *Journal of Manufacturing Processes*, *31*, 214–220. https://doi.org/10.1016/j.jmapro.2017.11.019.

Palanivel, R., Koshy Mathews, P., Murugan, N., & Dinaharan, I. (2012). Effect of tool rotational speed and pin profile on microstructure and tensile strength of dissimilar friction stir welded AA5083-H111 and AA6351-T6 aluminum alloys. *Materials and Design*, *40*, 7–16. https://doi.org/10.1016/j.matdes.2012.03.027.

Pankaj, P., Tiwari, A., Biswas, P., Rao, A. G., & Pal, S. (2020). Experimental studies on controlling of process parameters in dissimilar friction stir welding of DH36 shipbuilding steel-AISI 1008 steel. *Welding in the World, 64*(6), 963–986. https://doi.org/10.1007/s40194-020-00886-3.

Rezaee Hajideh, M., Farahani, M., Alavi, S. A. D., & Molla Ramezani, N. (2017). Investigation on the effects of tool geometry on the microstructure and the mechanical properties of dissimilar friction stir welded polyethylene and polypropylene sheets. *Journal of Manufacturing Processes, 26*, 269–279. https://doi.org/10.1016/j.jmapro.2017.02.018.

Santos, T. F. de A., Torres, E. A., & Ramirez, A. J. (2018). Friction stir welding of duplex stainless steels. *Welding International, 32*(2), 103–111. https://doi.org/10.1080/09507116.2017.1347323.

Saravanan, V., Rajakumar, S., Banerjee, N., & Amuthakkannan, R. (2016). Effect of shoulder diameter to pin diameter ratio on microstructure and mechanical properties of dissimilar friction stir welded AA2024-T6 and AA7075-T6 aluminum alloy joints. *International Journal of Advanced Manufacturing Technology, 87*(9–12), 3637–3645. https://doi.org/10.1007/s00170-016-8695-0.

Shah, L. H., Othman, N. H., & Gerlich, A. (2018). Review of research progress on aluminium-magnesium dissimilar friction stir welding. *Science and Technology of Welding and Joining, 23*(3), 256–270. https://doi.org/10.1080/13621718.2017.1370193.

Shanmuga Sundaram, N., & Murugan, N. (2010). Tensile behavior of dissimilar friction stir welded joints of aluminium alloys. *Materials and Design, 31*(9), 4184–4193. https://doi.org/10.1016/j.matdes.2010.04.035.

Shehabeldeen, T. A., Yin, Y., Ji, X., Shen, X., Zhang, Z., & Zhou, J. (2021). Investigation of the microstructure, mechanical properties and fracture mechanisms of dissimilar friction stir welded aluminium/titanium joints. *Journal of Materials Research and Technology, 11*, 507–518. https://doi.org/10.1016/j.jmrt.2021.01.026.

Shi, H., Chen, K., Liang, Z., Dong, F., Yu, T., Dong, X., Zhang, L., & Shan, A. (2017). Intermetallic compounds in the banded structure and their effect on mechanical properties of Al/Mg dissimilar friction stir welding joints. *Journal of Materials Science and Technology, 33*(4), 359–366. https://doi.org/10.1016/j.jmst.2016.05.006.

Singh, R., Kumar, V., Feo, L., & Fraternali, F. (2016). Experimental investigations for mechanical and metallurgical properties of friction stir welded recycled dissimilar polymer materials with metal powder reinforcement. *Composites Part B: Engineering, 103*, 90–97. https://doi.org/10.1016/j.compositesb.2016.08.005.

Singh, V. P., Patel, S. K., Ranjan, A., & Kuriachen, B. (2020). Recent research progress in solid state friction-stir welding of aluminium-magnesium alloys: A critical review. *Journal of Materials Research and Technology, 9*(3), 6217–6256. https://doi.org/10.1016/j.jmrt.2020.01.008.

Sinhmar, S., & Dwivedi, D. K. (2018). A study on corrosion behavior of friction stir welded and tungsten inert gas welded AA2014 aluminium alloy. *Corrosion Science, 133*, 25–35. https://doi.org/10.1016/j.corsci.2018.01.012.

Sinhmar, S., & Dwivedi, D. K. (2019). Effect of weld thermal cycle on metallurgical and corrosion behavior of friction stir weld joint of AA2014 aluminium alloy. *Journal of Manufacturing Processes, 37*(December 2018), 305–320. https://doi.org/10.1016/j.jmapro.2018.12.001.

Sinhmar, S., & Dwivedi, D. K. (2020a). Art of friction stir welding to produce weld joint without rotation of shoulder with narrow heat- affected zone and high corrosion resistance. *Science and Technology of Welding and Joining, 25*(6), 490–495. https://doi.org/10.1080/13621718.2020.1746512.

Sinhmar, S., & Dwivedi, D. K. (2020b). Mechanical behavior of FSW joint welded by a novel designed stationary shoulder tool. *Journal of Materials Processing Technology, 277*, 116482. https://doi.org/10.1016/j.jmatprotec.2019.116482.

Tanaka, T., Morishige, T., & Hirata, T. (2009). Comprehensive analysis of joint strength for dissimilar friction stir welds of mild steel to aluminum alloys. *Scripta Materialia*, *61*(7), 756–759. https://doi.org/10.1016/j.scriptamat.2009.06.022.

Thapliyal, S., & Dwivedi, D. K. (2018). Sliding wear behavior solid lubricant based Ni-Al-Bronze surface composite developed by friction stir processing. *Transactions of the Indian Institute of Metals*, *71*(5), 1193–1210. https://doi.org/10.1007/s12666-017-1255-y.

Thapliyal, T., & Dwivedi, D. K. (2020). Fatigue performance of friction stir welded Al2024 alloy in a different corrosive environment. *Materialwissenschaft Und Werkstofftechnik*, *51*(2), 174–180. https://doi.org/10.1002/mawe.201800171.

Verma, M., & Saha, P. (2023). Effect of micro-grooves featured tool and their depths on dissimilar micro-friction stir welding (μFSW) of aluminum alloys: A study of process responses and weld characteristics. *Materials Characterization*, *196*. https://doi.org/10.1016/j.matchar.2022.112614.

Wang, H., Wang, K., Wang, W., Huang, L., Peng, P., & Yu, H. (2019). Microstructure and mechanical properties of dissimilar friction stir welded type 304 austenitic stainless steel to Q235 low carbon steel. *Materials Characterization*, *155*, 109803. https://doi.org/10.1016/j.matchar.2019.109803.

Wang, T., Sidhar, H., Mishra, R. S., Hovanski, Y., Upadhyay, P., & Carlson, B. (2019). Effect of hook characteristics on the fracture behaviour of dissimilar friction stir welded aluminium alloy and mild steel sheets. *Science and Technology of Welding and Joining*, *24*(2), 178–184. https://doi.org/10.1080/13621718.2018.1503801.

Watanabe, T., Takayama, H., & Yanagisawa, A. (2006). Joining of aluminum alloy to steel by friction stir welding. *Journal of Materials Processing Technology*, *178*(1–3), 342–349. https://doi.org/10.1016/j.jmatprotec.2006.04.117.

West, P., Shunmugasamy, V. C., Usman, C. A., Karaman, I., & Mansoor, B. (2021). Part II.: Dissimilar friction stir welding of nickel titanium shape memory alloy to stainless steel - microstructure, mechanical and corrosion behavior. *Journal of Advanced Joining Processes*, *4*. https://doi.org/10.1016/j.jajp.2021.100072.

Yan, Y., Zhang, D. T., Qiu, C., & Zhang, W. (2010). Dissimilar friction stir welding between 5052 aluminum alloy and AZ31 magnesium alloy. *Transactions of Nonferrous Metals Society of China (English Edition)*, *20*(SUPPL. 2), 619–623. https://doi.org/10.1016/S1003-6326(10)60550-X.

Zhang, C., Huang, G., Cao, Y., Zhu, Y., Li, W., Wang, X., & Liu, Q. (2019). Microstructure and mechanical properties of dissimilar friction stir welded AA2024-7075 joints: Influence of joining material direction. *Materials Science and Engineering A*, *766*, 138368. https://doi.org/10.1016/j.msea.2019.138368.

8 Joining of Metallic Materials Using Microwave Hybrid Heating

Kadapa Vijaya Bhaskar Reddy,
K. V. Hari Shankar, and Gudipadu Venkatesh

8.1 INTRODUCTION

Microwaves are designated between infrared radiation and radio waves in the electromagnetic spectrum, and they span the range of wavelengths between 10^3 and 10^6 nm. Being electromagnetic waves, microwaves can travel at the speed of light and do not require any medium for transmission. They have a magnetic field and an electric field propagating in mutually orthogonal planes, and they oscillate normal to the direction of wave propagation, as shown in Figure 8.1. Both of these fields are vector fields since, at any given point in time, their magnitude and direction need to be specified. Microwaves can be produced using microwave generators such as magnetrons, gyrotrons, klystrons, etc. In most industrial microwave applicators and domestic microwave applicators, microwaves are produced using magnetrons due to their compactness, high efficiency, and low cost, and the typical frequency used is 2.45 GHz.

8.2 FUNDAMENTALS OF MICROWAVE THEORY

During microwave processing, materials are subjected to electric and magnetic fields, which result in losses inside the material and create a heating effect. Magnetic losses do not have a considerable influence on heating compared to dielectric losses.

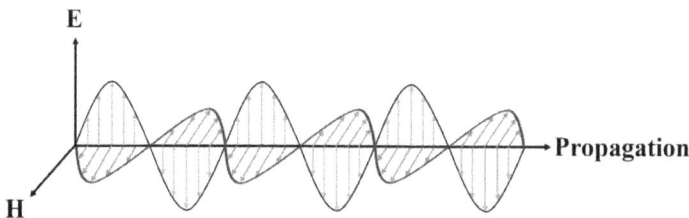

FIGURE 8.1 An electromagnetic wave (Horikoshi et al. 2017).

DOI: 10.1201/9781003327769-8

Dielectric losses are attributed to the polarization effect, which involves the redistribution of charges after interaction with a continuously changing electric field. So, polarization is very significant in microwave processing because it enhances the absorption of microwave energy and leads to heating. According to its origin, polarization may be classified as follows:

a. Electronic polarization: When an external electric field is applied to a dielectric material, the charges inside the atoms get redistributed, and the electrons get displaced away from the positive nucleus. This induces a dipole moment, which causes polarization. It is depicted in Figure 8.2.
b. Ionic polarization: The positive and negative ions inside the crystal lattice get relatively displaced under the influence of an electric field, which generates a net dipole moment and contributes to polarization, as illustrated in Figure 8.3.
c. Maxwell–Wagner polarization: In this type of polarization, interfaces within the material, such as grain boundaries and defect regions, become sites where the free charges accumulate under the action of an external electric field, as depicted in Figure 8.4. Hence, it is also known as interfacial polarization (Gupta and Eugene 2011).
d. Dipolar polarization: In materials having permanent dipole moments, under the application of an external electric field, these dipoles get re-aligned parallel to the electric field, thus generating a net dipole moment. When the electric field alternates, the dipoles rotate with a phase difference between field orientation and dipole orientation. This causes a random collision between the molecules and leads to heating. It is depicted in Figure 8.5.

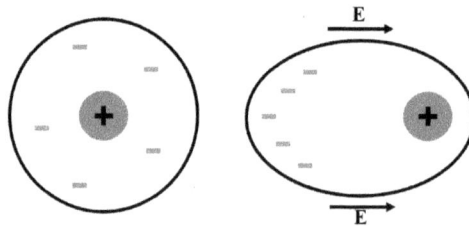

FIGURE 8.2 Electronic polarization (Uchino 2018).

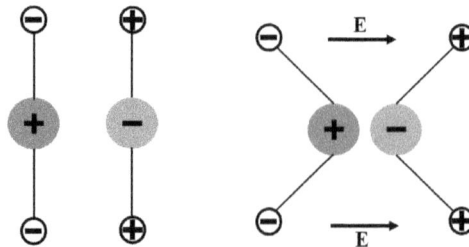

FIGURE 8.3 Ionic polarization (Gupta and Eugene 2011).

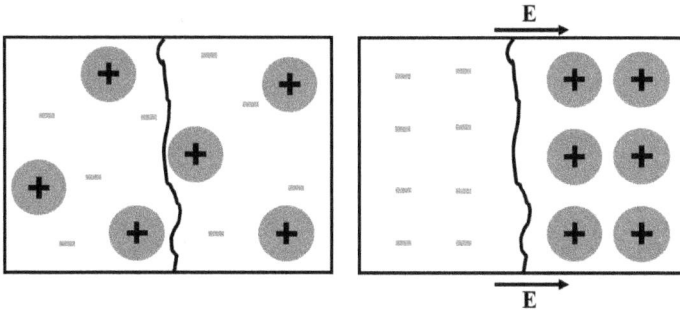

FIGURE 8.4 Maxwell–Wagner polarization.

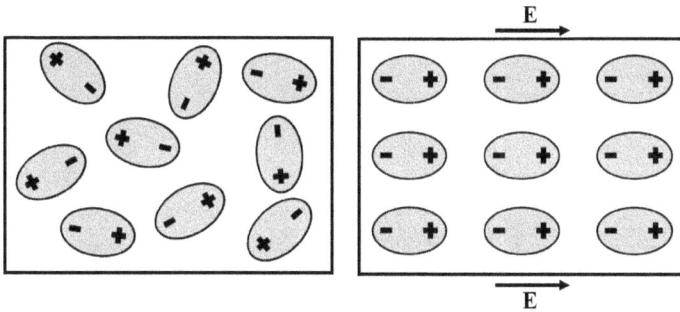

FIGURE 8.5 Dipolar polarization (Horikoshi et al. 2017).

During microwave irradiation of materials, heat may be generated inside the material at an atomic level. This is determined by the type of material and the absorption behavior exhibited by it upon microwave irradiation, as shown in Figure 8.6. Based on this, materials are classified into four types, as follows:

i. Transparent: Materials such as quartz and Teflon allow microwaves to pass through them without any absorption. They are low-loss insulators with total penetration.

ii. Opaque: All bulk metals are considered in this category as they are conductors with total reflection and negligible absorption of microwaves due to their low skin depth.

iii. Absorber: Dielectric materials such as water and silicon carbide are high-loss insulators, which involve total absorption of microwaves depending on the dielectric properties of the material.

iv. Partial absorbers: Composite materials come under this category because these advanced materials involve the absorption of microwaves with partial penetration. They are low-loss insulators with localized energy conversion. It is especially useful for selective and hybrid heating.

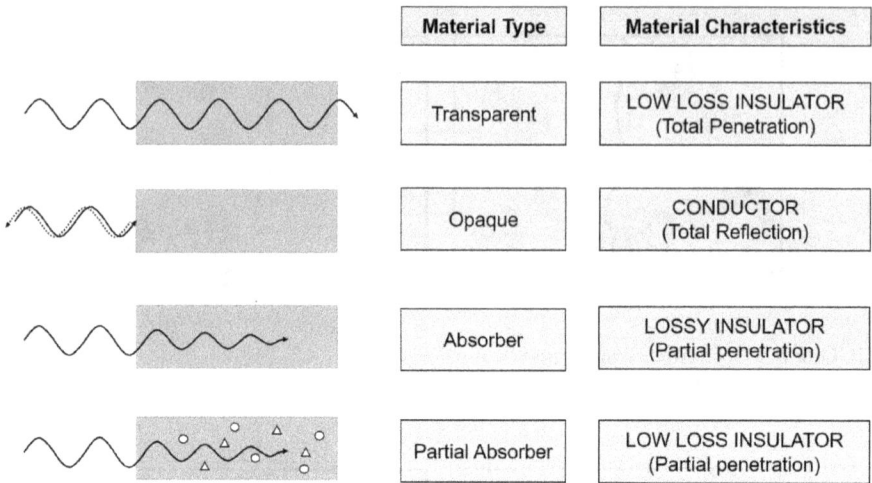

FIGURE 8.6 Material absorption characteristics (Vasudev et al. 2019).

Depending on microwave interaction with the material, the microwave energy may be converted to heat energy. This interaction depends on the material properties, such as dielectric and magnetic properties, because these properties determine the ability of the material to effectively couple with the microwaves.

8.2.1 PERMITTIVITY AND PERMEABILITY

Permittivity, also known as the dielectric constant, is the material property that represents the ability of a material to get polarized under the influence of an external electric field. The absolute permittivity of a material is given by Eq. 8.1 (Gupta and Eugene 2011)

$$\varepsilon' = \varepsilon_0 \varepsilon_r' \tag{8.1}$$

where $\varepsilon_0 = 8.854 \times 10^{-12}$ F/m is the vacuum permittivity and ε_r' is the relative permittivity.

The complex permittivity of a material accounts for the absorption and storage of electrical energy inside the material. It is given by Eq. 8.2 (Rawat et al. 2022)

$$\varepsilon^* = \varepsilon' - j\varepsilon'' \tag{8.2}$$

where ε'' is the dielectric loss factor and $j = \sqrt{-1}$ is the imaginary unit. The extent of penetration and absorption of microwaves inside a material is represented by its permittivity, ε', and its energy storage capacity is represented by the dielectric loss factor, ε''. The dielectric loss factor accounts for various losses occurring due to the resistance offered to translational and rotational motions of electrons, ions, and dipoles, which arise due to the electric field generation within the material.

Loss tangent $\tan \delta$ is the property of the material that represents the extent of the transformation of microwave energy into heat energy. It is given by the ratio of dielectric loss factor and permittivity, as shown in Eq. 8.3 (Tamang and Aravindan 2019)

$$\tan \delta = \frac{\varepsilon''}{\varepsilon'} = \frac{\text{Energy lost per cycle}}{\text{Energy stored per cycle}} \tag{8.3}$$

where δ is the angle between the resultant permittivity and its real part in the vector diagram as shown in Figure 8.7.

The impact of magnetic fields on the absorption of microwaves and heating is generally ignored since they are dominated by electric fields. However, a magnetic property called permeability (μ') has a significant influence on the skin depth of the material. For better penetration of microwaves, more skin depth is required, and this is possible only when the permeability is low.

Permeability (μ') accounts for the effect of the magnetic field on the material and is given by Eq. 8.4 (Gupta and Eugene 2011)

$$\mu' = \mu_0 \mu_r' \tag{8.4}$$

where $\mu_0 = 4\pi \times 10^{-7} \text{H/m}$ is the vacuum permeability and $\mu_r' =$ relative permeability.

The complex permeability, μ^*, is given by Eq. 8.5 (Rawat et al. 2022)

$$\mu^* = \mu' - j\mu'' \tag{8.5}$$

where $\mu'' =$ magnetic loss factor. The relaxation and resonance occurring due to the magnetic field are considered by the magnetic loss factor.

8.2.2 MAXWELL'S EQUATIONS

When a material is subjected to an electric field, it causes polarization of charges within the material. But when the electric field alternates rapidly, the charges are not able to change their polarization as fast as the electric field. This causes agitation, and

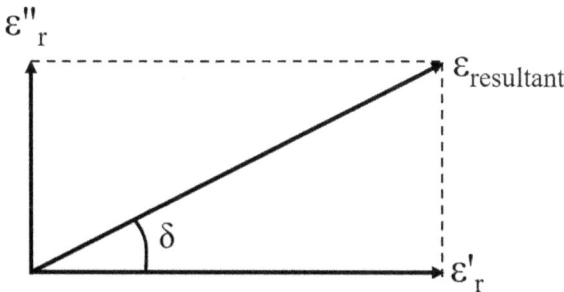

FIGURE 8.7 Vector diagram for resultant permittivity (Vasudev et al. 2019).

then heating occurs within the material. The heating is influenced by the power input and duration of the microwave radiation.

After Faraday discovered electromagnetic induction, Maxwell speculated on the existence of electromagnetic waves and gave his electromagnetic theory, which consisted of 20 equations. These equations were later reduced to only four equations by Heaviside, and they are famously known as Maxwell's equations, given in Eq. 8.6(a)–(d) (Gupta and Eugene 2011)

$$\textbf{Faraday's Law} \ : \ \nabla \times E = -\frac{\partial B}{\partial t} \tag{8.6a}$$

$$\textbf{Ampere's Law} \ : \ \nabla \times H = J + \frac{\partial D}{\partial t} \tag{8.6b}$$

$$\textbf{Gauss's Electric Law} \ : \ \nabla . D = q \tag{8.6c}$$

$$\textbf{Gauss's Magnetic Law} \ : \ \nabla . B = 0 \tag{8.6d}$$

where E = electric field intensity (V/m), H = magnetic field intensity (A/m), and $D = \varepsilon(\omega)E(t)$ is the electric displacement (N/V.m), $B = \mu(\omega)H(t)$ is the magnetic induction (T), $J = \sigma(\omega)E(t)$ is the flux of the electric current (A/m²), q is the electric charge density (C/m³), ω is the circular frequency (radians/s), σ is the electrical conductivity (S/m), ε is the permittivity (F/m), and μ is the permeability (H/m). Faraday's law describes the phenomenon of electromagnetic induction, which involves the interaction of a magnetic field with an electric circuit to generate an electromotive force. Ampere's law states that the intensity of the magnetic field is proportional to the electric current that produces that field. Gauss's electric law predicts that the net electric flux emanating from a closed region is equal to the total charge trapped within that region. Gauss's magnetic law indicates that the net magnetic flux coming out of a closed region is zero.

Maxwell's equations can be used to estimate the distribution of the electric and magnetic fields inside the microwave cavity, which would assist in finding the most suitable location of the experimental setup so that the processing time is reduced. Generally, if the setup is placed at locations where the field intensity is maximum, the material experiences enhanced coupling with microwaves, which results in faster processing.

Due to material–field interactions, microwave energy gets absorbed by the material, and power gets dissipated as heat. Assuming that power absorption is uniform throughout the volume, the power absorbed in a unit volume of the sample due to the electric field is given by Eq. 8.7 (Mishra and Sharma 2016a)

$$P = 2\pi f \varepsilon_0 \varepsilon'' |\mathbf{E}|^2 = 2\pi f \varepsilon_0 \varepsilon' \tan \delta |\mathbf{E}|^2 \tag{8.7}$$

where |E| is the absolute value of the electric field intensity. The absorbed power has a linear variation with the loss tangent, permittivity, and frequency, and it varies with the square of the electric field.

Similarly, the power absorbed in a unit volume of the sample due to the magnetic field is given by Eq. 8.8 (Gupta and Eugene 2011)

$$P = 2\pi f \mu_0 \mu'' |\mathbf{H}|^2 = 2\pi f \mu_0 \mu' \tan \delta_\mu |\mathbf{H}|^2 \tag{8.8}$$

where $|\mathbf{H}|$ is the absolute value of the magnetic field intensity. The absorbed power has a linear variation with the loss tangent, permeability, and frequency, and it varies with the square of the magnetic field.

Therefore, the total power absorbed per unit volume (W/m) of the sample is given by Eq. 8.9 (Gupta and Eugene 2011)

$$P = 2\pi f \varepsilon_0 \varepsilon'' E_{rms}^2 + 2\pi f \mu_0 \mu'' H_{rms}^2 \tag{8.9}$$

where E_{rms} is the root-mean-squared value of the electric field (V/m) and H_{rms} is the root-mean-squared value of the magnetic field (A/m). For diamagnetic materials, the value of permeability is very low, and thus the effect of the magnetic field on power absorption can be ignored. Therefore, Eq. 8.9 gets reduced to Eq. 8.10 (Gupta and Eugene 2011).

$$P = 2\pi f \varepsilon_0 \varepsilon'' E_{rms}^2 \tag{8.10}$$

As the material gets heated and its temperature rises, the material properties get altered and therefore influence the power absorbed.

The extent of penetration of microwaves inside the sample material from its surface has a significant effect on the power absorbed by the material. This depth of penetration from the surface is called Penetration Depth or Skin Depth depending on the type of material. The penetration of microwaves is different for different materials. While non-metals allow significant penetration of microwaves, for bulk metals, this penetration is negligible at room temperature. Hence, the term 'Penetration Depth' is used for non-metals, and 'Skin Depth' is used for metals. Penetration depth for non-metals is given in Eq. 8.11 (Gupta and Eugene 2011).

$$D_P = \frac{1}{\omega \sqrt{\left[0.5\mu_0\mu'\varepsilon_0\varepsilon' \left\{ \sqrt{1 + \left(\frac{\varepsilon''_{eff}}{\varepsilon'} \right)^2} - 1 \right\} \right]}} \tag{8.11}$$

Skin depth for metals is given by Eq. 8.12 (Mishra and Sharma 2016a)

$$D_S = \frac{1}{\sqrt{\pi f \mu' \sigma}} = 0.029 \left(\rho \lambda_0 \right)^{0.5} \tag{8.12}$$

where σ is the specific electrical conductance (S/m), ρ is the specific electrical resistance (Ω.m), and λ_0 (m) is the wavelength of the incident microwave. Therefore, it is clear that the absorption of power depends on electromagnetic variables and sample thickness.

8.2.3 LAMBERT'S LAW

It is difficult to determine the microwave power dissipation using Maxwell's equations unless the distribution of the electric field inside the microwave cavity is known. An alternate method to determine the power dissipation in the material during microwave processing without using the electric field distribution is given by Lambert's law. So, it could be useful in quickly obtaining the temperature distribution since the cumbersome procedure of electric field calculation is avoided. Lambert's law states that the power incident on the material is perpendicular to the surface of the sample, and the power gets dissipated exponentially as it progresses through the material. The power dissipated at a depth x from the surface of the sample is given by Eq. 8.13 (Mishra and Sharma 2016a)

$$P(x) = P_0 e^{-2\beta x} \tag{8.13}$$

where P_0 = power incident on the surface of the sample (W) and β = propagation constant (m^{-1}). The propagation constant depends on the loss tangent (tan δ), the radiation speed, and the frequency of the radiation.

For simplified power calculations, certain assumptions are associated with Lambert's law. These are:

1. The sample extends to infinity in one direction, i.e., it is semi-infinite.
2. The penetration of microwaves occurs in one direction only.
3. The standing wave effect is negligible.

Hence, the power calculation using Lambert's law, although faster, is not as accurate as the power calculation using Maxwell's equation. Still, Lambert's law is quite relevant because it provides a means for faster and more accurate prediction of temperature distribution. The obtained results are quite consistent with the experimental results. While using Maxwell's equations, the electric field needs to be determined initially, which involves dielectric properties that are very sensitive. So, even a slight error in the values of the dielectric properties causes a huge variation in the predicted temperature distribution. Hence, Maxwell's equations fail to maintain consistency with the experimental results.

8.3 HEATING MECHANISMS IN MICROWAVE PROCESSING

There are various mechanisms by which heat is generated inside the material after interaction with the microwaves, as shown in Figure 8.8. Depending on the type of material, these mechanisms may include agitation of electrons or dipoles, change in orientation of the domain wall, electron spin, etc. Magnetic materials are influenced by both electric and magnetic fields, whereas non-magnetic materials are only affected by the electric field.

8.3.1 FOR NON-MAGNETIC MATERIALS

Upon microwave interaction with non-magnetic materials, only the electric field component contributes to the heating effect. Depending on the conductivity of the material, the dominant loss mechanism for the heating effect may be conduction losses or dipolar

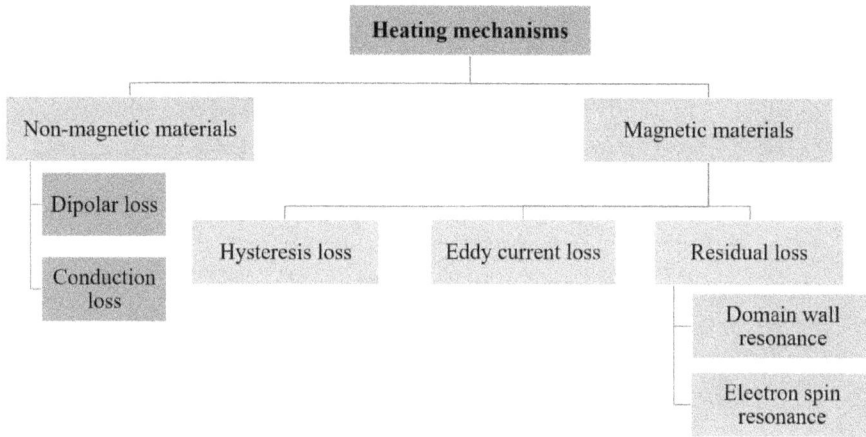

FIGURE 8.8 Heating mechanisms in microwave processing.

losses. In dielectric insulators, dipolar loss is dominant, whereas in non-magnetic conductors, conduction loss is the dominant mechanism for heat generation.

8.3.1.1 Dipolar Loss

When dielectric insulators are exposed to an external electric field, dipoles are generated. The dipoles change their orientation frequently to be aligned with the alternating electric field. This creates agitation of dipoles that is resisted by frictional, elastic, inertial, and molecular interaction forces, resulting in volumetric heating.

8.3.1.2 Conduction Loss

When non-magnetic conducting materials are subjected to an external electric field, the free electrons present in the material start to move in the direction of the electric field. It generates an electric current inside the material, which induces a magnetic field. The direction of the induced magnetic field is such that it creates a force against the direction of motion of the electrons. This resistance leads to the agitation of free electrons and creates a heating effect. When the applied electric field is alternating, the heating is volumetric.

8.3.2 FOR MAGNETIC MATERIALS

The heating effect of magnetic materials is influenced by both electric and magnetic fields. The electric field causes the electrons to move, whereas the magnetic field influences the electron spin, orientation of domains, etc. In magnetic materials, conduction losses are combined with other mechanisms such as eddy current loss, hysteresis loss, etc.

8.3.2.1 Hysteresis Loss

Hysteresis loss is found only in ferromagnetic materials. Due to the absence of hysteresis loss in paramagnetic and diamagnetic materials, they do not experience

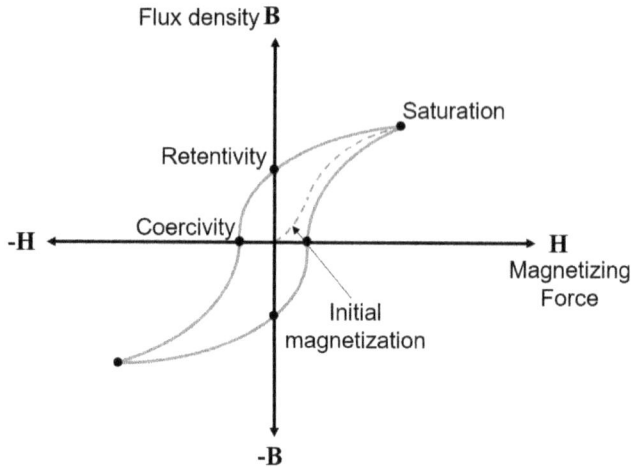

FIGURE 8.9   Hysteresis loop (msestudent.com n.d.).

the heating effect of microwaves. Within ferromagnetic materials, some magnetic domains have a direction of magnetization. In each domain, the atoms have their magnetic moments oriented in the same direction, which adds up to creating the magnetic moment of that domain. Initially, the direction of the magnetic moment of each domain is different from that of another domain. But when subjected to a magnetic field, these domain magnetic moments are aligned along the magnetic field direction. Due to disturbances in the orientation of domain magnetic moments, heating occurs. If an alternating magnetic field is applied, the heating is more uniform. The hysteresis loss is represented by the area enclosed inside the hysteresis loop, as shown in Figure 8.9. The more the area enclosed by the hysteresis loop, the greater the hysteresis loss.

8.3.2.2   Eddy Current Loss

Eddy currents have a significant effect on the heating of conductive materials like metals under the influence of an externally applied magnetic field. The material interacts with the magnetic field to induce an eddy current that opposes any variations in the applied magnetic field. If the externally applied magnetic field increases, the induced eddy current changes direction so that the net magnetic field reduces. Similarly, if the applied magnetic field decreases, the direction of the induced eddy current varies so that the net magnetic field increases. Thus, heat is generated due to variations in the direction of the induced eddy current. In the case of an alternating magnetic field, the heating is uniform. Eddy current loss reduces exponentially with an increase in skin depth. This is called the 'skin effect'.

8.3.2.3   Residual Losses

Although the exact reasons for residual losses are unknown, they are often attributed to resonance effects such as domain wall resonance and electron spin resonance (or FMR, i.e., Ferromagnetic Resonance). Residual losses occur due to domain wall displacement under the effect of an external electric field. It may also occur when the external electric field disturbs the spin of the electrons (Moulson and Herbert 1995). Domain wall resonance loss involves heat generation due to the extension and shrinkage of domain walls. Electron spin resonance loss involves heating due to some disturbance in the spin of the electrons.

a. Domain wall resonance loss: When a magnetic field is applied externally, some magnetic domains may have their magnetic moments aligned with the field. The domains in the same direction as those of the magnetic field expand while the other domains contract. These expansions and contractions of domains cause the shifting of domain walls, which is resisted by inertial and frictional effects. Due to this resistance, heat is generated. In the case of an alternating magnetic field, the heat generation is more uniform.

b. Electron spin loss (or FMR): Electron spin resonance is often called 'Ferromagnetic resonance' because it is predominant in ferromagnetic materials. Each magnetic domain in the material consists of electrons that have a certain direction of spin. The cumulative effect of the spin of the electrons generates a net magnetic moment, which is the domain magnetic moment. The anisotropic field on electrons produces an internal magnetic field as well. Initially, the magnetic moment is aligned with the internal magnetic field. But this alignment gets disturbed under the effect of an external magnetic field that is normal to the internal magnetic field. The material absorbs the magnetic energy, and the magnetic moment vector starts spinning around the internal magnetic field vector with an angle of precession. When the external magnetic field is deactivated, the absorbed energy is released as heat, and the magnetic moment vector spirals back into alignment with the internal magnetic field vector. This heating effect is uniform when the magnetic field is alternating.

8.4 MODES OF HEATING

8.4.1 CONVENTIONAL HEATING

In conventional heating, heat is transferred from a conventional heat source to the material, generally by conduction or convection. The heating rate is very low in conventional heating and is termed external heating because heat travels from outside to inside. Due to this, there is a temperature gradient within the material, which may lead to thermal residual stresses and inconsistencies in the properties of the material. For thermal equilibrium to be attained, a long processing time is required. Thus, a trade-off between processing time and product quality is inevitable when conventional heating is employed.

In conventional heating, energy in the form of heat is transferred, which is fundamentally different from microwave heating, in which energy conversion takes place from microwave energy to heat, leading to heat generation within the material itself. The efficiency of conventional heating increases with the thermal conductivity of the material.

8.4.2 Microwave Direct Heating (MDH)

In MDH, microwaves penetrate the material and interact with the molecules present inside it. The material is said to 'couple' with microwaves, which results in molecular agitation, leading to heat generation within the material. The heat generated inside the material travels out toward the surface. Hence, microwave heating is called 'inside-out heating' due to its reverse heating characteristics compared to conventional heating. There exists a thermal gradient in microwave heating that is the reverse of that of conventional heating.

MDH is limited to the processing of materials with high skin depth. Also, there is a limitation of the 'thermal runway' that is caused when there is an uncontrolled temperature rise. When the material has a low skin depth at room temperature, there is a need to increase the skin depth by increasing the temperature. This is done by initial heating of the material using susceptors.

8.4.3 Microwave Hybrid Heating (MHH)

In MDH, there is an issue of skin depth associated with it. Skin depth is the extent of penetration of microwaves inside the material from its surface. When the sample size is less than the skin depth, direct heating is capable of completely melting the sample because the heat gets generated from the core region and flows outward. But when the sample size is greater than the skin depth, the heat generation takes place only from the skin depth and then flows outwards because microwaves cannot penetrate the material beyond the skin depth. As a result, the core region experiences non-uniform heating, as depicted in Figure 8.10. Also, in MDH, the control of temperature becomes quite difficult.

To avoid these problems, MDH is combined with conventional heating. This is made possible using some special materials called susceptors. Susceptors such as silicon carbide, graphite, zirconia, charcoal, and alumina have the unique capability to absorb energy from microwaves and become heated up quickly. The heat energy is then transmitted to the sample material by conduction, i.e., conventional heating,

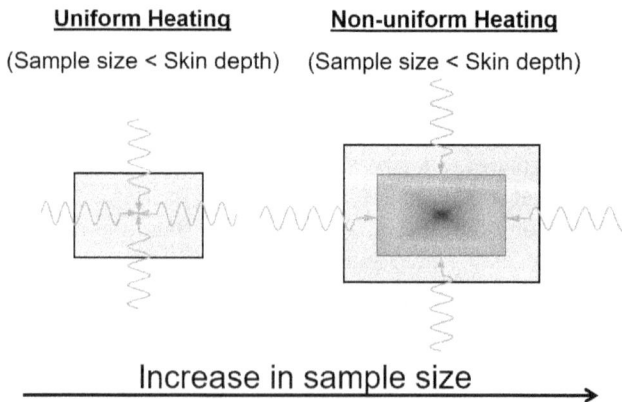

Uniform Heating   **Non-uniform Heating**

(Sample size < Skin depth)   (Sample size < Skin depth)

Increase in sample size

FIGURE 8.10   Absorption of microwave energy in a metallic sample (Mishra and Sharma 2016a).

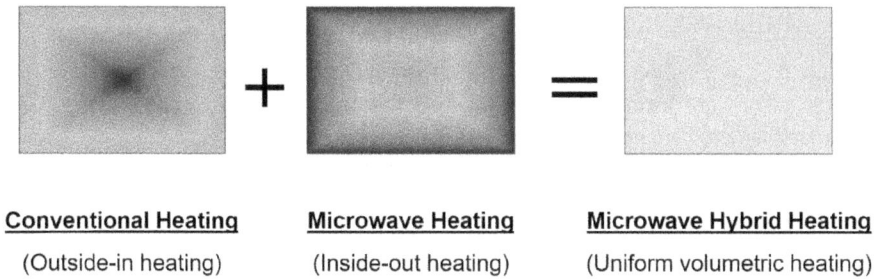

Conventional Heating   **Microwave Heating**   **Microwave Hybrid Heating**

(Outside-in heating)   (Inside-out heating)   (Uniform volumetric heating)

FIGURE 8.11 Uniform volumetric heating using microwave hybrid heating (Wong and Gupta 2015).

thus slowly increasing the temperature as well as the skin depth of the sample. When the sample temperature reaches its critical temperature, the skin depth of the material becomes equal to the sample size, and the entire sample starts to absorb microwave energy directly. Now, microwave heating begins, and the heat gets generated rapidly. So, the inside-out heating characteristic of MDH is combined with the outside-in characteristic of conventional heating, and the result is that MHH leads to uniform volumetric heating. This is illustrated in Figure 8.11. Another reason for this is that, due to the presence of susceptors, heat loss from the surface is reduced, which helps to maintain homogeneity in temperature. Uniform heating leads to a uniform microstructure in the sample material.

8.4.4 MICROWAVE-SELECTIVE HEATING (MSH)

There may be requirements where localized heating is desired in a part of the sample material. This is achieved using Microwave-Selective Heating (MSH), in which a certain region is selectively heated without disturbing the remaining part of the sample. A masking material is used to hide the regions of the sample where microwave exposure is not required. These regions will not undergo any changes in their properties. So, this technique provides better control over the properties of the material and its microstructure. It can be used to obtain unique microstructures and in FGM (Functionally Graded Material) synthesis.

8.4.5 SELECTIVE MICROWAVE HYBRID HEATING

Selective heating can also be achieved by a targeted supply of susceptor material over a certain region in the sample. The susceptor material is responsible for hybrid heating as it imparts heat to the sample by conventional heating until a critical temperature is attained, after which direct microwave heating begins. Thus, MHH and MSH can be combined to form a new technique called Selective Microwave Hybrid Heating. In Selective Microwave Hybrid Heating, the susceptor material is fed in the form of a powder by a vertical cavity feeder consisting of a vertical hole in a covering brick, such that the powder is filled inside the hole. Due to localized heating, the heat-affected zone and exposure time are reduced, and a better microstructure is obtained.

8.5 MICROWAVE JOINING

Srinath et al. worked on the joining of SS316 plates using MHH. Some intermetallics and carbides were formed, which enhanced the coupling of microwaves, resulting in a crack-free joint with good metallurgical bonding (Srinath, Sharma, and Kumar 2011a). Later, the same investigator further explored the joining of bulk copper with a Cu interlayer using MHH. The formation of oxides during the process enhanced microwave coupling, which resulted in reduced porosity as well as significant micro-hardness and strength (Srinath, Sharma, and Kumar 2011b).

In industry, various applications may require the joining of dissimilar metals. The joining of dissimilar metals is of great interest in research owing to the challenges faced during the process. Several attempts have been made to use MHH to join dissimilar metals. Srinath et al. investigated the bulk joining of SS316 to mild steel with nickel-based powder as the interlayer and charcoal powder as the susceptor, using microwave processing. Formation of various carbides, oxides, and intermetallics enhanced micro-wave coupling. Both ductile and brittle modes of failure were observed at the joint interface (Srinath, Sharma, and Kumar 2011c). Bansal et al. also conducted similar studies by using SS316 powder as the interlayer and SiC as the susceptor. The joint was found to have higher hardness, strength, and lower elongation compared to the joint prepared by Srinath et al (Bansal, Sharma, Kumar, et al. 2016). Inconel 718 powder was used as the interlayer for dissimilar metal joining of stainless steel (SS) with Inconel 718, and the microstructural investigation revealed the presence of the Laves phase in the base metal region adjacent to the fusion zone (Bansal, Sharma, Das, et al. 2016). Kumar et al. developed an SS304-SS316 joint with an interfacing powder of SS-316 using MHH. It was found that complete melting of the interface powder took place, which resulted in metallurgical bonding between powder and dissimilar workpieces (Kumar et al. 2020). Naik et al. conducted studies where they created dissimilar joints between MS and SS workpieces using TIG welding as well as microwave joining and then compared the joints formed by both techniques (Naik, Gadad, and Hebbale 2021).

Singh et al. developed microwave-processed cast iron joints using Ni-based pow-der slurry as the interlayer material. The subsequent characterization revealed the mutual diffusion of elements between the interlayer and base metal. The same author developed Hastelloy-X joints using empowering welding and cutting (EWAC) pow-der as the interlayer and performed mechanical and metallurgical characterization of the formed joints. Finer microstructure and metallurgical bonding between the powder and the workpieces were observed (Singh et al. 2017, 2019).

Badiger et al. carried out effective joining of Inconel-625 alloy through MHH. Later, mechanical characterization was performed, followed by microstructure examination of the fusion zone, which revealed that Laves phase was present at the grain boundaries. The study was extended by optimizing the process parameters using the Taguchi-based Grey model and ANOVA, which revealed that filler powder particle size was the most important control factor, followed by separator and sus-ceptor. Further analysis was done to understand the influence of power input on the metallurgical and mechanical properties of the joint formed (Badiger, Narendranath, and Srinath 2015; Badiger, Narendranath, and Srinath 2018; Badiger, Narendranath, and Srinath 2019; Badiger et al. 2019).

8.5.1 DEVELOPMENT OF STAINLESS-STEEL JOINTS USING MICROWAVE HYBRID HEATING

8.5.1.1 Numerical Simulation

8.5.1.1.1 Geometry

The microwave joining model was developed similar to the experimental setup using the COMSOL Multiphysics 5.5a software package, as shown in Figure 8.12. The workpieces, interlayer, and susceptor have been highlighted in the figure. The parameters used for modeling the geometry are listed in Table 8.1. A Dell laptop with an Intel i5 processor and 8GB of RAM was used for running the simulation. The time taken for computation was approximately 180 minutes. The material properties used in the simulation are listed in Table 8.2.

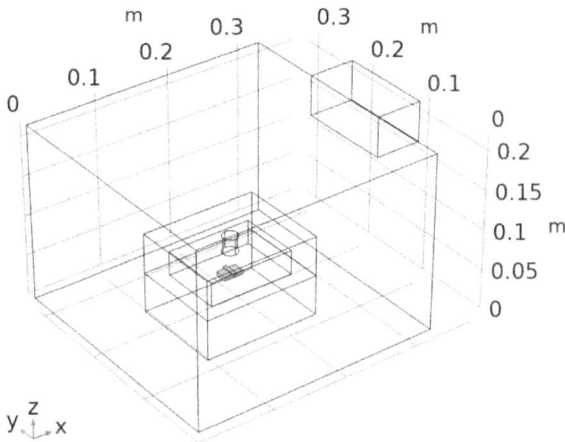

FIGURE 8.12 3D simulation model.

TABLE 8.1
Model Parameters Used for Simulation

Label	Dimension (mm)
Microwave cavity	$320 \times 320 \times 210$
Waveguide	$40 \times 80 \times 100$
Alumina cascade	$150 \times 120 \times 50$
Through-hole diameter	20
Graphite susceptor	$10 \times 15 \times 6$
EWAC interlayer	$1 \times 15 \times 5$
SS-316L plates	$25 \times 15 \times 5$

TABLE 8.2
Material Properties

Material	Density (ρ)	Thermal Conductivity (k)	Electrical Conductivity (σ)	Heat Capacity at Constant Pressure (C_p)	Relative Permittivity (ε)	Relative Permeability (μ)
Air (Tamang and Aravindan 2022)	-	-	0	-	1	1
Copper (Tamang and Aravindan 2022)	7990	400	5.998e7	385	1	1
Graphite (Tamang and Aravindan 2022)	2100	470	1700	830	15	1
EWAC powder (Matweb. online accessed March 2, 2022)	8900	76	1.33e-11	26	31-i×2.25	1
Stainless steel (Holian 1984)	8960	21.4	1.35e4	450	1	1
Alumina (Bhoi et al. 2022; Tamang and Aravindan 2019)	3900	27	1e-14	900	1	1

8.5.1.1.2 Boundary Conditions

Two modules were used for the simulation, namely, electromagnetic waves and heat transfer in solids. The following boundary conditions were applied to the developed model:

a. Port boundary condition: It was applied at the entrance to the rectangular waveguide as shown in Figure 8.13a, and transverse electric (TE$_{10}$) mode was considered with a 2.45 GHz frequency using Eq. 8.14 (Tamang and Aravindan 2022).

$$\text{Cut-off frequency } (f_c)_{mn} = \frac{C}{2}\sqrt{\frac{m^2}{a^2} - \frac{n^2}{b^2}} \tag{8.14}$$

b. Impedance boundary condition: It was applied on the walls of the microwave cavity and the waveguide as shown in Figure 8.13b to account for minute losses (skin effect) due to skin depth using Eq. 8.15 (Mishra and Sharma 2017b).

$$\sqrt{\frac{\mu_r \mu_0}{\varepsilon_r \varepsilon_0 - j\frac{\sigma}{\omega}}}\, n \times H + E - (n.E)n = (n.E_s)n - E_s \tag{8.15}$$

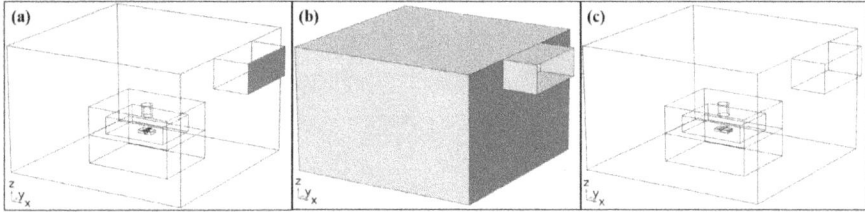

FIGURE 8.13 Boundary conditions: (a) port, (b) impedance, and (c) heat transfer in solids.

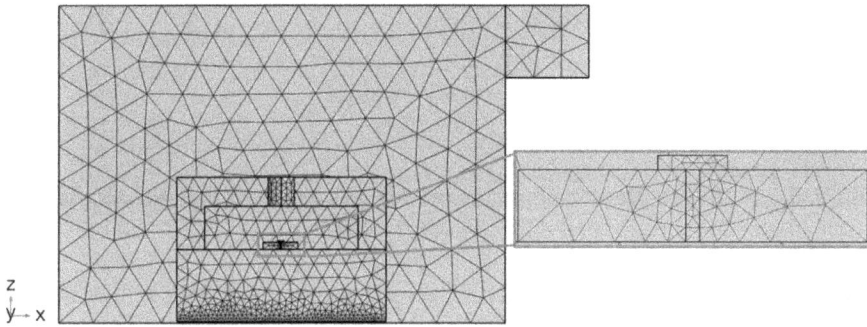

FIGURE 8.14 Mesh distribution of a microwave setup with an enlarged view of the interlayer.

   c. Heat transfer in solids: It was applied to SS-316L plates, the interlayer, and the susceptor using Eq. 8.16 (Rawat et al. 2022), as shown in Figure 8.13c.

$$\rho C_p \frac{\partial T}{\partial t} = \nabla \cdot (K\Delta T) + Q_{rms} \qquad (8.16)$$

8.5.1.1.3 Mesh

Physics-controlled mesh of size "extra fine" containing 5,13,541 elements was used for the simulation, as depicted in Figure 8.14. The elements were found to be mainly tetrahedral and then triangular. The enlarged view of the figure shows that the interlayer has very fine elements compared to the substrate, which is necessary for accurate results.

8.5.1.1.4 Results and Discussion

8.5.1.1.4.1 Electric Field Distribution   It is important to analyze the distribution of electric fields inside the microwave applicator, as it helps determine the appropriate location of the sample to ensure a shorter processing time. Positions with higher field intensity enable better microwave coupling of materials. Distribution of electric fields in the XY plane (E_z) is shown in Figure 8.15. The distortion in the electric field represents the conversion of microwave energy to heat energy. The locations of maximum and minimum electric fields represent the nodes and

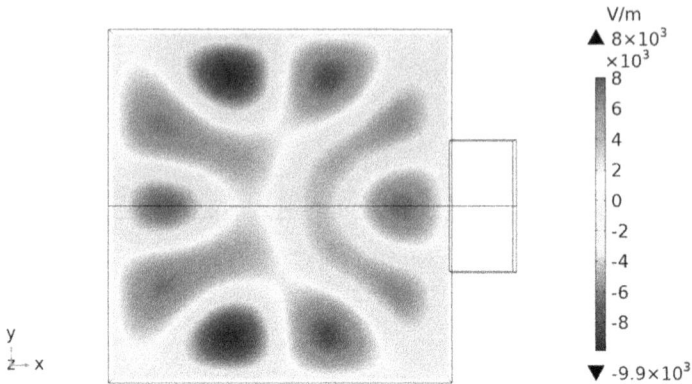

FIGURE 8.15   Electric field distribution in the XY plane.

anti-nodes in the propagation of microwaves. It was observed that the electric field distribution ranges from -9.9×10^3 to 8×10^3 (V/m).

8.5.1.1.4.2   Temperature Distribution   The temperature distribution at time instant 720 seconds is shown in Figure 8.16. It is seen that the melting point of EWAC powder, i.e., 1475°C, is obtained in the joint region. It is understood that the joining of interlayer powder and stainless-steel substrate occurs. The maximum temperature observed at the susceptor is 1480°C. It is consistent with the fact that heat is transferred from the graphite susceptor to the joining region.

8.5.1.2   Experiment

8.5.1.2.1   Materials and Methodology

Austenitic stainless-steel (SS-316L) plates having 25 mm × 15 mm × 5 mm dimensions and nickel-based (EWAC) interface powder having a spherical morphology of size ~40 μm were considered for microwave joining. Scanning Electron Micrographs of EWAC interface powder and base metal (SS-316L) are shown in Figure 8.17. Chemical composition of interface powder and base metal is shown in Table 8.3.

8.5.1.2.2   Development of Joint

A multimode microwave applicator operating at 2.45 GHz with 900 W power was considered for both experimental and simulation studies. The process parameters for the experiment are listed in Table 8.4. The experimental setup and its schematic diagram are illustrated in Figure 8.18a and b. EWAC interface powder was mixed with a chemically neutral resin to form an interface slurry. The prepared slurry was applied between SS-316L plates kept ~1mm apart as a butt joint, and a graphite susceptor was placed above the interlayer region. The entire setup was placed inside an alumina cascade with a hole in the top for focused heating, and the entire assembly was placed inside the microwave applicator. The procedure for melting the powder and forming

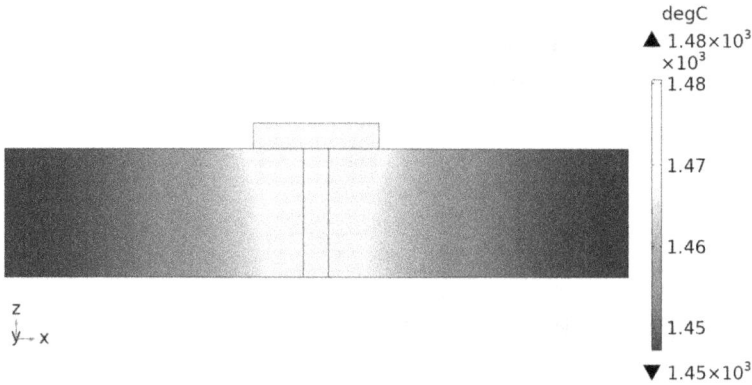

FIGURE 8.16 Temperature distribution after 720 seconds of processing.

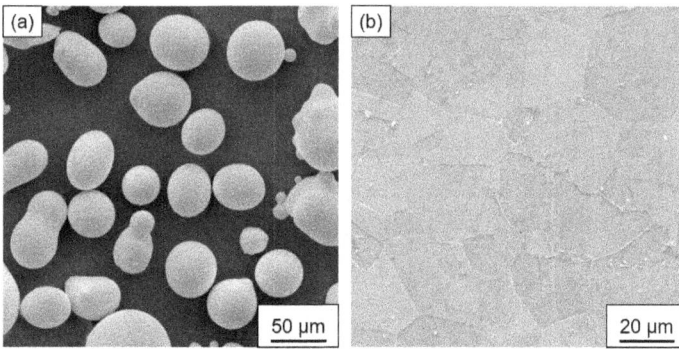

FIGURE 8.17 Microstructures of (a) nickel-based (EWAC) powder and (b) SS-316L plate.

TABLE 8.3
Chemical Composition of Materials Used

Element (Wt. %)	Fe	Ni	Cr	Mo	C	Mn	P	S	Si	N
SS-316L base metal	Bal.	12	18	3	0.03	2	0.045	0.03	0.75	0.1
Nickel-based powder	-	Bal.	0.2	-	-	-	-	-	2.8	-

the joint is depicted in Figure 8.19. The susceptor interacts with the microwaves, gets heated up, and then transfers the heat to the powder layer in its immediate contact. This rise in temperature (>400°C) of the slurry makes the resin evaporate. The skin depth of the top-most powder layer (i.e., first layer) increases, and the interaction of microwaves with powder particles is enhanced, as shown in Figure 8.19a. The temperature of the first layer rises rapidly until it melts, and heat gets transferred to

TABLE 8.4

Parameters Used for the Development of Microwave SS-316L Joints

Applicator	Microwave Applicator
Power	900 W
Frequency	2.45 GHz
Base metal	SS-316L
Interface powder	Nickel-based powder (~40 μm)
Susceptor	Graphite plate
Exposure	740 seconds

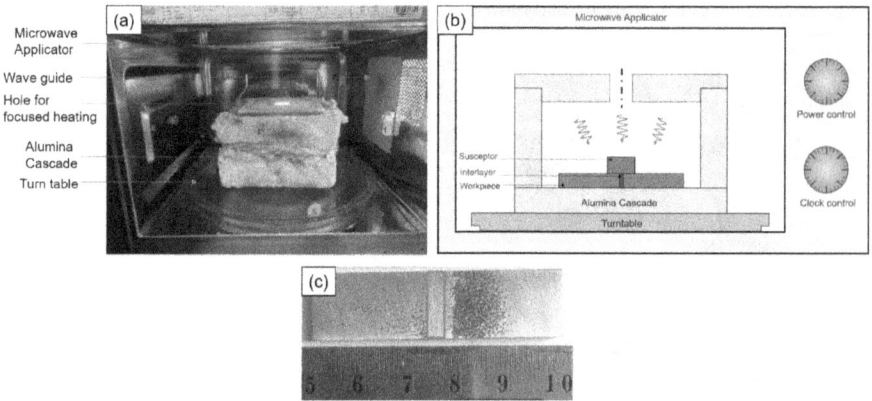

FIGURE 8.18   (a) Experimental setup, (b) schematic diagram, and (c) developed joint.

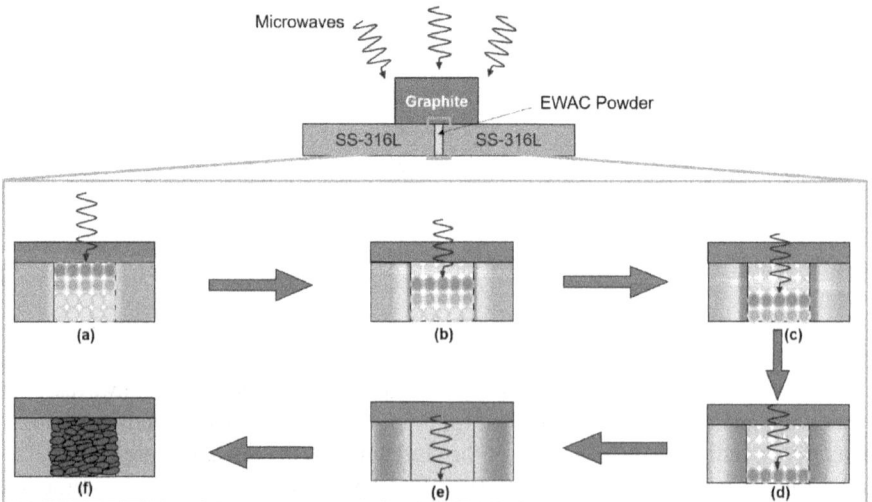

FIGURE 8.19   Melting of powder and formation of joint.

the powder layer beneath it (i.e., second layer). Then the skin depth of the second powder layer increases, and it experiences enhanced coupling with the microwaves, as depicted in Figure 8.19b. Its temperature rises, and then heat gets transferred to the next powder layer, and so on. This process continues as shown in Figure 8.19c and d until the entire powder melts and fills the voids left by the evaporated resin, as illustrated in Figure 8.19e. Simultaneously, the bulk metal pieces gradually undergo partial melting of interfacing surfaces, which leads to good metallurgical bonding with the interlayer. Thus, the joint is formed as depicted in Figure 8.19f. Several experiments were conducted to determine the optimum joining time, and it was found that 740 seconds was the time taken to form the bulk joint shown in Figure 8.18c.

8.5.1.2.3 Microstructural Analysis

The developed joint was cut across and ground with silicon carbide papers. The polished sample was etched with 1.5 mL of hydrogen peroxide, 30 mL of HCl, and 10 mL of distilled water for 30 seconds to reveal the microstructure. The microstructure of the EWAC-based joint obtained from a Scanning Electron Microscope is shown in Figure 8.20a–c. From Figure 8.20a, it is observed that the joint region is distinguishable and has wavy interfaces. This may be due to good metallurgical bonding between the bulk metal and interlayer. Columnar and inter-dendritic grains were observed in the joint region, along with the presence of a few pores, as shown in Figure 8.20b. The formation of carbides ($M_{23}C$) at the inter-dendritic and columnar grain boundaries is depicted in Figure 8.20c. Energy-dispersive X-ray spectroscopy analysis was performed to analyze the elemental composition in the grain interior and at the grain boundary in the joint region. The spectra obtained are depicted in Figure 8.20d and e. In the grain interior, iron (Fe) and nickel (Ni) were found to have

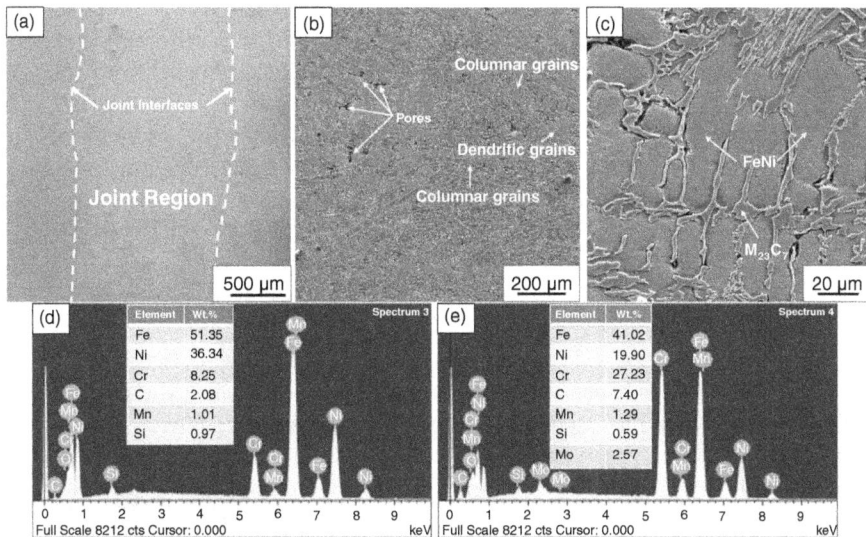

FIGURE 8.20 Micrographs of an EWAC-based joint at magnification – (a) 50×, (b) 200×, (c) 1000×, and EDS spectra of (d) grain interior (e) grain boundary.

higher weight percentages (wt. %) in composition. It is attributed to the formation of Fe-Ni intermetallics in the grain. At the grain boundaries, chromium (Cr) and carbon (C) have higher wt. % apart from Fe. At elevated temperatures, C has a very high affinity with Cr, which leads to the formation of $Cr_{23}C_7$ phase.

8.6 RECENT ADVANCES IN MICROWAVE PROCESSING OF METALLIC MATERIALS

8.6.1 MICROWAVE SINTERING

It is a common misconception that metals cannot be processed using microwaves, as an electron cloud present over them reflects the microwaves off the surface of metals. The successful sintering of metallic powders using microwaves was first reported in 1999 by Roy et al. The sintering of Fe-Cu-C and Fe-Ni-C systems was performed inside a microwave applicator having a controlled atmosphere with 1.4 kW power and 2.45 GHz operating frequency, and the total cycle time was approximately 90 minutes (Roy et al. 1999).

The microwave sintering of a Tungsten Heavy Alloy, W-Fe-Ni was performed by Upadhyaya et al. at 1500°C in a 2.45 GHz microwave furnace as well as in a conventional radiatively heated furnace, and then the sintering behavior was compared for both furnaces. It was found that the sintering time and power input for a microwave furnace were significantly lower as compared to a radiatively heated furnace. The mechanical properties obtained were also better using microwave sintering (Upadhyaya, Tiwari, and Mishra 2007).

Nagaraju et al. performed the sintering of SS-316L using microwave energy and studied the effect of sintering time and sintering temperature on the process and characteristics of the sintered sample (Nagaraju, Kumaran, and Srinivasa Rao 2019). Zhou et al. performed the microwave sintering of Mo-W-Cu alloys at low temperatures in a short period of time (~300 seconds). The results suggested that complete penetration of radiation inside the material was enabled by the microwave field (Zhou et al. 2021). Bhoi et al. carried out the simulation analysis of microwave hybrid sintering of aluminum and also performed experiments for the same. The results obtained from the experiments were consistent with the simulation results, as the error was less than 10% (Bhoi et al. 2022).

8.6.2 MICROWAVE CLADDING

Gupta and Sharma developed microwave cladding as a new processing technique that can be used to enhance the surface properties of samples. Microwave cladding of Austenitic stainless steel (SS-316) was carried out using EWAC powder in a 2.45 GHz microwave oven at a constant power input of 900 W. It was found that the substrate and clad partially diffused into each other, and a metallurgical bond was formed between them (Gupta and Sharma 2011).

The microwave cladding of WC-Co-based cermet over SS was performed by Zafar and Sharma. The average thickness of the clad was 1 mm. The clads were characterized, and the results revealed that partial diffusion and metallurgical bonding took place between the steel substrate and the cermet clad (Zafar and Sharma 2014). A similar study was performed on WC-Co-based clads over SS-316 and analyzed for their erosive wear and electrochemical corrosion behavior (Singh, Goyal, and Bansal 2021).

Singh and Zafar studied the time-temperature distribution in microwave cladding and the effect of power on the heating profile. It was found that the microwave input power had a strong influence on the sinter zone. The study was extended using modeling and simulation, and then the simulation results were compared with the experimental results. It was found that both the results were almost in agreement with each other, and the error was less than 5% (Singh and Zafar 2020, 2022).

Analysis of the control parameters and their effect on the responses in microwave cladding was performed using ANOVA to reduce slurry erosion in transportation systems. The important parameters considered were speed, angle of impact, and slurry exposure duration. It was found that the most dominant parameter was impact angle, and the level of significance was 0.05 (Vishwanatha et al. 2021).

8.6.3   MICROWAVE DRILLING

Microwave drilling of borosilicate glass was extensively studied by Gaurav Kumar et al. In this technique, a thin metallic concentrator is used to focus the microwave radiation into a narrow zone. The effect of dielectric fluid and concentrator material was studied to obtain better performance and minimum defects in the microwave drilling process. The influence of factors like the shape of the tool and its depth of immersion, feed rate, and machining gap on the quality of the hole produced by a graphite concentrator, was analyzed using modeling and simulation in COMSOL Multiphysics. It was found that the conical tip concentrator showed better performance compared to the cylindrical tip concentrator as the thermal energy was concentrated at the tip. In a later study, the authors analyzed the effect of input power and the velocity of the dielectric on the material removal rate using simulation in COMSOL Multiphysics software. Simulation and experimental analyses were also performed to minimize the defects during microwave drilling (Kumar and Sharma 2018, 2020; Kumar, Mishra, and Sharma 2020, 2021).

8.6.4   MICROWAVE CASTING

Mishra and Sharma investigated the in-situ microwave casting of aluminum alloy and found that an oxide layer was formed over the sample during the process, which acted as a susceptor and enhanced the heating process. Numerical simulation was done to analyze the melting characteristics, and the results were used to determine the effect of input power and setup location on the energy requirement and processing time. The temperature profiles obtained from simulation and experiment were found to be nearly coincidental, indicating consistency in the obtained results. It was revealed that the mold and susceptor materials have a great influence on the melting time and microstructure of the cast sample (Mishra and Sharma 2016b, 2017b, 2017a, 2018).

8.6.5 SIMULATION OF MICROWAVE PROCESSING

Singh and Zafar studied the time-temperature distribution in microwave cladding and the effect of power on the heating profile using experimental analysis as well as numerical simulation. It was found that experimental and simulation results were almost in agreement with each other, and the error was less than 5% (Singh and Zafar 2020, 2022). Tamang and Aravindan investigated the joining of copper with stainless steel (SS-316L) using microwaves, and an interfacial powder of copper was used for experimentation and simulation. Joining of dissimilar metals was simulated in COMSOL Multiphysics software using microwave energy as a heat source. The microstructure was analyzed based on the microwave exposure time, and it was found that upon increasing the exposure time, the porosity in the interlayer decreased (Tamang and Aravindan 2022).

Bhoi et al. performed the simulation analysis of microwave hybrid sintering of aluminum and carried out an experimental study as well. The experimental results were found to be consistent with the simulation results, as the difference was less than 10% (Bhoi et al. 2022). Microwave joining of SS304 samples was studied using two different interface powders, Inconel 718 and Nickel powder. Simulation of selective MHH was performed, and then a comparison was drawn to determine the best interface material (Rawat et al. 2022). Simulation and experimental studies were performed for MS pipe joining using selective MHH with nickel powder as the interlayer material. It was revealed that due to the presence of carbon in the susceptor material (charcoal), carbide precipitation took place in the joint zone (Thakur et al. 2022). Mishra and Sharma investigated the in-situ microwave casting of aluminum alloy and used numerical simulation to analyze the melting characteristics. The simulation results were used to determine the effect of input power and setup location on energy requirements and processing time (Mishra and Sharma 2017b).

COMSOL Multiphysics software was preferred as the analysis tool by most of the authors to study various microwave processing techniques.

8.7 SUMMARY

The chapter focused on MHH or processing for joining of metallic materials. Along with highlighting the characteristics, application, and mechanism of this technology, an experimental study is also reported. Properties such as permittivity, permeability, and loss tangent have been discussed, along with their significance in microwave processing. Absorption of power and penetration depth of microwaves, as given by standard equations such as Maxwell's equations and Lambert's law, have been explained. Heating mechanisms in different types of materials were explored, and it was observed that dipolar loss and conduction loss were the primary heating mechanisms for non-magnetic materials since they are only affected by electric fields. On the contrary, for magnetic materials, both electric and magnetic fields influence the heating mechanisms, resulting in conduction loss along with other losses such as eddy current loss, hysteresis loss, and residual loss. Different modes of heating were discussed, including conventional heating, microwave heating, MHH, microwave-selective heating, and selective MHH. Variants of microwave processing, such as microwave sintering, microwave cladding, microwave drilling, and microwave casting, are some futuristic, promising technologies that require some more detailed investigation to establish the field further.

8.8 FUTURE SCOPE OF MICROWAVE PROCESSING

8.8.1 CHALLENGES

1. Without the use of susceptors and complex tooling, thermal damage is likely to occur in microwave-processed materials due to their non-uniform-heating characteristics.
2. A rise in temperature during heating results in a change in material properties and alters the absorption of microwaves. These material properties need to be measured in real-time to accurately model the heating.
3. The experimental results obtained from microwave processing have poor repeatability.
4. The physics behind the phenomenon is still not well understood. Certain phenomena called 'Microwave effects' are still beyond the purview of researchers.
5. The existing microwave applicators have a standard size, which limits the size of the parts that can be produced. Customized microwave applicators may be designed, but they require a high investment.
6. The possibility of microwave leakage poses a hazard to the operator.

All these challenges limit the use of microwave processing for industrial applications. They need to be overcome, and hence, they also provide opportunities for future research and development in the area of microwave processing.

8.8.2 OPPORTUNITIES

Conventional processing of advanced materials is often very difficult, and this challenge led the researchers toward novel methods to process advanced materials. One of them is microwave processing. Further work can be done to have a better understanding of the process, which would lead to improved processing and maximized benefits. Some of the research problems that can be taken up are as follows:

1. Improvement in efficiency
   Currently, the efficiency of microwave processing in terms of energy utilization is quite low. It needs to be improved by enhancing the absorption of microwave energy by the material, which would result in better utilization of microwave energy. The microwave energy absorption can be increased by increasing the localized susceptor heating by applying a coated layer of a material with high dielectric loss over the sample material. Another measure to improve energy utilization is by enhancing the concentration of microwaves so that more energy is delivered per unit area of the sample

material. This can be achieved by developing improved systems to control the distribution of electric and magnetic fields.

2. A better understanding of the process

   Microwave processing and the governing principles in action need to be thoroughly understood by investigating the microwave-material interaction at the molecular level and the absorption of microwaves leading to a change in material properties. There is a need for a better theoretical understanding of the influence of electric and magnetic fields over microwave absorption. There must be sufficient experimental evidence to confirm the proposed theory. Modeling and simulation of the process form an effective tool to build a better understanding of the process. It would help to obtain results that are difficult to obtain experimentally, such as electric field distribution.

3. Improved susceptor design

   A proper susceptor material needs to be selected, and it should be designed in such a way that it increases the diffusion rate of microwave energy inside the material. This would enhance uniform-heating characteristics in MHH.

4. Microwave processing as a micromachining technique

   Microwaves can be used to effectively machine the workpieces at a micro-level. Microwave-assisted machining and microwave drilling need to be explored so that material removal is more precise and there is repeatability in the process.

5. Process optimization

   Optimization of processing parameters is necessary to ensure repeatability in the process. But it is not explored due to a lack of experimental data. Hence, large volumes of data need to be collected experimentally so that they can be used for optimization.

REFERENCES

Badiger, R. I., S. Narendranath, and M. S. Srinath. 2015. "Joining of Inconel-625 Alloy through Microwave Hybrid Heating and Its Characterization." *Journal of Manufacturing Processes* 18: 117–23. https://doi.org/10.1016/j.jmapro.2015.02.002.

Badiger, R. I., S. Narendranath, and M. S. Srinath. 2018. "Microstructure and Mechanical Properties of Inconel-625 Welded Joint Developed through Microwave Hybrid Heating." *Proceedings of the Institution of Mechanical Engineers, Part B: Journal of Engineering Manufacture* 232 (14): 2462–77. https://doi.org/10.1177/0954405417697350.

Badiger, R. I., S. Narendranath, and M. S. Srinath. 2019. "Optimization of Process Parameters by Taguchi Grey Relational Analysis in Joining Inconel-625 through Microwave Hybrid Heating." *Metallography, Microstructure, and Analysis* 8 (1): 92–108. https://doi.org/10.1007/s13632-018-0508-4.

Badiger, R. I., S. Narendranath, M. S. Srinath, and A. M. Hebbale. 2019. "Effect of Power Input on Metallurgical and Mechanical Characteristics of Inconel-625 Welded Joints Processed through Microwave Hybrid Heating." *Transactions of the Indian Institute of Metals* 72 (3): 811–24. https://doi.org/10.1007/s12666-018-1537-z.

Bansal, A., A. K. Sharma, S. Das, and P. Kumar. 2016. "On Microstructure and Strength Properties of Microwave Welded Inconel 718/ Stainless Steel (SS-316L)." *Proceedings of the Institution of Mechanical Engineers, Part L: Journal of Materials: Design and Applications* 230 (5): 939–48. https://doi.org/10.1177/1464420715589206.

Bansal, A., A. K. Sharma, P. Kumar, and S. Das. 2016. "Investigation on Microstructure and Mechanical Properties of the Dissimilar Weld between Mild Steel and Stainless Steel-316 Formed Using Microwave Energy." *Proceedings of the Institution of Mechanical Engineers, Part B: Journal of Engineering Manufacture* 230 (3): 439–48. https://doi.org/10.1177/0954405414558694.

Bhoi, N. K., D. K. Patel, H. Singh, S. Pratap, and P. K. Jain. 2022. "Multi-Physics Simulation Study of Microwave Hybrid Sintering of Aluminium and Mechanical Characteristics." *Proceedings of the Institution of Mechanical Engineers, Part E: Journal of Process Mechanical Engineering*, 1–11. https://doi.org/10.1177/09544089221074829.

Gupta, D., and A. K. Sharma. 2011. "Development and Microstructural Characterization of Microwave Cladding on Austenitic Stainless Steel." *Surface and Coatings Technology* 205 (21–22): 5147–55. https://doi.org/10.1016/j.surfcoat.2011.05.018.

Gupta, M., and W. W. L. Eugene. 2011. *Microwaves and Metals. Microwaves and Metals.* John Wiley & Sons (Asia) Pvt Ltd. https://doi.org/10.1002/9780470822746.

Horikoshi, S., R. F. Schiffmann, J. Fukushima, and N. Serpone. 2017. *Microwave Chemical and Materials Processing: A Tutorial.* https://doi.org/10.1007/978-981-10-6466-1.

Holian, K. 1984. *T-4 Handbook of Material Properties Data Bases: Equation of State.* edited by Kathleen S. Holian. Los Alamos National Laboratory. https://inis.iaea.org/search/search.aspx?orig_q=RN:17028984.

Kumar, G., R. R. Mishra, and A. K. Sharma. 2020. "On Finite Element Analysis of Material Removal Rate in Microwave Drilling of Borosilicate Glass." *Materials Today: Proceedings* 41: 759–64. https://doi.org/10.1016/j.matpr.2020.08.407.

Kumar, G., R. R. Mishra, and A. K. Sharma. 2021. "On Defect Minimization during Microwave Drilling of Borosilicate Glass at 2.45 GHz Using Flowing Dielectric and Optimized Input Power." *Journal of Thermal Science and Engineering Applications* 13 (3): 1–14. https://doi.org/10.1115/1.4048667.

Kumar, G., and A. K. Sharma. 2018. "Role of Dielectric Fluid and Concentrator Material in Microwave Drilling of Borosilicate Glass." *Journal of Manufacturing Processes* 33 (May): 184–93. https://doi.org/10.1016/j.jmapro.2018.05.010.

Kumar, G., and A. K. Sharma . 2020. "On Processing Strategy to Minimize Defects While Drilling Borosilicate Glass with Microwave Energy." *International Journal of Advanced Manufacturing Technology* 108 (11–12): 3517–36. https://doi.org/10.1007/s00170-020-05563-9.

Kumar, S., S. Sehgal, S. Singh, and A. K. Bagha. 2020. "Investigations on Material Characterization of Joints Produced Using Microwave Hybrid Heating." *Materials Today: Proceedings* 28: 1319–22. https://doi.org/10.1016/j.matpr.2020.04.588.

Mishra, R. R., and A. K. Sharma. 2016a. "Microwave-Material Interaction Phenomena: Heating Mechanisms, Challenges and Opportunities in Material Processing." *Composites Part A: Applied Science and Manufacturing* 81: 78–97. https://doi.org/10.1016/j.compositesa.2015.10.035.

Mishra, R. R., and A. K. Sharma. 2016b. "On Mechanism of In-Situ Microwave Casting of Aluminium Alloy 7039 and Cast Microstructure." *Materials and Design* 112: 97–106. https://doi.org/10.1016/j.matdes.2016.09.041.

Mishra, R. R., and A. K. Sharma. 2017a. "Effect of Susceptor and Mold Material on Microstructure of In-Situ Microwave Casts of Al-Zn-Mg Alloy." *Materials and Design* 131: 428–40. https://doi.org/10.1016/j.matdes.2017.06.038.

Mishra, R. R., and A. K. Sharma . 2017b. "On Melting Characteristics of Bulk Al-7039 Alloy during in-Situ Microwave Casting." *Applied Thermal Engineering* 111: 660–75. https://doi.org/10.1016/j.applthermaleng.2016.09.122.

Mishra, R. R., and A. K. Sharma . 2018. "Experimental Investigation on In-Situ Microwave Casting of Copper." *IOP Conference Series: Materials Science and Engineering* 346 (1). https://doi.org/10.1088/1757-899X/346/1/012052.

Moulson, A. J., Herbert, J. M. 1995. "Magnetic Ceramics." *Materials Characterization*. John Wiley & Sons, Inc. https://doi.org/10.1016/1044-5803(95)80089-1.

msestudent.com. "What Is Magnetic Hysteresis and Why Is It Important?". [Online]. Accessed March 10, 2022, https://msestudent.com/what-is-magnetic-hysteresis-and-why-is-it-important/.

Nagaraju, K. V.V., S. Kumaran, and T. S. Rao. 2019. "Microwave Sintering of 316L Stainless Steel: Influence of Sintering Temperature and Time." *Materials Today: Proceedings* 27: 2066–71. https://doi.org/10.1016/j.matpr.2019.09.062.

Naik, S. R., G. M. Gadad, and A. M. Hebbale. 2021. "Joining of Dissimilar Metals Using Microwave Hybrid Heating and Tungsten Inert Gas Welding - A Review." *Materials Today: Proceedings* 46: 2635–40. https://doi.org/10.1016/j.matpr.2021.02.322.

Matweb. "Overview of Materials for Nickel Alloy." [Online]. Accessed March 2, 2022. https://www.matweb.com/search/datasheet.aspx?matguid=8808b026f7c14d2f8d61f2d476aaeb26&n=1&ckck=1.

Rawat, S., R. Samyal, R. Bedi, and A. K. Bagha. 2022. "A Comparative Study of Interface Material through Selective Microwave Hybrid Heating for Joining Metal Plates." *Materials Today: Proceedings* 65: 3117–25. https://doi.org/10.1016/j.matpr.2022.05.346.

Roy, R., D. Agrawal, J. Cheng, and S. Gedevanlshvili. 1999. "Full Sintering of Powdered-Metal Bodies in a Microwave Field." *Nature* 399 (6737): 668–70. https://doi.org/10.1038/21390.

Singh, B., and S. Zafar. 2020. "Understanding Time-Temperature Characteristics in Microwave Cladding." *Manufacturing Letters* 25: 75–80. https://doi.org/10.1016/j.mfglet.2020.08.002.

Singh, B., and S. Zafar.. 2022. "Understanding Temperature Characteristics during Microwave Cladding through Process Modeling and Experimental Investigation." *CIRP Journal of Manufacturing Science and Technology* 37: 401–13. https://doi.org/10.1016/j.cirpj.2022.02.020.

Singh, P., D. K. Goyal, and A. Bansal. 2021. "Electrochemical Corrosion and Erosive Wear Behaviour of Microwave Processed WC-10Co4Cr Clad on SS-316." *Materials Today: Proceedings* 50: 1900–1905. https://doi.org/10.1016/j.matpr.2021.09.241.

Singh, S., R. Singh, D. Gupta, and V. Jain. 2017. "Preliminary Metallurgical and Mechanical Investigations of Microwave Processed Hastelloy Joints." *Journal of Manufacturing Science and Engineering, Transactions of the ASME* 139 (6): 1–5. https://doi.org/10.1115/1.4035370.

Singh, S., P. Singh, D. Gupta, V. Jain, R. Kumar, and S. Kaushal. 2019. "Development and Characterization of Microwave Processed Cast Iron Joint." *Engineering Science and Technology, an International Journal* 22 (2): 569–77. https://doi.org/10.1016/j.jestch.2018.10.012.

Srinath, M. S., A. K. Sharma, and P. Kumar. 2011a. "A Novel Route for Joining of Austenitic Stainless Steel (SS-316) Using Microwave Energy." *Proceedings of the Institution of Mechanical Engineers, Part B: Journal of Engineering Manufacture* 225 (7): 1083–91. https://doi.org/10.1177/2041297510393451.

Srinath, M. S., A. K. Sharma, and P. Kumar. 2011b. "A New Approach to Joining of Bulk Copper Using Microwave Energy." *Materials and Design* 32 (5): 2685–94. https://doi.org/10.1016/j.matdes.2011.01.023.

Srinath, M. S., A. K. Sharma, and P. Kumar . 2011c. "Investigation on Microstructural and Mechanical Properties of Microwave Processed Dissimilar Joints." *Journal of Manufacturing Processes* 13 (2): 141–46. https://doi.org/10.1016/j.jmapro.2011.03.001.

Tamang, S., and S. Aravindan. 2019. "3D Numerical Modelling of Microwave Heating of SiC Susceptor." *Applied Thermal Engineering* 162 (March): 114250. https://doi.org/10.1016/j. applthermaleng.2019.114250.

Tamang, S., and S. Aravindan. 2019. 2022. "Joining of Dissimilar Metals by Microwave Hybrid Heating: 3D Numerical Simulation and Experiment." *International Journal of Thermal Sciences* 172 (PA): 107281. https://doi.org/10.1016/j.ijthermalsci.2021.107281.

Thakur, A., R. Bedi, A. K. Bagha, and P. S. Rao. 2022. "Microstructural and Mechanical Properties of Mild Steel Pipes Joined Using Selective Microwave Hybrid Heating." *Journal of Manufacturing Processes* 82 (September): 848–59. https://doi.org/10.1016/j. jmapro.2022.08.024.

Uchino, K. 2018. *Ferroelectric Devices. Ferroelectric Devices.* 2nd edition. CRC Press. https:// doi.org/10.1201/b15852.

Upadhyaya, A., S. K. Tiwari, and P. Mishra. 2007. "Microwave Sintering of W-Ni-Fe Alloy." *Scripta Materialia* 56 (1): 5–8. https://doi.org/10.1016/j.scriptamat.2006.09.010.

Vasudev, H., G. Singh, A. Bansal, S. Vardhan, and L. Thakur. 2019. "Microwave Heating and Its Applications in Surface Engineering: A Review." *Materials Research Express*, 38. https://doi.org/https://doi.org/10.1088/2053-1591/ab3674.

Vishwanatha, J. S., A. M. Hebbale, N. Kumar, M. S. Srinath, and R. I. Badiger. 2021. "ANOVA Studies and Control Factors Effect Analysis of Cobalt Based Microwave Clad." *Materials Today: Proceedings* 46: 2409–13. https://doi.org/10.1016/j.matpr.2021.01.214.

Wong, W., and M. Gupta. 2015. "Using Microwave Energy to Synthesize Light Weight/Energy Saving Magnesium Based Materials: A Review." *Technologies* 3 (1): 1–18. https://doi. org/10.3390/technologies3010001.

Zafar, S., and A. K. Sharma. 2014. "Development and Characterisations of WC-12Co Microwave Clad." *Materials Characterization* 96: 241–48. https://doi.org/10.1016/j. matchar.2014.08.015.

Zhou, H., K. Feng, S. Ke, and Y. Liu. 2021. "Study on the Microwave Sintering of the Novel Mo-W-Cu Alloys." *Journal of Alloys and Compounds* 881: 160584. https://doi. org/10.1016/j.jallcom.2021.160584.

9 Hybrid Welding Technologies

Amrit Raj Paul and Manidipto Mukherjee

9.1 INTRODUCTION

Welding is a technique used for joining two similar or dissimilar materials, mainly metals or thermoplastics, by performing localized melting and allowing them to solidify, resulting in fusion (Weman 2003). Welding differs from lower-temperature metal-joining procedures like brazing and soldering in that welding does not melt the base metal. In addition to melting the base metal, a filler material is usually introduced to the joint to generate a pool of molten material (the weld pool) that cools to form a joint that can be stronger than the base metal depending on the welded design (butt, full penetration, fillet, etc.) and the parent metal (Lancaster 1999; Weman 2003). Pressure can be used alone or in combination with heat to create a welded joint. Welding also incorporates a shield (such as inert gases, fluxes, etc.) to prevent contamination or oxidation of the filler or molten metals. A gas flame (chemical), an electric arc (electrical), a laser, an electron beam, friction, and ultrasound friction are all examples of energy sources that can be utilized for welding (Weman 2003; Lancaster 1999). As the demand for reliable and economical joining methods grew during the early 20th century, welding technology improved swiftly. Following the wars, numerous advanced welding techniques have been introduced, along with manual methods such as shielded metal arc welding, which is now one of the most widely used, as well as semi-automatic and automatic processes such as gas metal arc welding, submerged arc welding, flux-cored arc welding (FCAW), and electroslag welding (Cary 2004; Lancaster 1999). In the second half of the century, developments proceeded with the creation of laser beam welding, electron beam welding, and friction stir welding (FSW). Despite the advanced welding techniques that have been developed in the context of improved weld quality and weld strength, some limitations are always associated with the welding technique, which is illustrated in Table 9.1 (Ahmed 2005; da Silva, El-Zein, and Martins 2020; Chaturvedi and Arungalai Vendan 2021; Norrish 2006; Cary 2004). In this context, hybrid welding nowadays is getting a lot of enchantment from enthusiastic researchers as it can maximize the advantages and mitigate the limitations of an individual welding process (Casalino et al. 2010). The term "hybrid welding" stands for a new welding technique developed by mixing two or more available welding techniques altogether to eradicate the limitations associated with the available welding techniques. In the 1980s, M. Steen first came up with a hybrid welding technique combining laser welding and gas tungsten arc welding (GTAW) altogether to minimize the

DOI: 10.1201/9781003327769-9

TABLE 9.1

Limitations of Available Welding Techniques

Welding Technique	Heat Source	Limitations
GMAW	Electric arc	• Atmospheric contamination. • Difficult to control the arc size. • Large HAZ. • Higher chance for lack of fusion due to low depth of penetration.
GTAW	Electric arc	• Atmospheric contamination. • Relatively slow welding process. • Difficult to weld highly oxidizable materials. • Limitation of the thickness of parent metal.
FSW	Mechanical friction	• Only suitable for flat surfaces. • Non-forgeable material cannot be welded.
Laser Welding	High-energy laser beam	• Difficult to weld reflective materials. • Low reinforcement. • Requires precise joint gap and position.
Electron Beam Welding	High-energy electron beam	• Limitation of weld size. • Low reinforcement. • Requires precise joint gap and position.

metallurgical difficulties and reflectivity-related issues of the laser welding technique (Steen 1980). Since then, lots of research articles have been done on the topic, i.e., hybrid welding. However, from that time on, only laser-arc hybrid welding has gained huge popularity in the science and technology domain, which is quite obvious due to its higher stability, less generation of residual stress, higher depth of penetration and large reinforcement, cleaner weld aesthetic, superior metallurgical and mechanical properties, and, most importantly, excellent adaptability between laser and arc (Acherjee 2018; Bunaziv et al. 2021; Wang, Liu, and Liu 2020; Huang et al. 2021; Borrisutthekul, Miyashita, and Mutoh 2005). On the contrary, a few articles also discussed the hybridization of laser welding with other welding techniques such as FSW (Munitz 2002; Padhy, Wu, and Gao 2015) and ultrasonic welding (USW) (Savu, Savu, and Sebes 2013; Guo et al. 2020). It has been seen that almost all of the developed hybrid welding technology deals with the introduction of laser welding techniques in other welding technologies, though there may be some other possibilities that give rise to hybridized welding techniques without the incorporation of laser welding. In this context, exploring other possibilities of developing hybrid welding techniques, it has been observed that the hybridization of welding techniques can be executed in two ways, i.e., power source hybridization and filler

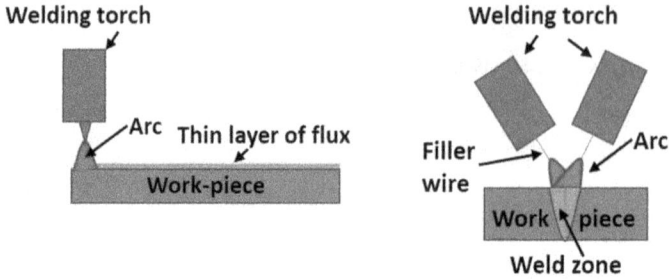

FIGURE 9.1 Schematic diagram of filler material hybridization by (a) adding powdered flux into the weld zone and (b) twining similar/dissimilar filler wire in the weld zone.

material hybridization. Power source hybridization stands for the introduction of one type of welding power source with another and totally different types of weld power source, such as laser-arc hybrid welding, laser-FSW hybrid welding, etc. Filler material hybridization includes the combination of different forms of filler material, such as powder fluxes or filler wires, as shown in Figure 9.1. Material hybridization is not a novel technique as it has been used frequently in activated flux tungsten inert gas welding (A-TIG) (Tseng and Hsu 2011; Sakthivel et al. 2011) process or dual/triple wire metal inert gas (MIG) welding process (Liu et al. 2015; Lin and Zhang 2020; Yang et al. 2021). However, it will be the first time this method is introduced as a material hybridization technique for welding processes. This chapter will elaboratively deal with both types of hybridization methodologies, primarily focusing on the pros and cons of the particular hybridized welding techniques. Moreover, the chapter will also try to shed some light on the possibility of other hybrid welding techniques as a part of future research interests.

9.2 POWER SOURCE HYBRIDIZATION

Power source hybridization is the process of combining two or more power sources for the purpose of joining the two materials. In this way, it is possible to counter the disadvantage of one welding process with the advantage of another. The historical development of power source hybridization is more prominently discussed in the literature. To date, 80% of the hybrid welding literature has discussed the laser-arc hybrid welding technique. There are only a few articles available that deal with the laser-FSW (Munitz 2002; Padhy, Wu, and Gao 2015) or laser-USW technique (Savu, Savu, and Sebes 2013; Guo et al. 2020). Moreover, the other aspects of hybridization, such as arc-FSW, USW-FSW, electric field/magnetic field-FSW (Padhy, Wu, and Gao 2015), and arc-USW (Krajewski et al. 2012), have also been discussed in a few research articles. The following subsection discusses all the available power source hybridization techniques one by one to give a clear idea of the effect of those hybridization methodologies on the weld quality and properties.

9.2.1 LASER-ARC HYBRID WELDING

Power source hybridization was first to come into the picture in the 1970s when M. Steen (1980) developed a gas metal arc welding (GMAW)-assisted laser welding setup that increased the welding velocity and depth of penetration with comparatively low laser power capacity. The hybridization effect demonstrated a notable improvement in welding speed, penetration depth, and process stability. Later, many research and development works have been carried out on the hybridization of laser and arc welding power sources, and clear benefits of arc augmentation have been noticed compared to the individual setups. The ability of the materials to bridge gaps and their reflectivity are no longer obstacles to successful welding, which is one of the many downsides of the individual arc and individual laser welding techniques, respectively. On the basis of various combinations of laser and arc welding systems, a variety of hybrid laser and arc welding systems have been developed in the past few decades. The most popular arc welding methods utilized in laser-arc hybrid welding techniques include GMAW, GTAW, and PAW (plasma arc welding), as shown in Figure 9.2, whereas CO_2 lasers and Nd:YAG lasers are the most widely used lasers (Bagger and Olsen 2003; Vaidya et al. 2006; Emmelmann, Kirchhoff, and Petri 2011; Shi 2005; Nilsson et al. 2003; Ishide, Tsubota, and Watanabe 2003). A comparative analysis of laser source hybridization with GMAW, GTAW, and PAW is given in Table 9.2.

Hybrid laser-arc welding is an appealing solution for several industries because of its numerous process benefits, including higher productivity and improved weld quality compared to laser welding or arc welding. In order to lower the cost per unit length of weld, the welding system's capital cost can be offset by offering faster welding speed and deeper penetration. A wide range of metals, including non-ferrous metals like titanium (Zhang et al. 2015), magnesium (Zeng et al. 2011; Tan et al. 2015), aluminum (Leo et al. 2015), and copper (Tan et al. 2015; Zhang et al. 2015), as well

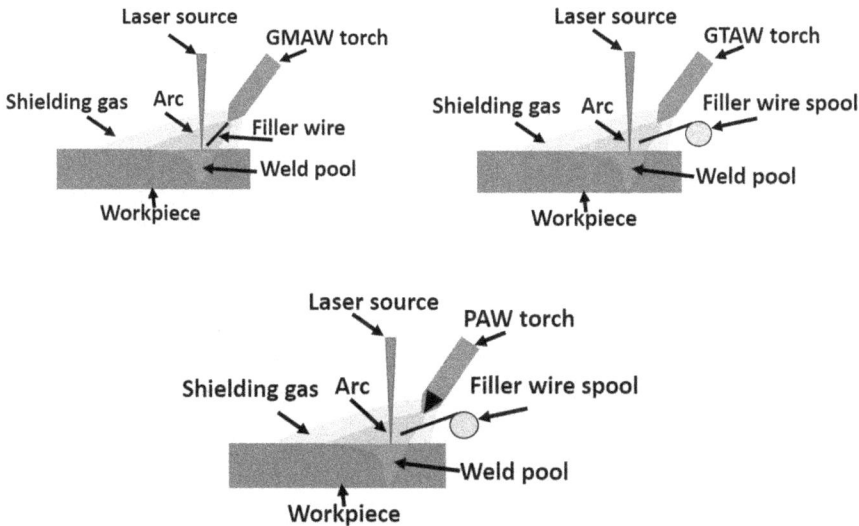

FIGURE 9.2 Schematic diagram of laser hybridization with (a) GMAW, (b) GTAW, and (c) PAW.

TABLE 9.2

Comparative Analysis of Different Laser-Arc Hybrid Welding Setups

Power Source	Type	Characteristics
Primary Power Source		
Laser	CO_2	• High wavelength: 10.6 μm • Difficult to weld highly reflective metals • Low weld pool stability • Average weld quality • Lower welding efficiency
	Nd:YAG	• Low wavelength: 1.064 μm • Easy to weld reflective material as the laser beam can be transported from source to workpiece through optical fibers • High weld pool stability • Superior weld quality • Higher welding efficiency
Secondary Power Source		
Arc	GMAW	• Larger volume of the melt pool • Easy to weld thicker sections with large root gap • Less arc stability • High heat input and high residual stress
	GTAW	• Smaller volume of the melt pool • Used for welding thin gauge materials • Low heat input and low residual stress • Contamination of non-consumable electrodes is very high
	PAW	• Lowest heat input • Highest arc stability • Highest welding speed and depth of penetration • Contamination of non-consumable electrodes is very high

as ferrous metals like steel and stainless steel (Hao et al. 2017; Zeng et al. 2011), can be joined together using the hybrid laser-arc welding process. Both thin sheet metal and thick metal plates can be welded using this technique. Because of the fast welding speed and deep penetration, big plates can be welded with fewer passes, which lessens weld distortion. The automotive and shipbuilding industries, where numerous metallic components are welded to construct light or heavy cars and ships, are where laser-arc hybrid welding is most commonly used. Other applications of laser-arc hybrid welding can be found in the power industry where tight wall panels and ribbed boiler pipes are welded together.

9.2.2 Laser-FSW Hybrid Welding

Laser-FSW hybrid welding is one of the most widely used hybrid welding techniques, primarily based on the FSW. The FSW process is generally summed up with certain

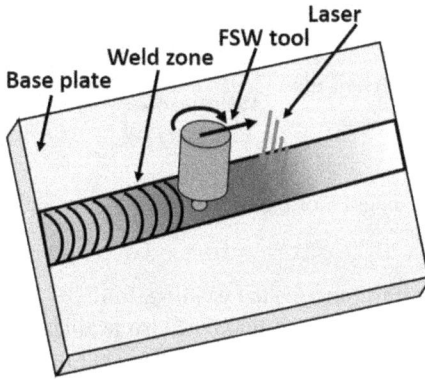

FIGURE 9.3 Schematic diagram of the laser-FSW hybrid welding process.

drawbacks, such as cavities due to lack of fusion (Padhy, Wu, and Gao 2015), frequent wear of tools (Padhy, Wu, and Gao 2015), poor mechanical properties (Kumar Rajak et al. 2020), kissing bond defects (Zhao et al. 2014), etc. The laser is used to preheat the base plate ahead of friction welding to minimize the defect associated with conventional FSW. Nd:YAG fiber lasers, diode lasers, and CO_2 lasers are the three most frequently utilized laser sources in the laser-FSW hybrid welding technique (Padhy, Wu, and Gao 2015). The focus of the laser is generally kept closer to the FSW tool, in the range of 20–40 mm, at a 45° inclination with the FSW tool axis. A schematic diagram of laser-FSW hybrid welding is shown in Figure 9.3. The resistance of the material to tool penetration and forward motion is reduced by the use of laser preheating. As a result, there is less need to exert strong forces on the tool and the workpieces, causing a reduction in the mechanical force in the range of 21%–43% (Padhy, Wu, and Gao 2015). The reduction in mechanical force in the plunging and movement of the FSW tool reduces its wear significantly. Apart from the concerns of tool wear, preheating helps the tool properly mix the materials at the retreating and advancing sides, eradicating the formation of lack of fusion cavity and cracks (Chang et al. 2011). Moreover, the improvement in mechanical properties of the weld zone, such as hardness, tensile strength, endurance strength, etc., has also been noticed in the laser-FSW hybrid welding due to increased levels of material plasticization and thorough grain refinements at the weld zone.

Despite improved mechanical and metallurgical properties, high flowability of material at the weld zone, and reduced tool forces, this method is associated with certain limitations, such as the reflectivity of material to the laser and, obviously, the extremely high cost of the laser-FSW setup. This can be avoided by incorporating other preheating techniques to replace the laser source in the setup.

9.2.3 LASER-USW HYBRID WELDING

The USW process is most prominently applied for welding small components at high production rates. High-frequency vibrations between 20 and 40 kHz are used to complete the welding process. It is a kind of spot welding in which mechanical vibrations

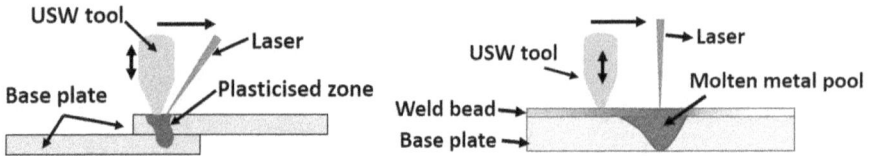

FIGURE 9.4 Schematic diagram of (a) laser-assisted hybrid USW and (b) USW-assisted hybrid laser welding.

and intermolecular bonding generate the welding. Initially, it was employed for polymers, but technological development makes it also available for metals, specifically thin foils of metals. However, low plasticity at the weld interface results in a lack of fusion and cracks at the weld interface, hampering the welding efficiency. Therefore, a preheating system may be required, which could impart adequate plasticity at the weld interface to incorporate the perfect weld between the metals. However, during the welding of electronic circuits, the bulk heating of the weld interface may alter the electronic properties of the circuit parts closer to the bulk heating regions. Therefore, localized heating of the interface with the help of a laser could be a better option. A schematic diagram of laser-USW hybrid welding is shown in Figure 9.4a. By using the thermo-compression action of the material, laser-USW hybrid welding can be used to create welds specifically for electrical connections (Savu, Savu, and Sebes 2013). The base materials are heated during preheating to temperatures that make them softer and more susceptible to plastic deformation during pressure applications that are unique to the laser-USW hybrid welding process. The temperature where the focal point contacts the base material rises to 100°C–200°C (Savu, Savu, and Sebes 2013). In the context of foundation materials for the welding process, this temperature has no adverse effects on the parts that will be welded or the nearby elements. By controlling the process parameters to incorporate a slow cooling rate at the weld interface, the microstructural changes are negligible, which leads to lowering the interfacial brittleness and diminishing the crack formation.

The above description validates the assistance of lasers in the USW process. However, the USW can also be used as a secondary welding unit during the laser welding technique. To eradicate welding porosity and poor welding formability, Lei et al. (2018) hybridized the conventional laser welding technique by incorporating the USW technique into it. Figure 9.4b shows the schematic representation of the USW-assisted hybrid laser welding technique. The USW is applied post to the laser welding, which enables it to refine the grain size and diminish the pores in the weld zone. As a result, the metallurgical and mechanical properties at the weld zone improve (Lei et al. 2018).

9.2.4 OTHER HYBRID WELDING TECHNIQUES

The arc-FSW hybrid welding technique is further subcategorized into two sections based upon the type of arc used, i.e., plasma arc (Kumar Yaduwanshi, Pal, and Bag 2014; Yaduwanshi, Bag, and Pal 2014) or GTAW arc (Bang et al. 2012, 2013). However, the plasma arc gives an upper hand over the GTAW arc in terms of arc

Arc zone   Weld line   Arc zone   Weld line

FSW tool   FSW tool   Harder material

Weld zone   Weld zone   Softer material

Similar base plate   Dissimilar base plate

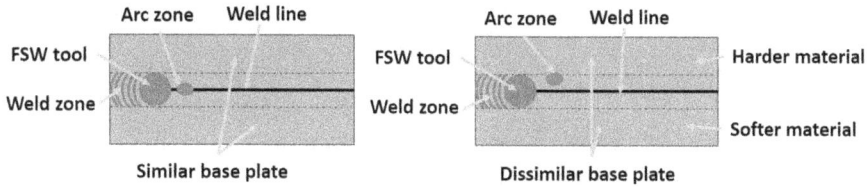

FIGURE 9.5 Schematic diagram of the arc-FSW hybrid welding process (a) for similar material and (b) for dissimilar material.

stability and control, energy concentration, and equipment cost. In both cases, the arc is transferred onto the workpiece for the sake of preheating the parent material ahead of FSW. For similar welding, the arc is focused on the weld line, whereas for dissimilar welding, the arc is focused on the material with a higher melting point, so that the difference between the melting points of the harder and softer materials is compensated. The schematic diagram of arc-FSW hybrid welding for similar and dissimilar welding is shown in Figure 9.5a and b, respectively.

It has also been noticed that, due to the broader arc dome compared to the plasma arc, the GTAW arc is mainly used for the welding of dissimilar materials. The wider arc dome may deteriorate the mechanical properties in the heat-affected zones of similar materials, which is not desirable. Whereas for dissimilar materials, focusing the GTAW arc over the high melting point material does not bring significant changes in the mechanical properties of the high melting point material. Therefore, the use of the GTAW arc for dissimilar material welding is equitable. However, both the arcs could bring grain refinement, improved tensile strength, and suppressed void formation at the weld interface due to increased plasticization and material flowability (Bang et al. 2012; Yaduwanshi, Bag, and Pal 2014; Kumar Yaduwanshi, Pal, and Bag 2014; Bang et al. 2013). Despite the improved mechanical and metallurgical properties, control over the heat-affected zone with respect to the weld line and FSW tool-arc distance is still a challenge that may be part of future research.

In the previous section, it was already discussed that it is difficult to use arc to assist the USW due to the wider arc dome and heat-affected zone. However, the use of USW to assist arc welding is possible. According to studies conducted so far, the structure and characteristics of welded joints are obviously impacted by mechanical vibrations. The vibrations alter the structure in both the melted zone and the heat-affected zone, regardless of the parent metal or the welding technique (Krajewski 2016). In this context, Krajewski et al. (2012) performed a study on the USW-assisted arc welding process and noticed a significant improvement in porosity suppression and weld strength. The schematic diagram of the arc-USW hybrid welding setup is shown in Figure 9.6.

Similarly, the USW technique can also be utilized to assist FSW as a post-processing unit to refine the grains and suppress the voids in the weld zone. However, when combined with a static load, high-frequency vibrations like ultrasonic waves can significantly soften metallic materials without producing a lot of heat (Siddiq and el Sayed 2012; Rusinko 2011). The effect of ultrasonic energy is comparable to that of thermal softening, and, according to experimental findings, it takes 10^7 times less ultrasonic

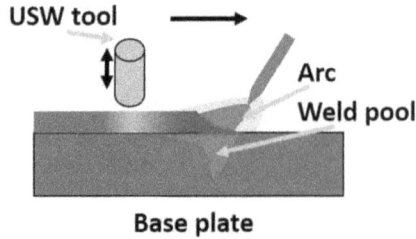

FIGURE 9.6   Schematic diagram of an arc-USW hybrid welding setup.

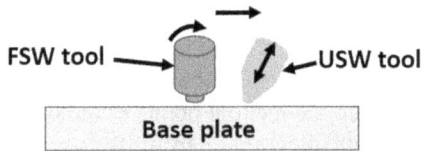

FIGURE 9.7   Schematic diagram of USW-FSW hybrid welding setup.

energy than thermal energy to produce the same degree of softening (Hung and Lin 2013; Shi, Wu, and Liu 2015; Rusinko 2011). This could be explained by the fact that the ultrasonic energy is preferentially absorbed by the defects in the grain, not in the defect-free regions, whereas the thermal energy is uniformly distributed throughout the material by conduction (Shi, Wu, and Liu 2015; Rusinko 2011). At Shandong University (Jinan, China), Wu et al. (Liu, Wu, and Padhy 2015a, 2015b) created a novel USW-FSW hybrid welding system. In this system, the weld center line of the workpieces receives ultrasonic energy at a 45° angle. The transmitting sonotrode moves 20 mm in front of the FSW tool during the welding process. Figure 9.7 shows the schematic diagram for this system. This appears to be the most effective way to transmit ultrasonic energy so far because it is applied exactly into the weld zone. The high-frequency vibration at the weld center line causes softening of the material, resulting in a reduction in axial force during plunging and welding. The axial forces are further reduced by increasing the amplitude of ultrasonic vibration. However, a decrease in axial forces is a result of more heat input at the weld interface, which may widen the plastically deformed zone at the weld region. Despite this, significant improvements in tensile strength and elongation have been noticed in the welded material (Liu, Wu, and Padhy 2015a). Additionally, the growth of voids and tunnel flaws is suppressed by ultrasonic vibrations (Shi, Wu, and Liu 2015). These findings demonstrate that the use of ultrasonic energy in the FSW of aluminum alloys is advantageous because it increases material flow and expands the plastically deformed region.

In this section, the available power source hybridized welding setups are discussed. It is noticed that the power source hybridization significantly coped with the shortcomings of the individual welding process in terms of metallurgical and mechanical properties. Moreover, the power source hybridization lowers the requirement for high energy in an individual welding process and also economizes the whole welding processing cost to a certain extent. However, the above-mentioned processes could only improve the mechanical and metallurgical properties of the weld zone through grain refinement technique/strain hardening techniques. For the sake of further improvement of weld properties, solid solution strengthening or precipitation hardening is required, which involves in-situ alloying of the weld region. This in-situ alloying can be processed by filler material hybridization, which is discussed in the next section.

9.3 MATERIAL HYBRIDIZATION

Material hybridization can be done by incorporating two different types of filler material in the weld zone. This technique can only be possible in the fusion welding process, such as laser welding, arc welding, etc. The availability of turbulent molten pools during fusion welding (Mukherjee et al. 2015) mechanically mixes the additionally added materials in the weld bead. This is sometimes also called the in-situ alloying of the weld bead during the welding process (Wang, Chen, and Yu 2000). Previously, this material hybridization during the welding process was done for the GTAW process as an activated GTAW process. During the process, a layer of some surface-active agents such as oxides, chlorides, or sulfides (Kumar, Ahmad, and Singh 2019) is sprayed over the weld zone to increase the depth of penetration of welding. The material hybridization is not directly related to advancing the weld depth of penetration. In fact, it is more focused on increasing the strength of the weld through solid solution strengthening/precipitation hardening. In this process, hard particles such as SiC, Al_2O_3, etc. are mixed in the weld pool, and, as a result, the solidified weld bead becomes a dispersed composite (Wang, Chen, and Yu 2000), which will, obviously, have much higher strength than the base plate. The mixing of these oxides/carbides particles during the welding process can be done in two distinctive ways: by applying a layer of powder to the surface to be welded and by feeding the powder directly into the weld pool, as shown in Figure 9.8a and b. Apart from mixing the non-metallic refractory compounds in the weldpool, in-situ alloying by mixing different metals/alloys can also be performed, in contrast with the

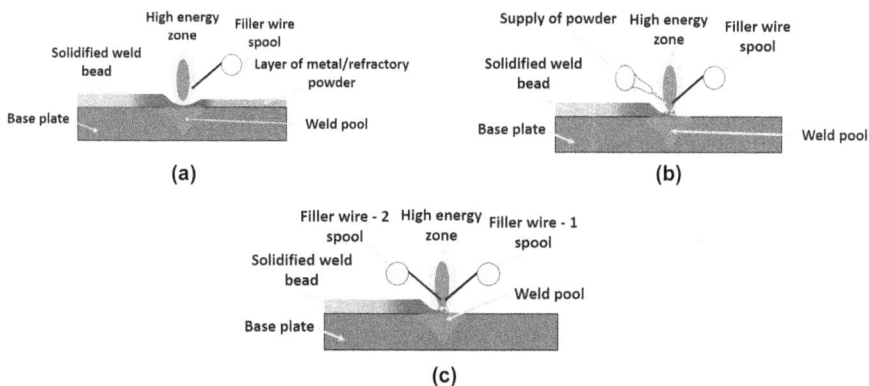

FIGURE 9.8 Schematic diagram of hybrid welding with (a) a layer of metal/ceramic powder, (b) direct feeding of metal/ceramic powder in the high-energy zone, and (c) a multi-wire system.

improvement of the mechanical properties of weld beads. The feeding of alloys into the base metal's weldpool can be done again in three ways: by applying a layer of metal/alloy powder to the surface to be welded (Figure 9.8a), by feeding the metal powder directly into the weldpool (Figure 9.8b), and by feeding the multiple filler wires in the high-energy zone (Figure 9.8c) during the welding process. However, the material hybridization of welding is generally found to be employed for coating purposes. The fusion of such hard ceramic particles advances the wear and corrosion properties of the surfaces. It has been found that the addition of carbide particles (Ureña et al. 2001; Lin and Lin 2011; Mir et al. 2022) in the weld pool acts as a grain refiner and a pinning agent for the dislocation lines, which in turn significantly improves the hardness and strength of the deposited weld bead. However, not being surface-active agents, carbides do not affect the bonding between the deposited beads and the base metal. On the other hand, the addition of oxides in the weld bead (Pouriamanesh et al. 2016; Lee et al. 2015; Ahmed et al. 2020) enhances the depth of penetration and dilution between the weld bead and the base metal, which certainly increases the joint strength between the weld bead and the base metal. However, the carbides are harder than the oxides, so the improvement in mechanical properties is greater in the case of carbides than in the case of the oxides. The layer of ceramic/metal powders will be difficult to apply directly to the weld surfaces, during the welding, the powders may be blown away by the arc forces, etc. (forces acting at the high-energy zone). Therefore, the powders should be applied in the form of a slurry, which will be prepared by thoroughly mixing acetone/alcohol with the ceramic powders. This slurry will then be applied to the weld surfaces, and after letting it dry completely, the welding process should be started. However, it will be difficult to uniformly apply the slurry throughout the weld surface, so an automatic system will be required that ensures the uniform distribution of slurry throughout the weld surfaces.

Nonetheless, the manufacturing process of ceramics (carbides or oxides) particles is very costly and could be overestimated for many industrial applications. Therefore, instead of using powdered filler materials, wire can be used extensively. However, the production of ceramic wires is not possible, which could be hybridized with wired filler material by directly feeding wire and powder into the arc. But, with the advent of FCAW techniques, wire-powder filler material hybridization has become much easier than before. FCAW is nothing but an ordinary arc welding technique that uses a hollow metal wire filled with fluxes that cover the weld bead during the welding process and avoid its atmospheric interaction, just like shielded metal arc welding. Thus, the shielding gas becomes an optional accessory for the FCAW technique. With the change in technology and time, the shielding flux of the hollow metal wire is replaced with ceramic and metal powders to incorporate the in-situ alloying of the weldment during the welding process. Such changes to FCAW made the shielding gas a compulsory agent during the welding. However, the cost of powder production made the filler wire costlier than the solid metal wire. Therefore, if wires are available for the specific material, then it is recommended to use a multi-wire feeding system during the welding process. However, with the increase in the number of wires, the feeding system becomes more complex, and the cost of the welding again increases. Therefore, for more than three alloying elements, the metal-cored filler wire should be used for welding. In addition to the

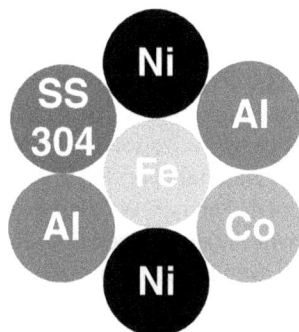

FIGURE 9.9 Schematic diagram of the cable configuration of Al-Co-Cr-Fe-Ni HEA used for WAAM process.

welding process, metal-cored filler wires are now extensively used for the additive manufacturing of high-entropy alloys (HEAs) (Pańcikiewicz 2021). The base metal of HEA is kept in the form of hollow wire, whereas the other alloying elements are filled inside the hollow wire in powder form. During the welding/deposition process, the wire and powder melt and are mixed together to develop the desired elemental configuration. Apart from metal-cored wires, combined cable wire (CCW) systems have also been found in trends for the development of HEAs. The CCW system is somehow different from the multi-wire system. The CCW uses a core wire (made of base elements), around which the other wires of alloying elements are torsionally wound to form a single cable with multiple strands (Shen, Kong, and Chen 2021), as shown in Figure 9.9. The figure depicts the cable configuration of Al-Co-Cr-Fe-Ni HEA used for the WAAM process (Shen, Kong, and Chen 2021). The diameter of individual wires is chosen in such a way that it produces exactly the same compositional configuration upon melting.

9.4 SUMMARY

This chapter elaborately discussed the different aspects of hybrid welding techniques. The hybridization of welding is done to improve the weld strength and quality to a significant level and to assure that failure should not occur through the weld. The hybridization of welding can be done in two ways: power source hybridization and material hybridization. The power source hybridization mixes up two or more types of completely different power sources altogether to eradicate the limitations of individual power sources and improve the weld quality. In general, laser-based hybrid welding are primarily available in the science and technology domain, but other power source hybridizations, such as Arc-FSW, Arc-USW, FSW-USW, etc., can also be done to improve the weld quality. Apart from this, in material hybridization, the different types and forms of filler materials are mixed together in the weld pool to incorporate in-situ alloying of the weldment to improvise the weld strength. In such a way, two major materials hybridization methods are discussed in this chapter: Wire+Powder material hybridization and Wire+Wire material hybridization. The Wire+Powder material hybridization can be done in three ways: depositing wired material on the powdered bed, feeding powder and wire in the same high-energy zone (arc), and using flux/metal-cored wire for welding. Similarly, the Wire+Wire material hybridization can be incorporated in two distinctive manners: by feeding different filler wires in the same high-energy zone and by using CCW made of different wire strands. Power source hybridization is a bit of an old technique and can be a costlier deal in comparison to material hybridization. However, material hybridization, such as flux/metal-cored wires and CCWs, is quite new in the field but relatively cheaper than power source hybridization. Therefore, material hybridization should be given more focus to be well established in the near future.

REFERENCES

Acherjee, Bappa. 2018. "Hybrid Laser Arc Welding: State-of-Art Review." *Optics and Laser Technology*. doi:10.1016/j.optlastec.2017.09.038.

Ahmed, Khalique Ejaz, B. M. Nagesh, B. S. Raju, and D. N. Drakshayani. 2020. "Studies on the Effect of Welding Parameters for Friction Stir Welded AA6082 Reinforced with Aluminium Oxide." In *Materials Today: Proceedings*. Vol. 20. doi:10.1016/j.matpr.2019.10.059.

Ahmed, Nasir. 2005. *New Developments in Advanced Welding*. doi:10.1533/9781845690892.

Bagger, Claus, and Flemming Ove Olsen. 2003. "Comparison of Plasma, Metal Inactive Gas (MIG) and Tungsten Inactive Gas (TIG) Processes for Laser Hybrid Welding." In *ICALEO 2003 – 22nd International Congress on Applications of Laser and Electro-Optics, Congress Proceedings*. doi:10.2351/1.5060041.

Bang, Han Sur, Hee Seon Bang, Geun Hong Jeon, Ik Hyun Oh, and Chan Seung Ro. 2012. "Gas Tungsten Arc Welding Assisted Hybrid Friction Stir Welding of Dissimilar Materials Al6061-T6 Aluminum Alloy and STS304 Stainless Steel." *Materials and Design* 37 (May): 48–55. doi:10.1016/j.matdes.2011.12.018.

Bang, Han Sur, Hee Seon Bang, Hyun Jong Song, and Sung Min Joo. 2013. "Joint Properties of Dissimilar Al6061-T6 Aluminum Alloy/Ti-6%Al-4%V Titanium Alloy by Gas Tungsten Arc Welding Assisted Hybrid Friction Stir Welding." *Materials and Design* 51: 544–51. doi:10.1016/j.matdes.2013.04.057.

Borrisutthekul, Rattana, Yukio Miyashita, and Yoshiharu Mutoh. 2005. "Dissimilar Material Laser Welding between Magnesium Alloy AZ31B and Aluminum Alloy A5052-O." *Science and Technology of Advanced Materials* 6 (2). doi:10.1016/j.stam.2004.11.014.

Bunaziv, Ivan, Odd M. Akselsen, Xiaobo Ren, Bård Nyhus, and Magnus Eriksson. 2021. "Laser Beam and Laser-Arc Hybrid Welding of Aluminium Alloys." *Metals*. doi:10.3390/met11081150.

Cary, Horward B. 2004. "Modern Welding Technology 5/e." *Industrial Robot: An International Journal* 31 (4). doi:10.1108/ir.2004.31.4.379.3.

Casalino, Giuseppe, Umberto Dal Maso, Andrea Angelastro, and Sabina Luisa Campanelli. 2010. "Hybrid Laser Welding: A Review." In *DAAAM International Scientific Book 2010*. DAAAM International Vienna. doi:10.2507/daaam.scibook.2010.38.

Chang, Woong Seong, Sudhakar R. Rajesh, Chang Keun Chun, and Heung Ju Kim. 2011. "Microstructure and Mechanical Properties of Hybrid Laser-Friction Stir Welding between AA6061-T6 Al Alloy and AZ31 Mg Alloy." *Journal of Materials Science and Technology* 27 (3). doi:10.1016/S1005-0302(11)60049-2.

Chaturvedi, Mukti, and S. Arungalai Vendan. 2021. *Advanced Welding Techniques*. Advanced Welding Techniques. doi:10.1007/978-981-33-6621-3.

Emmelmann, Claus, Marc Kirchhoff, and Nikolai Petri. 2011. "Development of Plasma-Laser-Hybrid Welding Process." *Physics Procedia*. doi:10.1016/j.phpro.2011.03.025.

Guo, Hengtong, Dengming Zhang, Yaobang Zhao, Zhenglong Lei, Qian Li, Yuchen Yang, and Yanli Xu. 2020. "Study on the Formation of Welding, Microstructure, and Properties of 5A06 Aluminum Alloy by Ultrasonic Laser-Assisted Filler Welding." *Zhongguo Kexue Jishu Kexue/Scientia Sinica Technologica* 50 (11). doi:10.1360/SST-2019-0407.

Hao, Kangda, Chen Zhang, Xiaoyan Zeng, and Ming Gao. 2017. "Effect of Heat Input on Weld Microstructure and Toughness of Laser-Arc Hybrid Welding of Martensitic Stainless Steel." *Journal of Materials Processing Technology* 245. doi:10.1016/j.jmatprotec.2017.02.007.

Huang, Hanxuan, Peilei Zhang, Hua Yan, Zhengjun Liu, Zhishui Yu, Di Wu, Haichuan Shi, and Yingtao Tian. 2021. "Research on Weld Formation Mechanism of Laser-MIG Arc Hybrid Welding with Butt Gap." *Optics and Laser Technology* 133. doi:10.1016/j.optlastec.2020.106530.

Hung, Jung Chung, and Chih Chia Lin. 2013. "Investigations on the Material Property Changes of Ultrasonic-Vibration Assisted Aluminum Alloy Upsetting." *Materials and Design* 45: 412–20. doi:10.1016/j.matdes.2012.07.021.

Ishide, Takashi, Shuho Tsubota, and Masao Watanabe. 2003. "Latest MIG, TIG Arc-YAG Laser Hybrid Welding Systems for Various Welding Products." In *First International Symposium on High-Power Laser Macroprocessing*. Vol. 4831. doi:10.1117/12.497771.

Krajewski, Arkadiusz. 2016. "Mechanical Vibrations in Welding Processes." *Welding International* 30 (1): 27–32. doi:10.1080/09507119.2014.937609.

Krajewski, Arkadiusz, Wladyslaw Wlosinski, Tomasz Chmielewski, and Pawel Kolodziejczak. 2012. "Ultrasonic-Vibration Assisted Arc-Welding of Aluminum Alloys." *Bulletin of the Polish Academy of Sciences: Technical Sciences* 60 (4): 841–52. doi:10.2478/v10175-012-0098-2.

Kumar, Hemant, Gulshad Nawaz Ahmad, and Nirmal Kumar Singh. 2019. "Activated Flux TIG Welding of Inconel 718 Super Alloy in Presence of Tri-Component Flux." *Materials and Manufacturing Processes* 34 (2): 216–23. doi:10.1080/10426914.2018.1532581.

Kumar Rajak, Dipen, Durgesh D. Pagar, Pradeep L. Menezes, and Arameh Eyvazian. 2020. "Friction-Based Welding Processes: Friction Welding and Friction Stir Welding." *Journal of Adhesion Science and Technology*. doi:10.1080/01694243.2020.1780719.

Kumar Yaduwanshi, Deepak, Sukhomay Pal, and Swarup Bag. 2014. "Effect of Preheating on Mechanical Properties of Hybrid Friction Stir Welded Dissimilar Joint." *In Proceedings of 5th International & 26th All India Manufacturing Technology, Design and Research Conference*.

Lancaster, John Frederick. 1999. *Metallurgy of Welding*. 6th edition, Woodhead Publishing Limited, England.

Lee, Jung Gu, Jin Ju Park, Min Ku Lee, Chang Kyu Rhee, Tae Kyu Kim, Alexey Spirin, Vasiliy Krutikov, and Sergey Paranin. 2015. "End Closure Joining of Ferritic-Martensitic and Oxide-Dispersion Strengthened Steel Cladding Tubes by Magnetic Pulse Welding." *Metallurgical and Materials Transactions A: Physical Metallurgy and Materials Science* 46 (7). doi:10.1007/s11661-015-2905-5.

Lei, Zhenglong, Jiang Bi, Peng Li, Tao Guo, Yaobang Zhao, and Dengming Zhang. 2018. "Analysis on Welding Characteristics of Ultrasonic Assisted Laser Welding of AZ31B Magnesium Alloy." *Optics and Laser Technology* 105 (September): 15–22. doi:10.1016/j.optlastec.2018.02.050.

Leo, Paola, Gilda Renna, Giuseppe Casalino, and Abdul Ghani Olabi. 2015. "Effect of Power Distribution on the Weld Quality during Hybrid Laser Welding of an Al-Mg Alloy." *Optics and Laser Technology* 73. doi:10.1016/j.optlastec.2015.04.021.

Lin, Xue, and Zhisheng Zhang. 2020. "The Study of Industrial Design: The Transformation of Welder Based on Tandem Double-Wire MIG Welding." In *IOP Conference Series: Materials Science and Engineering*. Vol. 711. doi:10.1088/1757-899X/711/1/012011.

Lin, Yuan Ching, and Yu Chi Lin. 2011. "Microstructure and Tribological Performance of Ti-6Al-4V Cladding with SiC Powder." *Surface and Coatings Technology* 205 (23–24). doi:10.1016/j.surfcoat.2011.09.001.

Liu, Xiaochao, Chuansong Wu, and Girish Kumar Padhy. 2015a. "Improved Weld Macrosection, Microstructure and Mechanical Properties of 2024Al-T4 Butt Joints in Ultrasonic Vibration Enhanced Friction Stir Welding." *Science and Technology of Welding and Joining* 20 (4): 345–52. doi:10.1179/1362171815Y.0000000021.

Liu, Xiaochao, Chuansong Wu, and Girish Kumar Padhy. 2015b. "Characterization of Plastic Deformation and Material Flow in Ultrasonic Vibration Enhanced Friction Stir Welding." *Scripta Materialia* 102 (June): 95–98. doi:10.1016/j.scriptamat.2015.02.022.

Liu, Yong Qiang, Huan Li, Li Jun Yang, Kai Zheng, and Ying Gao. 2015. "Arc Spectrum Diagnostic and Heat Coupling Mechanism Analysis of Double Wire Pulsed MIG Welding." *Guang Pu Xue Yu Guang Pu Fen Xi/Spectroscopy and Spectral Analysis* 35 (1). doi:10.3964/j.issn.1000-0593(2015)01-0005-05.

Mir, Fayaz Ahmad, Noor Zaman Khan, Arshad Noor Siddiquee, and Saad Parvez. 2022. "Joining of Aluminium Matrix Composites Using Friction Stir Welding: A Review." *Proceedings of the Institution of Mechanical Engineers, Part L: Journal of Materials: Design and Applications*. doi:10.1177/14644207211069619.

Mukherjee, Manidipto, Saptarshi Saha, Tapan Kumar Pal, and Prasanta Kanjilal. 2015. "Influence of Modes of Metal Transfer on Grain Structure and Direction of Grain Growth in Low Nickel Austenitic Stainless Steel Weld Metals." *Materials Characterization* 102: 9–18. doi:10.1016/j.matchar.2015.02.009.

Munitz, Abraham. 2002. "Laser-Assisted Friction Stir Welding." https://www.researchgate.net/publication/284801953.

Nilsson, Klas, Sebastian Heimbs, Hans Engström, and Alexander F. H. Kaplan. 2003. "Parameter Influence in CO_2-Laser/MIG Hybrid Welding." *56th Annual Assembly of International Institute of Welding*, Romania.

Norrish, John. 2006. *Advanced Welding Processes*. doi:10.1533/9781845691707.

Padhy, Girish K., Chuansong Wu, and Shan Gao. 2015. "Auxiliary Energy Assisted Friction Stir Welding - Status Review." *Science and Technology of Welding and Joining*. doi:10.1179/1362171815Y.0000000048.

Pańcikiewicz, Krzysztof. 2021. "Preliminary Process and Microstructure Examination of Flux-Cored Wire Arc Additive Manufactured 18ni-12co-4mo-Ti Maraging Steel." *Materials* 14 (21). doi:10.3390/ma14216725.

Pouriamanesh, Rasoul, Kamran Dehghani, Rudolf Vallant, Norbert Enzinger, and Metallurgical Engineering. 2016. "Effect of Friction Stir Welding on Microstructure and Properties of Micro-TiO_2 Doped HSLA Steel." In *41th CWS International Conference Welding 2016*, Croatia.

Rusinko, Andrew 2011. "Analytical Description of Ultrasonic Hardening and Softening." *Ultrasonics* 51 (6): 709–14. doi:10.1016/j.ultras.2011.02.003.

Sakthivel, T., M. Vasudevan, Kinkar Laha, Padmanabhan Parameswaran, Kovi S. Chandravathi, M. D. Mathew, and A. K. Bhaduri. 2011. "Comparison of Creep Rupture Behaviour of Type 316L(N) Austenitic Stainless Steel Joints Welded by TIG and Activated TIG Welding Processes." *Materials Science and Engineering A* 528 (22–23). doi:10.1016/j.msea.2011.05.052.

Savu, Ionel Danut., Sorin Vasile Savu, and Grigore Sebes. 2013. "Preheating and Heat Addition by LASER Beam in Hybrid LASER-Ultrasonic Welding." *Journal of Thermal Analysis and Calorimetry* 111 (2): 1221–29. doi:10.1007/s10973-012-2449-5.

Shen, Qingkai, Xiangdong Kong, and Xizhang Chen. 2021. "Fabrication of Bulk Al-Co-Cr-Fe-Ni High-Entropy Alloy Using Combined Cable Wire Arc Additive Manufacturing (CCW-AAM): Microstructure and Mechanical Properties." *Journal of Materials Science and Technology* 74 (May): 136–42. doi:10.1016/j.jmst.2020.10.037.

Shi, Lei, Chuansong Wu, and Xiaochao Liu. 2015. "Modeling the Effects of Ultrasonic Vibration on Friction Stir Welding." *Journal of Materials Processing Technology* 222: 91–102. doi:10.1016/j.jmatprotec.2015.03.002.

Shi, Steve. 2005. "Laser and Hybrid Laser MAG Welding of Thick Section." *International Forum on Welding Technologies in Energy Engineering*, Shanghai.

Siddiq, Amir, and Tamer el Sayed. 2012. "Ultrasonic-Assisted Manufacturing Processes: Variational Model and Numerical Simulations." *Ultrasonics* 52 (4): 521–29. doi:10.1016/j.ultras.2011.11.004.

Silva, Lucas F. M. da, Mohamad S. El-Zein, and Paulo Martins. 2020. *Advanced Joining Processes: Welding, Plastic Deformation, and Adhesion. Advanced Joining Processes: Welding, Plastic Deformation, and Adhesion.* doi:10.1016/B978-0-12-820787-1.01001-9.

Steen, William M. 1980. "Arc Augmented Laser Processing of Materials." *Journal of Applied Physics* 51 (11). doi:10.1063/1.327560.

Tan, Caiwang, Wenxiong He, Xiangtao Gong, Liqun Li, and Jicai Feng. 2015. "Influence of Laser Power on Microstructure and Mechanical Properties of Fiber Laser-Tungsten Inert Gas Hybrid Welded Mg/Cu Dissimilar Joints." *Materials and Design* 78. doi:10.1016/j.matdes.2015.04.022.

Tseng, Kuang Hung, and Chih Yu Hsu. 2011. "Performance of Activated TIG Process in Austenitic Stainless Steel Welds." *Journal of Materials Processing Technology* 211 (3). doi:10.1016/j.jmatprotec.2010.11.003.

Ureña, Alejandro, Paula Rodrigo, Luis Gil, M. D. Escalera, and Juan L. Baldonedo. 2001. "Interfacial Reactions in an Al-Cu-Mg (2009)/SiCw Composite during Liquid Processing. Part II: Arc Welding." *Journal of Materials Science* 36 (2). doi:10.1023/A:1004832713790.

Vaidya, W. V., Kandavelmani Angamuthu, Mustafa Koçak, R. Grube, and Jens Hackius. 2006. "Strength and Fatigue Resistance of Laser-Mig Hybrid Butt Welds of an Airframe Aluminium Alloy AA6013." In *Welding in the World*. Vol. 50. doi:10.1007/BF03263465.

Wang, Huaming, Yelin Chen, and LigenYu. 2000. 'In-Situ' Weld-Alloying/Laser Beam Welding of SiCp/6061Al MMC." *Materials Science and Engineering A* 293: 1–6. www.elsevier.com/locate/msea.

Wang, Hongyang, Xiaohong Liu, and Liming Liu. 2020. "Research on Laser-TIG Hybrid Welding of 6061-T6 Aluminum Alloys Joint and Post Heat Treatment." *Metals* 10 (1). doi:10.3390/met10010130.

Weman, Klas. 2003. *Welding Processes Handbook*. 2nd edition, Woodhead Publication, United Kingdom.

Yaduwanshi, Deepak Kumar, Swarup Bag, and Sourav Pal. 2014. "Effect of Preheating in Hybrid Friction Stir Welding of Aluminum Alloy." *Journal of Materials Engineering and Performance* 23 (10): 3794–3803. doi:10.1007/s11665-014-1170-x.

Yang, Ke, Fei Wang, Hongbing Liu, Peng Wang, Chuanguang Luo, Zhishui Yu, Lijun Yang, and Huan Li. 2021. "Double-pulse Triple-wire Mig Welding of 6082-t6 Aluminum Alloy: Process Characteristics and Joint Performances." *Metals* 11 (9). doi:10.3390/met11091388.

Zeng, Zhi, Xunbo Li, Yugang Miao, Gang Wu, and Zijun Zhao. 2011. "Numerical and Experiment Analysis of Residual Stress on Magnesium Alloy and Steel Butt Joint by Hybrid Laser-TIG Welding." *Computational Materials Science* 50 (5). doi:10.1016/j.commatsci.2011.01.011.

Zhang, Kezhao, Zhenglong Lei, Yanbin Chen, Ming Liu, and Yang Liu. 2015. "Microstructure Characteristics and Mechanical Properties of Laser-TIG Hybrid Welded Dissimilar Joints of Ti-22Al-27Nb and TA15." *Optics and Laser Technology* 73. doi:10.1016/j. optlastec.2015.04.028.

Zhang, Lin Jie, Jie Ning, Xing Jun Zhang, Gui Feng Zhang, and Jian Xun Zhang. 2015. "Single Pass Hybrid Laser-MIG Welding of 4-mm Thick Copper without Preheating." *Materials and Design* 74. doi:10.1016/j.matdes.2015.02.027.

Zhao, Yong, Lilong Zhou, Qingzhao Wang, Keng Yan, and Jiasheng Zou. 2014. "Defects and Tensile Properties of 6013 Aluminum Alloy T-Joints by Friction Stir Welding." *Materials and Design* 57. doi:10.1016/j.matdes.2013.12.021.

10 Clinching
A Deformation-Based Advanced Joining Technique

V. Kumar and J. P. Misra

10.1 INTRODUCTION

With the worldwide call for sustainable development and reducing greenhouse gas emissions, the automotive sector has been looking for various fuel-efficient technologies. Reducing the weight of aircraft or automobiles is one of the most practical and sought-after methods of enhancing fuel economy and lowering emissions. Weight reduction without affecting the safety and performance of the vehicle is of utmost importance. Weight may be reduced by employing lightweight materials in vehicle and aircraft bodywork. As a result, more and more new body shapes and advanced materials are being used to make car bodies. Composite material, for example, is frequently utilized in aircraft bodies due to its low weight-to-strength ratio. The appropriate material is chosen based on the convenience of the production process and the cost involved (Mayyas et al., 2012; Peng et al., 2022; Yuce et al., 2014).

Every year, there is a higher demand for building materials with better strength and performance that can meet more stringent standards. The dynamic development, selection, and implementation of such an advanced lightweight, high-strength material creates an entirely new technical circumstance for the automotive sector. Technology and science have made the most advancements, and many new materials are being developed. The challenge for today's engineers is to assemble car parts from varying novel materials. These elements specify the assembly process's distinctiveness. With certainty, it can be stated that mass production of automated, environmentally friendly motor vehicles is what the modern world needs. The basis of contemporary production philosophy is that items should be produced with as little negative influence on the environment as possible. Technologies for "white" (or "sustainable") assembly are now being explored. New processes and products bring about specific design and operational modifications. Automobile manufacturers are changing how they employ joining technologies to create components out of unique materials due to mounting pressure on ecological and economic issues. Therefore, it becomes necessary to create new advanced material-joining technologies due to such advanced materials.

Moreover, conventional welding and joining techniques do not produce sound weld joints with advanced materials, mostly during welding with dissimilar materials. For example, Al alloys and steel alloy welding are among the most extensively utilized industrial processes. But due to the inferior wetting property of Al, the distinction in physical and chemical characteristics, the high difference in melting point of Al and steel alloys, and the appearance of brittle intermetallic compounds led to the formation of poor and weak weld joints by the conventional welding process. As a result, researchers worldwide are improving and developing traditional welding and joining processes (Bach et al., 2005; Mehta, 2017).

Advantages of advanced welding and joining techniques include increased productivity and quality, a significant reduction in component cost, as well as the cost associated with the manufacturing process, conducive joint properties, more control over process parameters, successful welding of complex geometrical objects, the ability to handle a wide variety of materials, and more flexibility in material selection. Moreover, these are more mechanized and automated than conventional welding processes and produce fewer welding defects. Advanced joining techniques have a far lower adverse impact on parent material characteristics and less distortion.

10.1.1 CLINCHING

Due to the movement toward sustainable manufacturing and lightweight materials, clinched joints have substantially increased utilization in engineering industries (Eshtayeh et al., 2016). In the automobile sector, pressure welding fell out of favor toward laser-beam welding and joining by plastic deformation, including clinching, around 2010, as shown in Figure 10.1.

The clinching idea was first referenced in a patent in 1897. However, it didn't start being employed in industry until the late 20th century. In the modern sheet metal processing business, clinching is a cutting-edge joining technique used to connect sheets

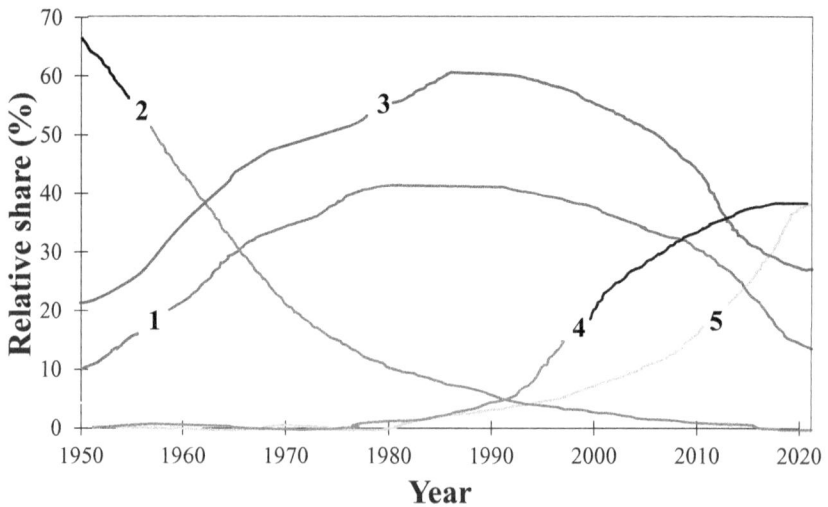

1- MIG Welding; 2- Gas brazing 3- Resistance welding 4- Clinching 5- Laser-beam welding

FIGURE 10.1 Contribution of different joining practices employed for the manufacturing of automobile bodies.

with thickness varying from 0.1 to 12 mm. Clinching is a mixture of deep drawing and shaping techniques in which thin metallic sheets are connected locally without additional parts. Deep drawing induces two-dimensional broadening of sheets by creating a confined hollow chamber with a decrease in sheet thickness. At the joint bottom, the total sheet thickness is lowered by about 60% of its original thickness (Varis, 2002). Compression causes material to move radially and fill die grooves. It is a cold-forming procedure that uses a punch, blank holder, and dies set to shape a tiny portion of the sheet metal pieces being joined together. This whole clinching process generally happens in five stages, as described in Figure 10.2a (Mucha, 2017). The sheet metal blanks are clamped or forced against the die surface before forming the joint. The purpose of the clamping is to guarantee proper metal flow during the forming procedure. During the joining operation, the blanks are pushed against the punch using a hydraulic machine at the joint-forming location (Chastel and Passemard, 2014). The sheets are gradually pushed toward the die imprint until the bottom material contacts the surface. To achieve clinching without local incision, it is crucial that the diameter of the punch be roughly 65%–70% of the diameter of the die. The worksheets are distorted plastically and progressively inserted into the die cavity. Cold plasticity deformation occurs between the connecting sheets. After the cavity has been completely filled, the sheet imprint bottom is firmly pressed.

At this point, a lock is created by carefully structuring the sheet material so that the components may be clinched together. Therefore, the two worksheets are mechanically linked by producing an interlock between the bottom and top materials by continuously reducing the thickness of the sheets, and metallic sheet materials are extruded into the die grooves (He, 2017; Peng et al., 2020). Cohesion is thus founded on form and force. Thus, force locking, material locking, and "S" shape locking all contribute to the clinched joint's strength, as shown in Figure 10.2b. Split dies with moving lamellas or a ring groove in the die aid in the creation of interlocks. The quantity of interlock produced governs the strength of the bonded sheets. Further, it is also impacted by structural changes in the material of the components that have been clinched, in addition to the geometry and form of the lock. Thus, undercut, bottom, and neck thickness are three primary joint characteristics, as depicted in Figure 10.2c. The neck thickness and clinch lock significantly affect the connection's strength. A minor clinch lock causes low joint strength by drawing the top material away from the bottom worksheet (due to weak interlocking). A narrow neck might induce an upper sheet fracture. Increasing both parameters would enhance the mechanically clinched joint's strength.

Further, clinching involves plasticizing the joining partners to create geometric interlocking, and it often has a round or square form. Nevertheless, it is preferred to keep the more ductile sheet on the punch side since the top material experiences significant plastic deformation during clinching. In other words, the thickest and strongest sheet is often positioned toward the punch side. For instance, the most typical method for clinching metal and composite parts involves pushing the metallic part into the composite part to create a clinched joint. A die is used to position the composite component, and the metal component is put on top.

However, pre-forming the die-sided sheet is one way to improve the interlock (Abe et al., 2018). Heating the connecting partners before clinching might also increase

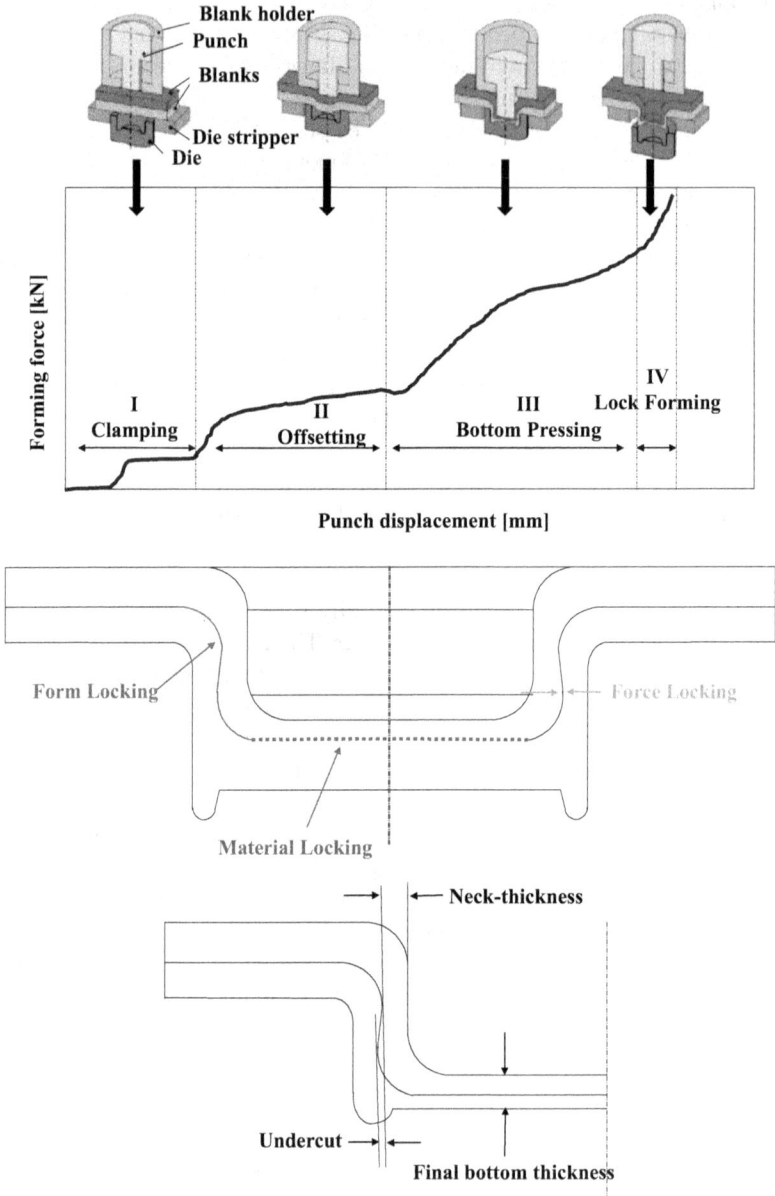

FIGURE 10.2 (a) Sequential steps in the formation of a clinched joint (Mucha, 2017), (b) joining mechanism in clinching, and (c) characteristics of a clinched joint (He, 2017).

forming limitations. For AA7075 T6, a quick heat treatment at 250°C for three seconds can increase the joinability (Jäckel et al., 2017). The concept of clinching at high temperatures was used for joining magnesium (Hahn et al., 2005) and hardenable 6xxx aluminum (Lambiase, 2015. The flow stress decreases, and the forming limit rises due to the elevated part temperature. As a result, more significant plastic stresses are possible.

The advantages of clinching are as follows:

- There is no need for extra component addition. Because of this, the procedure is quick, inexpensive, lightweight, and highly automated.
- The method is made simpler because thermoplastic matrices do not require surface preparation or predrilled holes.
- No metallurgical changes in the junction, no heating, no fume generation, and green technology are a few other advantages.
- Low investment cost: no expensive electrical installations, cooling, or exhaust devices.
- Minimal operating costs include cheap power consumption, no power use when the tool is idle, low prices for worn components due to the tool's extended lifespan, and no additional power use through the use of exhausting devices.

However, the following disadvantages are associated with the clinching technique:

- Once the joint is clinched, it cannot be disassembled without causing component damage.
- Clinching does not apply to all materials since the process relies on the linked materials' ductility. Because it is a cold-forming technique, it cannot combine brittle materials.
- Because the punch side material experiences significant plastic deformation during clinching, more ductile work pieces should be kept on top.
- If the hole is punched rather than pre-cut, the clinching process may result in little damage to composite materials (dragging, delaminations, and cracking).
- If a substrate consisting of steel or aluminum alloys is utilized simultaneously with a CFRP composite, galvanic corrosion may happen during the joining process.

10.1.2 Types of Clinching

The automotive industry uses composites, ultra- and advanced HSS, sophisticated polymers, Mg, and Al alloys to increase performance and energy efficiency while lowering CO_2 secretions. Nevertheless, due to their varied properties, connecting them is a challenge in the automobile sector. Researchers have suggested a variety of unique clinching procedures to join these materials. There are two groups into which these developed clinching techniques fall:

a. Modified clinching techniques: The modified clinching methods use specialized tools to control the material flow, resulting in high-performance joints. In other words, enhanced clinching tools enhance material flow in the die cavity.

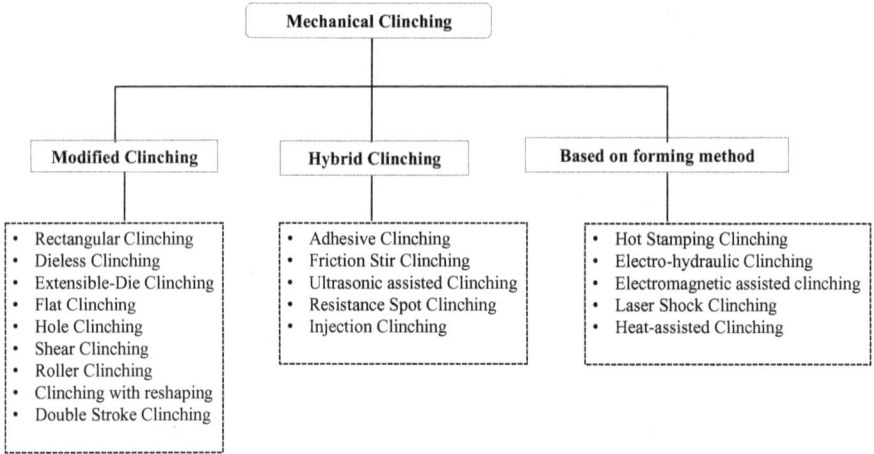

FIGURE 10.3   Types of clinching.

    b. Hybrid clinching techniques: Hybrid clinching is a mixture of different join-
       ing processes and mechanical clinching that is more sophisticated, energy-
       intensive, and suited to connecting specific sheet materials.
    c. Based on forming method: However, based on the energy applied to form
       the sheet materials, different types of clinching have also been developed,
       as mentioned in Figure 10.3.

10.2 VARIANTS OF CLINCHING

10.2.1 Flat Clinching

The formation of protrusions at the joint is one of the disadvantages of clinching
technology. Such protrusion height limits the application of clinched joints in visible
and operating parts. Therefore, a new method called flat clinching has been developed, which produces joints without protrusion height by employing a flat die. The flat
clinching technology was patented by Chemnitz University of Technology (Carboni
et al., 2006; Lee et al., 2010). Figure 10.4a illustrates the flat clinching procedure.
Compared to the traditional clinching technique, the flat clinching procedure forms
the junction by flowing the joining partners upwardly and radially. The blank holder
directs the sheet to flow upwardly as well as controls the material to flow radially. The
method of flat clinching requires more substantial punch power to create a joint.

    Planar anvils are utilized in place of fixed or extendable dies. Thus, a one-step
forming procedure yields a fully flat, one-sided junction that may be used for both
functional surfaces and viewable portions. Numerous elements and their interactions
impact how the mechanical interlock is formed. Figure 10.4b depicts the parameters
of a flat-clinched joint. The diameter and power of the punch were suggested by
Neugebauer et al. (2008) as the two main influential parameters of flat clinching. By
using numerical evaluations, Gerstmann and Awiszus (2014) proposed the friction

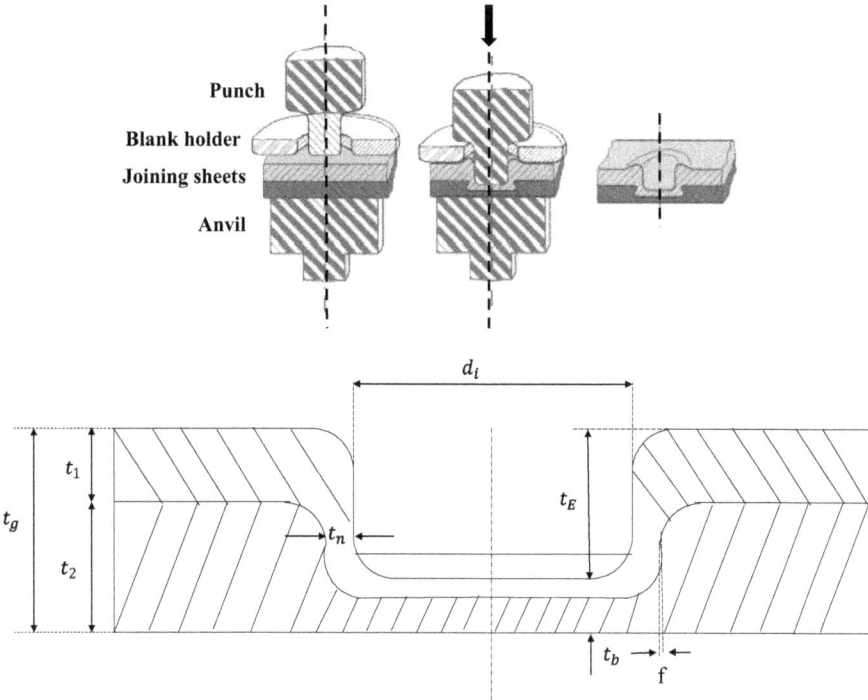

t_n- neck thickness; t_b- bottom thickness; t_E- penetration depth; f - interlocking;

d_i- inside diameter; t_1- thickness part 1-punch side; t_2- thickness part 2 - anvil side;

t_g- total thickness of parts

FIGURE 10.4   (a) Process of flat clinching and (b) characteristics of flat clinching.

between sheets, material selection, and clinching tool shape as the most influencing factors. Chen et al. (2017) investigated the interaction between the flat clinching procedure and the material-forming methods using experimental and computational tools. The performance of the joint could be enhanced by enlarging the neck thickness and increasing the punch's diameter. Neck thickness may also be enhanced by the increasing blank holder force; however, the length of the interlock shortens. Additionally, increasing the forming power might enhance the dependability of the joining technique. An effective mechanical interlock might be made using a forming force between 60 and 90 kN.

10.2.2   HOLE CLINCHING

Materials with widely varying mechanical characteristics and thicknesses may be joined via hole clinching. More notably, hole clinching may link steel sheets with different degrees of ductility and strength (Lee et al., 2014; Messler, 2000).

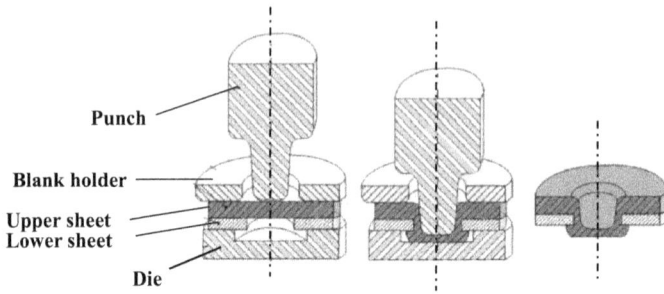

FIGURE 10.5   Process of hole clinching.

Before clinching, a punched bottom sheet is necessary, as shown in Figure 10.5. Punch compresses the top material. The top worksheet is dented and disseminated through the bottom worksheet's hole. The hole-clinching technique has been studied and developed by several researchers. Factors affecting the hole-clinched joint have been described below:

- Alignment of the hole in the bottom sheet and the center of the clinching tools influences the neck thickness, the strength of the connection, and the development of the undercut.
- The punch's corner radius affects how the interlock forms.

The efficiency of this clinching process is also decreased by the requirement for per-forated sheets before clinching. To address problems like efficiency and alignment, the hole clinching tools need to be further upgraded.

10.2.3   DIE-LESS CLINCHING

Since a flat-bottom anvil is utilized, die-less clinching is simple to heat. Additionally, it has a wider acceptance for little discrepancies in punch and counter-tool concen-tricity. In contrast to traditional clinching, die-less clinching does not require a bot-tom die. Neugebauer et al. (2008) described it as a method of creating connections that does not involve the use of any intermediary tools or the removal of any raw materials. Figure 10.6 depicts the die-less clinching procedure. The punch firmly crushes the sheet. As the punch moves, the sheet's local material is displaced and flows backward. The sheets rise as a result of the sheets' material flow.

For connecting magnesium sheets, die-less clinching works ideally. Less than 1 second might be needed to complete the pre-heating process for connecting the sheets. It implies that the die-less clinching method may quickly and reliably con-nect magnesium sheets. A different study has been performed using DEFORM-2D to optimize the punch's parameter. The findings demonstrate a negative association between neck thickness and punch edge radius. However, neck thickness and punch diameter are positively correlated.

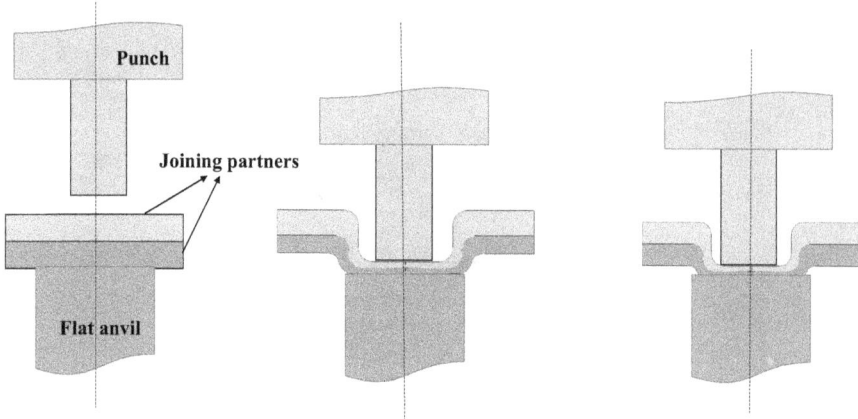

FIGURE 10.6 Process principle of die-less clinching.

FIGURE 10.7 Rectangular clinching tool (Lambiase, 2015a,b).

10.2.4 RECTANGULAR CLINCHING

The geometric design of the tools makes rectangular clinching distinct from traditional clinching, as shown in Figure 10.7. The clinching of rectangular shapes can provide high interlock while reducing the joining force. The rectangular junction is thinner than the circular joint and more than 1.7 times as strong as the circular joint. Failure of a rectangular clinched joint is a combination of pulled-out and neck fractures (Zhao et al., 2014). With less connecting force, rectangular clinching produces a significant interlock with stronger strength than split joints.

However, as illustrated in Figure 10.8, the rectangle clinching tools are further modified by including a square punch and a square die (Abe et al., 2018). Ultra-high-strength steel sheets may be joined using these techniques with strength that is more than twice as great as traditional clinched joints.

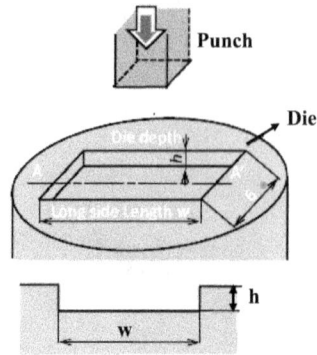

FIGURE 10.8   Modified rectangular clinching (square clinching) (Abe et al., 2018a,b).

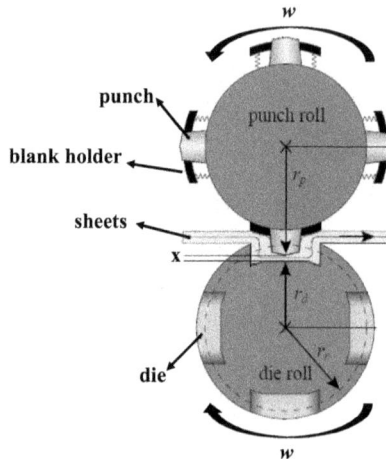

FIGURE 10.9   Representation of roller clinching (Hiller and Volk, 2015).

The fundamental benefit of square clinching is that the punching effort is comparatively low. The connecting partners tend to break during clinching, which might affect the connection's effectiveness.

10.2.5   ROLLER CLINCHING

Roller clinching is an efficient way to connect metal sheets. The use of revolving rollers makes the clinching process continuous, as demonstrated in Figure 10.9. Punch and die rollers are the primary equipment utilized in this operation. Metal sheets are continually fed between the rollers as the mold and punch roller revolve differently at a specific angular speed. The clinching procedure starts when the punch and die spin into a particular location. Compressing sheets with the spring-suspended holder stops sheets from being extruded beyond the die cavity.

The capacity to quickly attach metallic material is a benefit of roller clinching. However, due to the unique kinematics of the process, the roller-clinched junction has an unsymmetrical configuration, as shown in Figure 10.10. A few enhanced techniques have been created to enhance the roller-clinched joint's development.

Additionally, it was discovered that an increase in rolling radius, rather than larger stripping pressures, impacted the geometrical profile of the roller-clinched joints. Unbuttoning and "neck fracture" are two of the joints' failure mechanisms when shear loads are applied at a 0° angle. When the shear force direction is 180°, the joint often fails due to a neck fracture, and during a pullout load test, the joint mostly fails through "unbuttoning."

10.2.6 LASER SHOCK CLINCHING

A cutting-edge technique for combining identical and different metal foils is laser shock clinching, as illustrated in Figure 10.11. It can attach metal sheets that range in

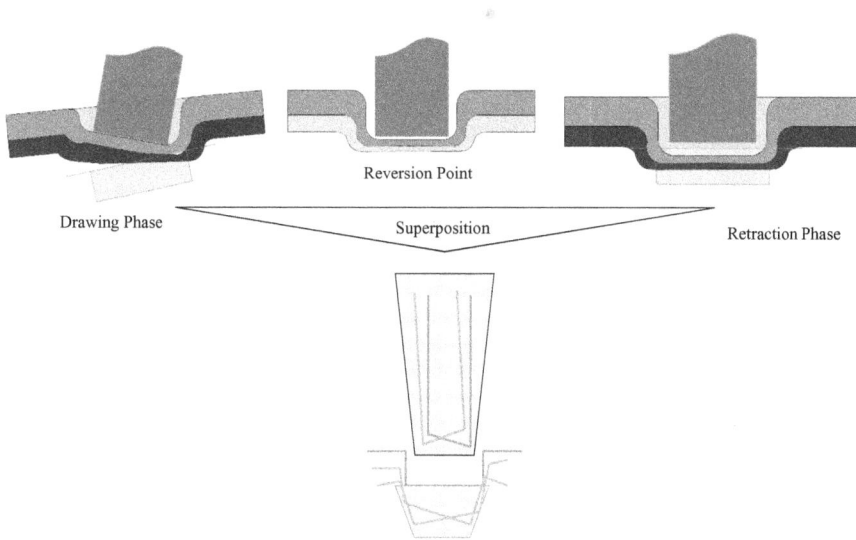

FIGURE 10.10 Roller clinching process steps with overlapping horizontal tool movement (Hiller and Volk, 2015).

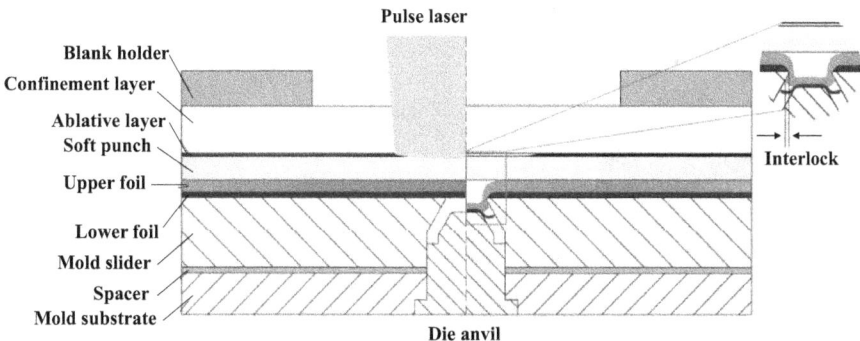

FIGURE 10.11 Schematic illustration of LSC.

thickness from 20 to 300 µm. The equipment used in laser shock clinching consists of a blank holder, an ablative layer, a mold substrate, a confinement layer, metal foils, a soft punch, a spacer, a mold slider, and a die anvil. The high-energy laser beam is first focused on the ablative layer through the confinement layer. Localized high-pressure, high-temperature plasma is formed from the ablative layer. A laser-supported detonation wave is created as the laser energy is introduced into the plasma. The soft punch is then placed on the higher sheets due to the shock wave's propagation through the hyper-elastic soft punch. After the top sheet pours into the mold to produce a clinched junction, the bottom sheet is detached.

Connecting annealed Cu/Al foils needs a maximum of three laser pulses. The interlock may be more extensive, using several laser pulses and a moderate degree of laser intensity. The top foil's minimal thickness decreased while the joints' interlocking strengthened as laser energy increased (Wang et al., 2016). Wang et al. (2016) studied the significance of the thickness of the ablative layer on sheet material flow. If the ablative layer is less than 80 µm thick, more ablative layer thickness leads to a suitable amount of material flow.

In contrast, a thicker ablative layer prevents enough material from flowing when the ablative thickness is more than 80 µm. Miniature sheets with a thickness typically between 20 and 300 µm are appropriate for connecting using laser shock clinching. Laser shock, which is easier to manage, provides punching power. But because so many impact overlays are required, installing the sheets might be challenging.

10.2.7 HYDRO CLINCHING

"Hydro clinching" refers to a novel process incorporating the benefits of clinching and hydroforming. Some advantages include high precision, fewer processing stages, adaptability to complicated tube-workpiece joining, and suitability for connecting dissimilar metal sheets. Neugebauer et al. (2005, 2008) suggested the hydro-clinching technique.

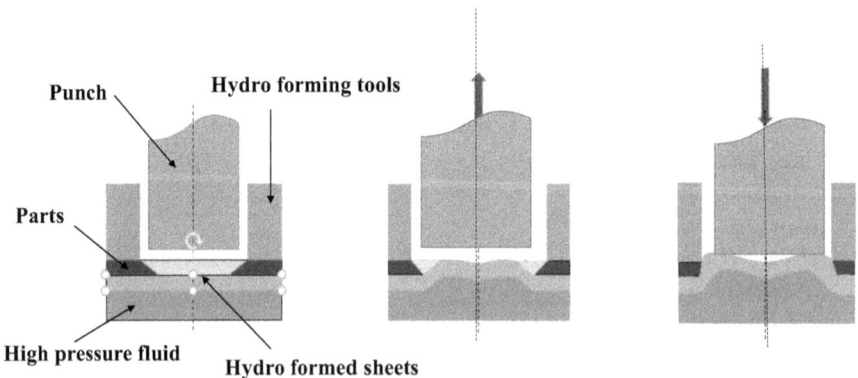

FIGURE 10.12   The hydro-clinching operational sequence (Neugebauer et al., 2008).

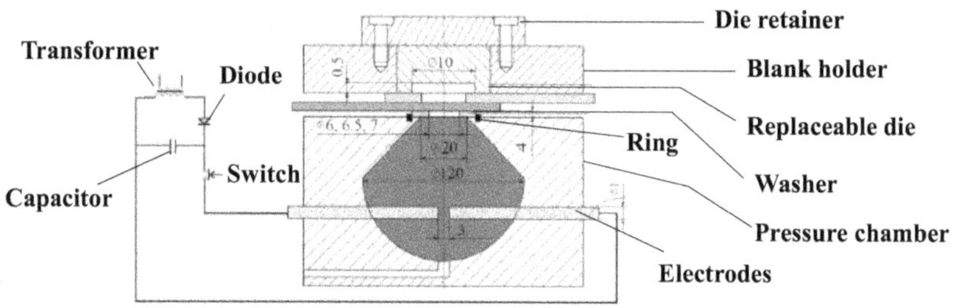

FIGURE 10.13 Diagram and the actual configuration of the electro-hydraulic hole clinching system (Babalo et al., 2018).

The geometry of the hole chamfers, the joining partner's flow, and the liquid force affect how tightly the hydro-clinched connection interlocks. Babalo et al. (2018) invented the novel hydraulic-clinching technique known as electro-hydraulic clinching. Electrical energy is used in this technique to create liquid shock waves that form the sheets. The water-filled pressure chamber is seen in Figure 10.13. The pressure chamber's upper hole is covered with sheets. Unlike traditional clinching, which uses a punch, the replacement die is fastened above and is substituted by a liquid shock wave. By abruptly discharging the electrode, a shock wave may be produced. The shock wave works on the lower material to create a joint by moving liquid water.

Compared to traditional clinched joints, electro-hydraulic clinched joints are nearly twice as effective. By enlarging the pre-cut diameter, the available area of the top sheet may be diminished. Joint interlock and neck thickness are affected by the placement of the die cavity axis and the shock origination site. The two hydro-clinching systems generate high-pressure liquid in distinct ways. The first is supplied by an external hydraulic pump, while electric pulses generate the second. It is also capable of joining tube materials. Electro-hydraulic hole clinching is very efficient and has a minimal energy footprint. Both of these techniques combine hole clinching with different technologies. Nevertheless, none of these options addresses the faults with hole clinching, such as stringent requirements for the concurrency of the holed materials and lower die.

10.2.8 INJECTION CLINCHING

For hybrid constructions made of thermoplastics and metals or thermoset polymers, ICJ is a revolutionary spot-joining technique. The method is based on adhesive bonding, staking, and injection molding. A thermoplastic element (often a cylindrical stud) is put into a through hole (cavity) in a metallic/thermoset component, heated, and deformed to make joints. By employing a piece of the structure to create a rivet, weight is saved and mechanical dependability is increased. The injection clinching technology consists of four distinct procedures, as shown in Figure 10.14. The initial phase is to position the tools toward the top layer. The heating unit then begins heating the polymeric stud. In the last heating phase, the punch applies forging pressure

(a) tool approach

(b) heating

(c) forging

(d) tool separation

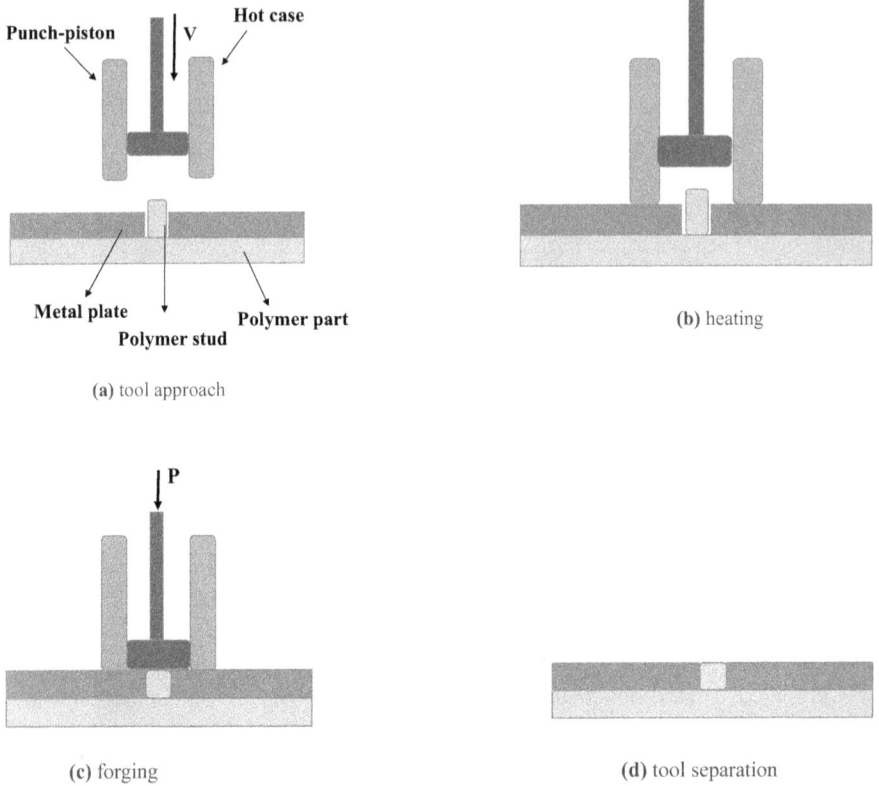

FIGURE 10.14 Steps of the injection clinching process: (a) tool approach, (b) heating, (c) forging, and (d) tool separation (Abibe et al., 2013).

to the softened/molten polymeric stud. The polymer studs are pressed into the holes of the metal dies. After heating, the pressure is sustained to avoid polymer easing. Finally, the punch is withdrawn, and following air cooling, the joint is cemented. The working temperature, the heating duration, and the injection rate primarily control injection clinching joining procedures. The injection clinching parameters, such as the joining force, have an immediate impact on the viscosity of the polymer. In addition, the researchers discovered that the polymer stud exhibited virtuous cavity-filling characteristics at 300°C. However, the damaged stud had an immense void, and the polymer experienced significant mass loss. The junction strength was between 34.9% and 88.5% of the strength of the foundation sheets. The major failure mechanisms of the junction were rivet withdrawal and net tension failure. The causes of joint failure are proportional to the amount of water present in the polymeric component (Abibe et al., 2013).

Additionally, Abibe et al. (2013) developed a novel polymer-metal joining process that uses friction heating rather than resistance heating. They analyzed the native characteristics and microstructure of the 6082-T6 aluminum alloy and polyetherimide junction. The microstructure and characteristics of the polyetherimide were altered by the thermomechanical treatment during the process. The polymer can undergo thermal degradation, chain reorientation, and recrystallization, depending on strain rates, temperatures, and the polymer. The hollow design of friction injection clinching joints boosts the lap shear and cross tension strengths by 18% and 21%, respectively (Abibe et al., 2016).

Injection clinching is an effective technique for attaching polymers to metallic sheets. But the optimal heating temperature (300 °C) and cooling period are both long. Other heating and cooling techniques need to be stressed to enhance the efficacy of the process.

10.2.9 FRICTION STIR CLINCHING

Friction stir clinching is a revolutionary method incorporating friction welding with clinching. As illustrated in Figure 10.15, it incorporates a revolving punch, which is used to heat the material and increase the sheet material's formability. It is accomplished similarly to hole clinching. First, polymer and metal sheets are pierced with holes. Then, the punch revolves at a specified rate and crushes the metallic material to the desired thickness. The clinched connection is established when the punch returns to its starting location. Such techniques can decrease joining pressure and improve the formability of sheet material. In addition, friction-assisted clinching can guarantee the safety of connecting highly reflective materials. Since the heating zone is limited to the joint site, the heating process parameters may be readily controlled and monitored. The primary benefit is that it can retain the maximum temperature at the contact site, which helps optimize localized sheet material flow. However, how the heat is distributed during the clinching process is not quite apparent.

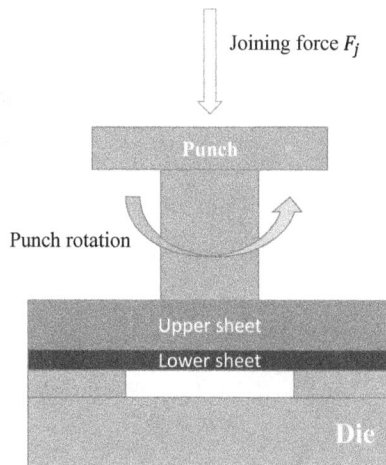

FIGURE 10.15 Process of friction-assisted clinching.

10.2.10   LASER-ASSISTED CLINCHING

The characteristics of materials are improved by both friction- and laser-assisted heating processes. Using a laser, sheets are heated during clinching. Because this process does not straighten the joining partners, the heated part may be better regulated. The laser beam heats the sheet by passing through the punch hole, as depicted in Figure 10.16.

The LAC methods produce a clinched joint with a slight bump in the middle (see Figure 10.17), which depends on the punch's area. Moreover, the bump can be eliminated by incorporating optical fibers within the punch's bore. The gaps between the sheets are bigger in the clinched joints, which are produced via LAC (see Figure 10.17). No research has been done to demonstrate the process of its

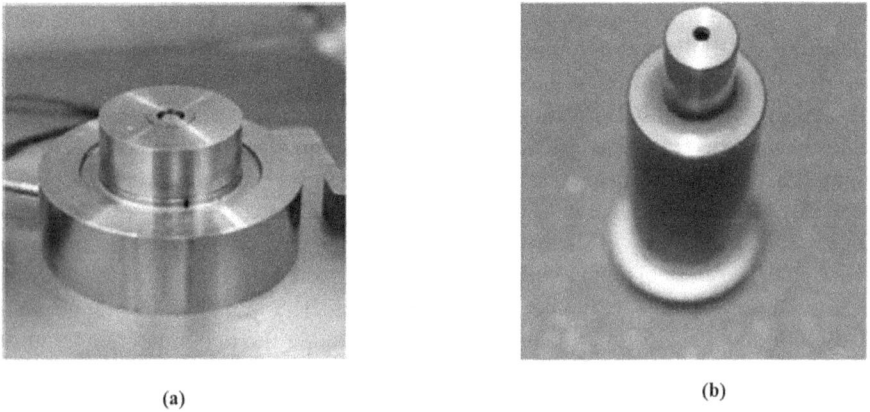

(a)   (b)

FIGURE 10.16   LAC tools: (a) modified die and (b) modified punch (Reich et al., 2014).

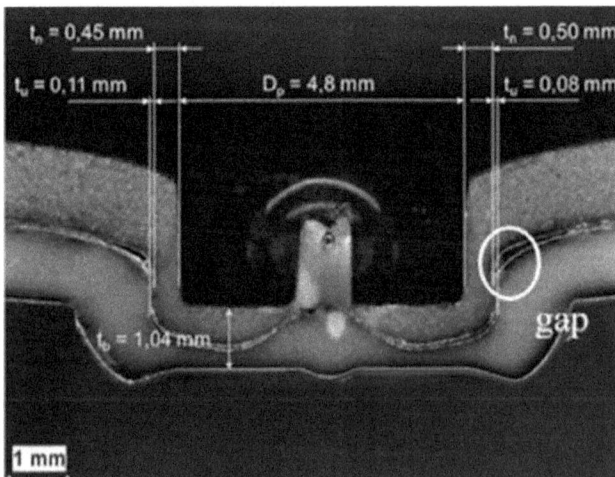

FIGURE 10.17   The sectional view of a laser-assisted clinched joint (Reich et al., 2014).

formation. However, it may be brought on by unequal temperature spreading or varied plate hardness. To remove the effect of the gap on the joint properties, more research should be done on the mechanical performance of laser-assisted clinched joints.

More significantly, laser heating affects various parts of the top sheets in different ways. The measuring area comprises the bottom area, throat area, heat-affected region, and base material area. The neck area has harder rock than the bottom, which has softer rock. The distribution of top sheet hardness matches the distribution of heat input. To connect high-hardness steels, the clinching procedure has to reduce yield strength and improve ductility.

10.2.11 SHEAR CLINCHING

Shear clinching is another form of clinching technology developed by Busse et al. (2010). Such a type of clinching technique expands the capabilities of clinching to join poor-formability sheets. Two key advantages of shear clinching over clinching with a pre-hole are the removal of positioning and the pre-hole process. The shear-clinching die, as seen in Figure 10.18, is made up of a mobile anvil and multiple lamellae that are also mobile but have inner edges with less than 0.5 mm radii. The punch setup consists of a blank holder, an inside punch, and an outside punch.

Additionally, the outer punch is spring-loaded and has an exterior contour that can be either hemispherical or cone-shaped. A specified disconnected inner punch pierces the upper workpiece while the lowermost material is shear-cut throughout the operation. This results in a form-fitting, negative joint. In other words, the top material is forced into the cut-out hole while the bottom sheet is indirectly sheared. As a result, even extremely strong materials, such as hot-stamped 22MnB5, may be clinched in one step. Shear clinching has also been found suitable for joining three sheets (Wiesenmayer and Merklein, 2021). However, the mechanical characteristics of the top connecting partner also restrict the procedure. The mechanical aspects of the bottom sheet have to fall within the range of TS = 359 MPa, UE = 34.5%, and TS = 1202 MPa, UE = 10.2% for shear clinching (Busse et al., 2010). That implies that shear clinching can link standard dual-phase steels or press-hardened steels.

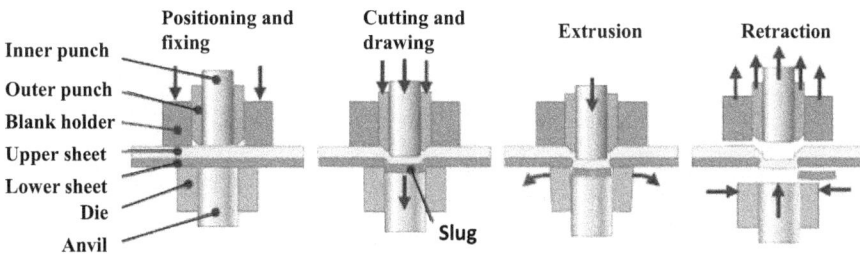

FIGURE 10.18   Illustration of shear clinching process (Busse et al., 2010).

10.2.12 FIXED AND EXTENSIBLE DIE CLINCHING

Both fixed and extensible dies are used in clinching technology with specific diameter punches, as shown in Figure 10.19. The bottom dies of fixed dies have a groove with a defined size. In contrast, the bottom of the dies on extensible dies features a sector slider that may move outwards in response to the distortion and drawing of the sheets during clinching.

The tool for extendable die clinching is contrasted in Figure 10.20. Pushing the material toward the die cavity creates an interlock during the fixed die-clinching operation. In contrast, several movable sectors make up the extendable die. Compared to a fixed die, a superior interlock is achieved when the material is dispersed radially instead of toward the die groove. Additionally, extendable die-clinched joints exhibit differing geometrical and mechanical characteristics. The punch, blank holder, sliding sectors, and fixed die are built of high-strength steel components to accomplish the intended durability. Periodically, the rubber spring has to be changed.

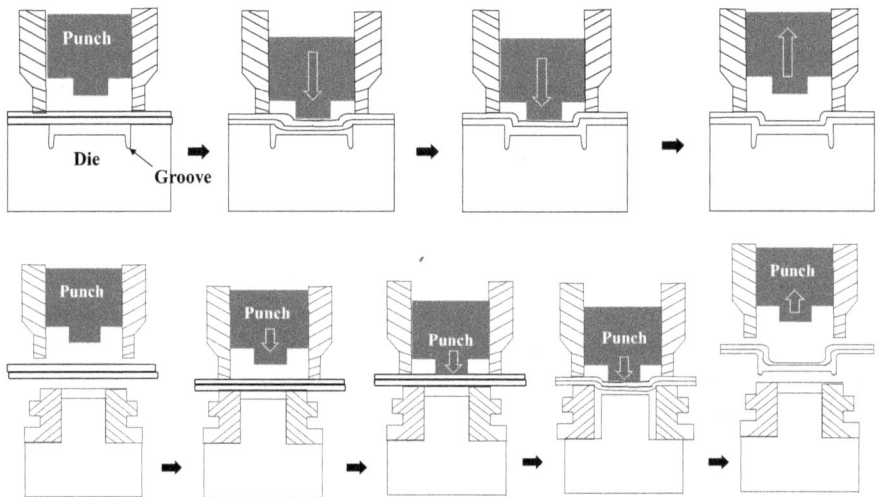

FIGURE 10.19 Clinching with (a) fixed die (Lin and Lo, 2016) and (b) extensible die (He, 2017).

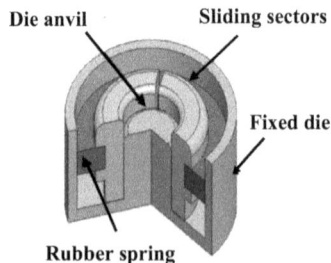

FIGURE 10.20 Schematic of the extensible die-clinching tool (He et al., 2014).

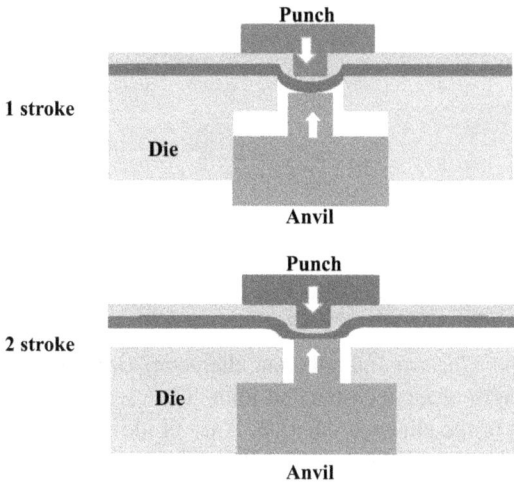

FIGURE 10.21 Double-stroke clinching by ATTEROX (Sadowski et al., 2015).

10.2.13 DOUBLE-STROKE CLINCHING

ATTEROX (ATTEROX Tools S.A., Renens/Lausanne, Switzerland) has presented a new clinching method called double-stroke clinching (Figure 10.21). With the action of a punch, the sheets to be connected are driven into a stiff die. The punch is active in the first stroke, creating a performance of the joining partners. But because the anvil in the die's base is merely secured by a spring, the punch may be free to move. The anvil is mechanically locked during the second stroke. Further, preform is pressed between the punch and the anvil outside the stiff die, generating a rivet-like connection (Sadowski et al., 2015).

10.3 CLINCHING-BASED HYBRID JOINING

Hybrid joining uses many joining techniques to produce joints with desirable characteristics beyond those possible with a single joining method. In addition to clinching, various methods, like adhesive bonding, riveting, etc., may enhance joints' mechanical qualities.

10.3.1 CLINCH BONDING

Clinch bonding is a novel hybrid joining technique. The clinching and adhesive bonding techniques are combined in the adhesive clinching technique. According to the order in which the adhesive clinching occurs, there are three primary categories for the adhesive clinching process. The first kind involves applying glue to one of the sheets before joining them together. The second approach, in contrast, involves first connecting the sheets before using low-viscosity glue in the clinched junction. In the third approach, the sheets are clinched after the adhesive has dried, as illustrated

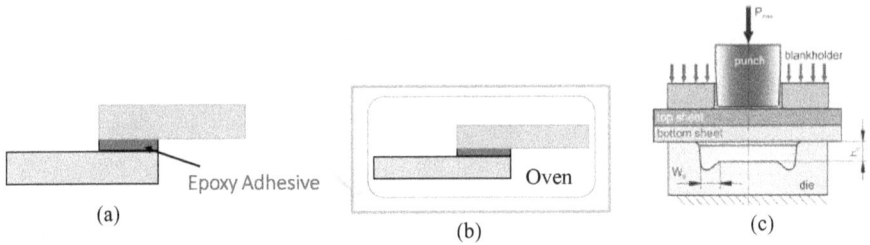

FIGURE 10.22   (a–c) Clinch bonding process.

in Figure 10.22. The adhesive and adherent characteristics play a substantial role in enhancing the strength of a clinch-bonded joint. The use of flexible adhesive would reduce the strength of the clinched junction. A novel idea for a mixed adhesive joint has been proposed to address this problem. The mixed adhesive combines two types of glue: brittle and ductile. The combined glue can strengthen joints, especially when exposed to temperatures between 50°C and 200°C. All adhesive-clinched joints have energy absorption values that follow a normal distribution, and they perform better than traditional joints in terms of load carrying and energy absorption. Balawender et al. (2011) also found that clinching before adhesive curing produces better outcomes than adhesive curing first. The strength and corrosion resistance of the joint may be enhanced by clinch bonding. Glue clinching requires a lot of time to ensure that the adhesive cures, and temperature significantly impacts how stable the joint is after clinching. To some extent, the joint's quality is determined by the adhesive's quality.

10.3.2   Resistance Spot Clinching (RSC)

RSC is a hybrid technique that integrates resistance spot welding and clinching. RSC can address typical resistance spot welding issues, such as electrode deterioration, welding flaws, and base metal softening. In Figure 10.23, the RSC procedure is described. There are four steps to clinching a resistance position in one cycle. The first stage involves fixing the workpieces in the middle of the top and bottom dies and sandwiching a narrow processing tape between the joining partner and mold. Punching down to the bottom mold occurs in the second phase. A tiny impression is created in the connecting region at this stage when a temporary low-level current flows over the joining partners. In the third step, the heat produced by a high-level current melts the connection region of the metal sheet. Detachment is the last step, in which the top mold retracts to its starting place. Furthermore, the processing tape breaks away from the joining region.

Metallurgical interconnection is the primary connecting method for RSC. During the RSC procedure, no interlock is created. The welding zone is subdued and contrasts subtly with the columnar dendritic site. The high temperature of lower copper molds sufficiently quenches the bottom of this region. Therefore, the columnar grains in the fusion area are found to be fully grown (Zhang et al., 2017). The microstructure of the joints made with the RSC process displays more uniformly dispersed dendritic zones than resistance spot-welded joints. The author concluded that RSC

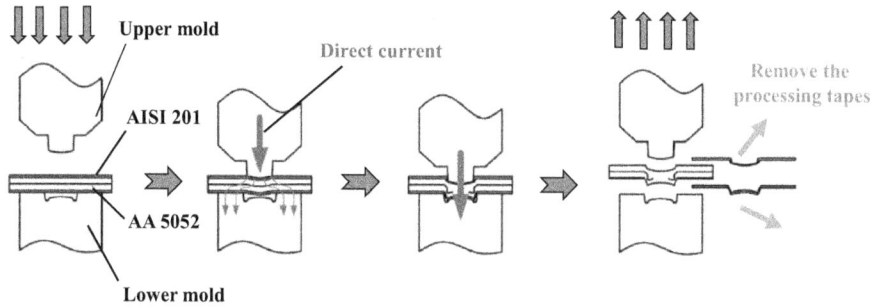

FIGURE 10.23   Resistance spot clinching process (Zhang et al., 2017).

joints had superior mechanical qualities than resistance spot welding joints because they have had good microstructural features. The author also discovered that RSC joints had superior load-bearing capability even at lower heat inputs than conventional resistance spot welding joints.

RSC joints experience significant deformation later than resistance spot welding joints do. RSC joints have higher tensile-shear load-bearing properties than resistance spot welding joints because they have a lower distortion concentration.

10.4   FACTORS AFFECTING CLINCHED JOINT FORMATION

Several elements impact the creation of a clinch, such as the interlocking of the joining partners, the thinning of the top and rear workpiece, and the decrease in thickness of the bottom work. Many research studies have regarded these elements as the critical determinants of process joinability. Additionally, the geometric characteristics of the tool, such as the configuration of the punch and die, influence these variables. Benabderrahmane and Ali (2013) investigated the impact of tool form, material characteristics, and friction on the clinching technique. The mechanical or geometric factors of the tool on the capability of the joint exerted a significant impact. Their perfect design will enable the optimal creation of a sound clinch joint with satisfactory mechanical performance. Figure 10.24 illustrates the impact of friction, punch diameter, and die height on punch load. In the process of clinching, friction is a crucial factor. The impact of friction on the varying thickness of the material and the amount of effort required is notable because it naturally affects the point's ultimate shape. Figure 10.24a depicts the progression of effort. Initially, the advancement of the load exhibits a predictable rise proportional to the punch's displacement. The effort required for a coefficient of zero friction increases to 40,000N. When the friction effect is accounted for, however, we observe an increase in effort until 50,000N.

The shape of the utilized tools still governs the performance of clinched connections. In addition, the punch corner radius, die groove, and die depths significantly impact the joining and the layer thickness. Figure 10.24b and c illustrates the impact of punch diameter and matrix depth on punch effort. It can be observed that the effort of the punch rises as its diameter grows. On the other hand, as the depth of the die increases, the effort of the punch reduces. Indeed, the rise in punch corner

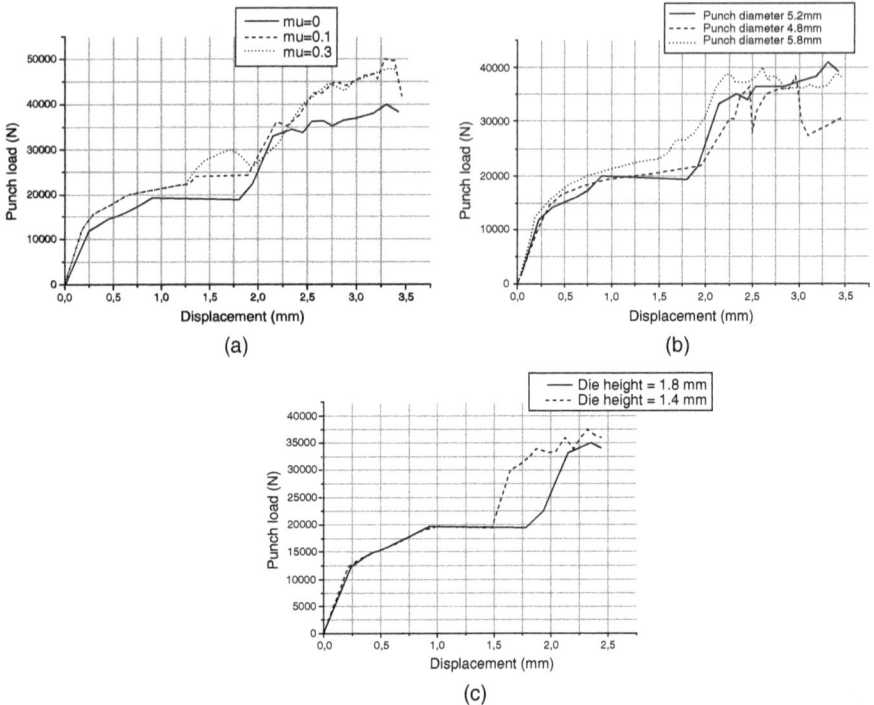

FIGURE 10.24 Load-displacement curves for different (a) friction coefficients, (b) punch diameters, and (c) die heights (Benabderrahmane and Ali 2013).

radius increased the bottom layer thickness, which is connected to the minimal plastic deformation of the top material.

Moreover, the two sheets did not interlock due to the wide corner radius (Abe et al., 2009). In addition, the undercut length and neck thickness of the top material have a substantial effect on the joint qualities. Moreover, proportional relationships were discovered between die radii and neck thickness and an inverse correlation with the mechanical interlock length of the clinched junction (Lee et al., 2010). Conclusively, the joint interlock, neck thickness, bottom thickness, bottom die, die depth, groove depth, groove fillet radius, groove width, and draft angle are identified as the essential geometrical features of a clinched connection.

10.5 MECHANICAL AND METALLURGICAL CHARACTERISTICS OF CLINCHED JOINTS

Understanding the mechanical characteristics of clinched connections is crucial for engineering industries. Numerous computational and experimental studies on clinched junctions' static and dynamic performance have been conducted. To measure the joint's strength, however, several aspects and areas are to be considered, including static strength, dynamic strength, and crash resistance. It is dependent on several collective factors, including the joining partner (type and thickness), the

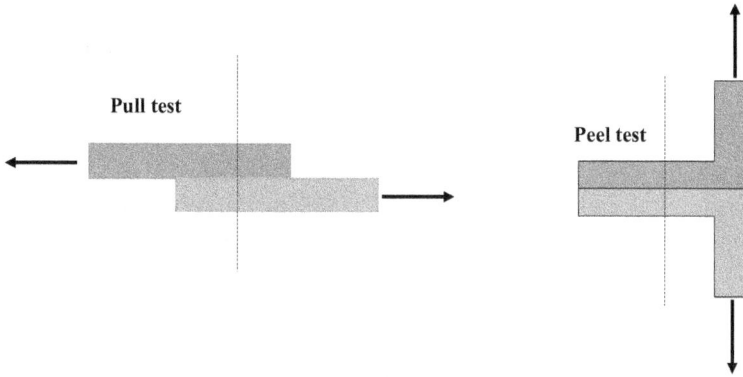

FIGURE 10.25 Test methods for a clinched joint.

joining direction (thin into thick, hard into soft), the diameter of the clinching point, and the shape of the clinching point. There are typically two loading scenarios in mechanical joining: static (rigid body in equilibrium) or dynamic (cyclic loads). Additionally, there are two tendencies for looking at joints: empirically, by loading the joints and seeing the mechanical reaction, and mathematically, by computing the mechanical behavior using precise techniques.

Tensile tests were conducted for two different configurations (longitudinal and transverse) to estimate the UTS during the static loading of the clinched joints. According to Carboni et al. (2006), the direction of the clinching has no impact on the clinched joint's ultimate strength. In general, two experimental techniques are used to determine the strength of a clinch joint: "pull" (tensile-shear) and "peel" (peel-tension), as shown in Figure 10.25. The pull approach usually produces more strength for a clinch joint than the peel method. A tensile test is used to evaluate both approaches. The collective shearing strength is assessed using the pull technique, whereas the axial strength is assessed using the peel mode.

A clinch joint's strength mainly depends on four key elements:

- Material properties
- Thickness
- Clinch point size
- Surface properties of the substance

The choice of connected materials significantly impacts how strong the clinched joint is. The mechanical characteristics of the combined materials should be considered while designing the geometry of the forming tools. Figures 10.26 displays the shear strength of clinch joints created using the same tool setup but for pairs of different materials (Sadowski et al., 2015). As is evident, two distinct values of shearing strength are produced for the same pair of materials. The strength of the junction is stronger when the thicker, more durable sheet is facing the punch and the thinner, more delicate sheet is facing the die. According to Abe et al. (2009), the quantity of interlock diminishes as the bottom sheet's strength rises because of the lower sheet's high flow stress.

The joint's quality improves as the sheets' total thickness increases. Like other mechanical joining techniques, a clinch joint's bigger diameter corresponds to a stronger joint. The surface condition of the material also affects the joint strength. A dry surface results in a stronger junction than an oiled or greased one. Nevertheless, such consequences are very modest in the case of steel sheets while being quite significant in the case of aluminum sheets. Various finite element analysis (FEA) analyses reveal that the joint's lateral wall is the most distorted area. This side wall area dramatically impacts the strength of the clinched joint. Due to the significant amount of strain hardening that happens here, the joint's mechanical strength is increased.

10.6 FAILURE MODES OF CLINCHED JOINT

Failure of a clinch joint can occur in a variety of ways. Sometimes the joint opens in response to the sheets' focused deformation. Most of the time, it happens because the joining partners don't lock together. Another method of failure is generated due to significant material weakening in the neck region of the top sheet, causing a fracture at that place. It occurs due to the material's limited ductility. The third type of failure is an amalgamation of the first two. In this mechanism, cracks at the end are formed owing to the tensile stress produced. However, the groove depth may be decreased to prevent it.

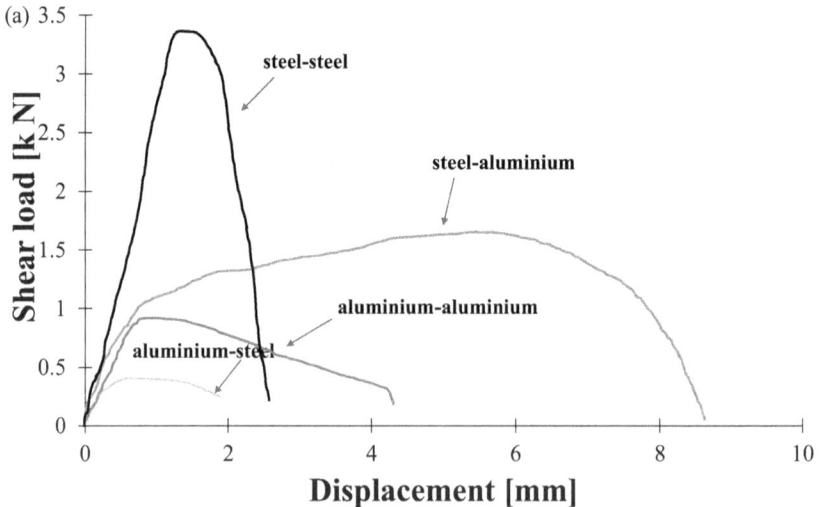

FIGURE 10.26 (a–c) Typical clinched joint shear measurements for various materials (Sadowski et al., 2015).

(Continued)

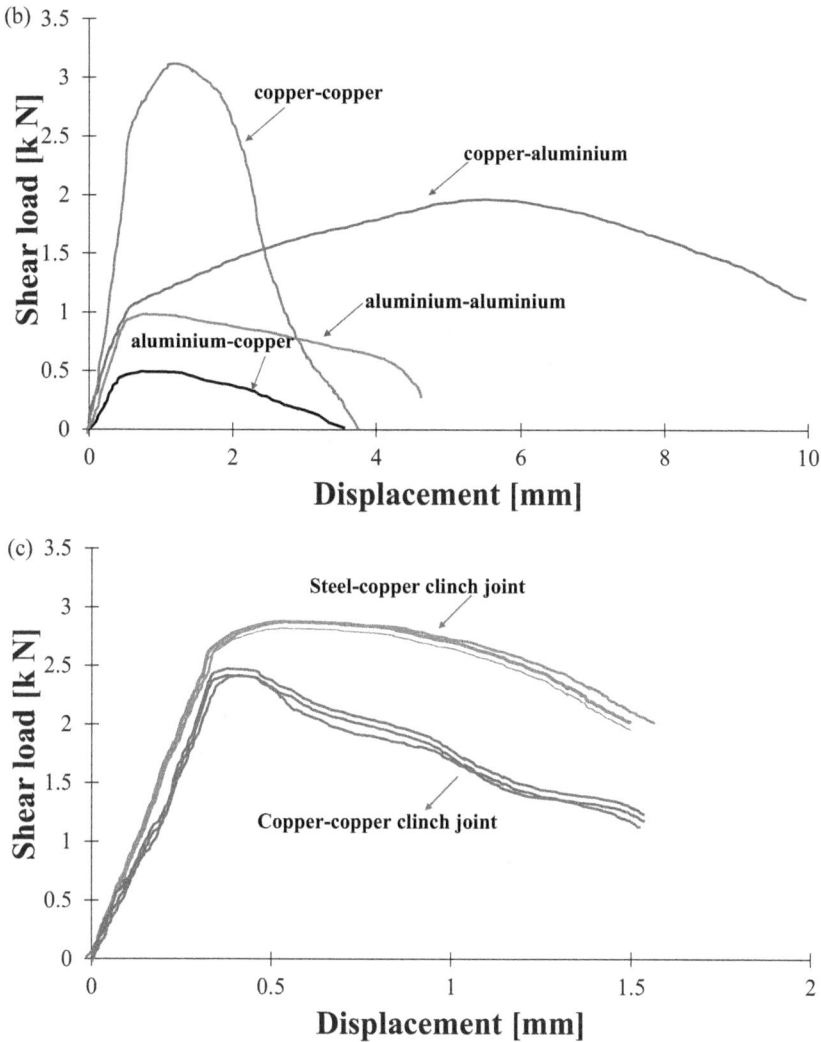

FIGURE 10.26 (*Continued*) (a–c) Typical clinched joint shear measurements for various materials (Sadowski et al., 2015).

As seen in Figure 10.27, the forming of the top material in the neck may not be enough to create the joint lock, or it may reach a material-limiting strain. This might result in either material breaking off in the neck or opening as a press stud, leading to two possible mechanisms of clinch joint failure (Sadowski et al., 2015). These two flaws in the clinched joint are due to excessive top sheet thinning (small neck thickness) and minor interface folding (small joint lock).

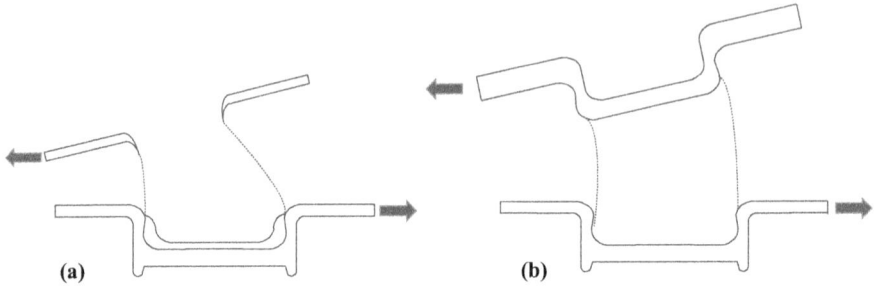

FIGURE 10.27 Failure mechanism of clinched joints (a) neck fracture mode and (b) press-stud fastener mode (Sadowski et al., 2015).

10.7 NUMERICAL MODELING OF THE CLINCHING TECHNIQUE

Modeling approach arose from the necessity to address intricate structural analysis issues. Since clinching is regarded as a challenging cold metalworking procedure, a sufficiently effective FEA simulation necessitates a thorough understanding of several factors, such as the joining partners and friction properties. The literature on the FEA of joining technique is evaluated in terms of the commercial program employed, the element employed, the workpiece thickness, and the type. Employing FEA to improve joining parameters will result in a higher rate of successful joint fabrication. Further, FEM will make it possible to conduct several experiments and tests that would otherwise be too costly or time-consuming to do in the real world. The application of FEA would minimize time and costs while improving the quality, strength, and production of joints.

10.7.1 Modeling

Clinching is a complicated metal elastoplastic deformation process that involves three nonlinearities: shape, material, and contact. Several FEM software programs were utilized to simulate the clinching procedure, including ABAQUS, Deform 2D/3D, LS-DYNA, ANSYS, MARC, Simufact, Forge, and FEAP. Despite considerable variances, these software programs adhere to the same principle. When using FEM, the standard procedures of geometry modeling, material modeling, specifying boundary conditions, contact modeling, meshing, and computation must be considered. Usually, the FEM utilized in clinching may be classified as static, dynamic, implicit, or explicit approaches. Most scholars agree on certain modeling conventions, including the simplicity of the two-dimensional axisymmetric model to reduce computer time, the stiff or elastoplastic model for the tools, disregarding the anisotropy of joining partners, and disregarding the spring back after clinching.

10.7.2 Meshing

Proper mesh density looks particularly crucial for FEA of clinching. Extremely wide grids would compromise simulation precision, whereas extremely fine grids would

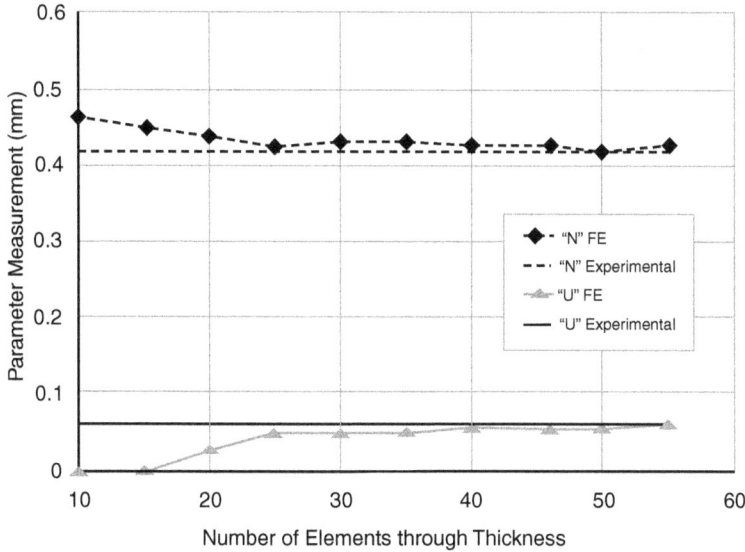

FIGURE 10.28   Element size impacts neck thickness N and interlock depth U (Atia and Jain, 2018).

be long-delayed. In general, mesh convergence analysis was used to identify the selected mesh size. The mesh was progressively refined to undertake this analysis until there was barely a difference between the two succeeding models. Atia and Jain (2018) employed differing element dimensions in the modeling of die-less clinching. Modeling was found to be incapable of predicting the interlock when the mesh size is large, up to one-fifteenth of the material thickness. However, when the mesh size reduces from one-fifteenth to one-fortieth of the material thickness, the modeled interlock increases and is accordant with the experiment with an inaccuracy of less than three percent (Figure 10.28). The ideal mesh aspect ratio of 1.3 was learned to boost computing efficacy without considerably compromising computational precision.

Moreover, a suitable mesh arrangement is essential to assure simulation accuracy and reduce computational time. Lambiase and Di Ilio (2016) and Atia and Jain (2018) employed a grid division approach in which fine mesh was used in regions of high strain and coarse mesh was utilized elsewhere (Figure 10.29).

Roux and Bouchard (2013) employed a more appropriate meshing approach, utilizing the Forge 2009 program. Here, 2× refinement was applied to the border of both sheets, and the isotropic mesh size was 0.1 mm for the global sheet. This mesh refinement approach improves contact computation precision, particularly when considerable curvature is present.

Similarly, dense meshes are given to tool materials and contact regions, considering tools as line elastomers and joining partners as elastic-plastic bodies (Figure 10.30). Gerstmann and Awiszus (2014) incorporated an improved meshing operation in the punch's corner radius and the associated anvil-sided materials in their simulation of the flat clinching of aluminum and polystyrene material.

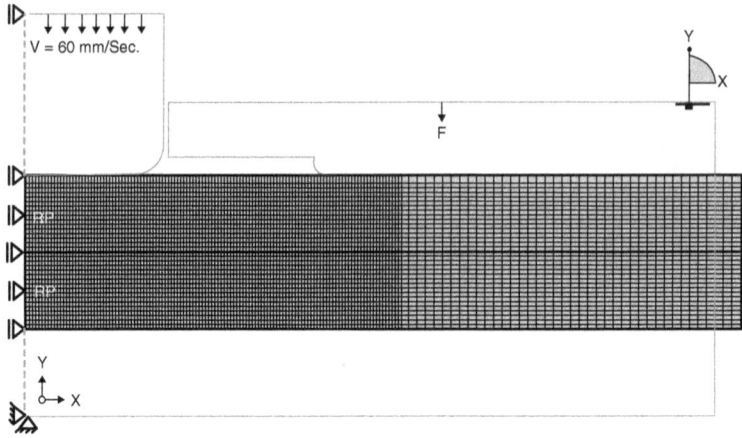

FIGURE 10.29   Fine mess in the deformation zone (Atia and Jain, 2018).

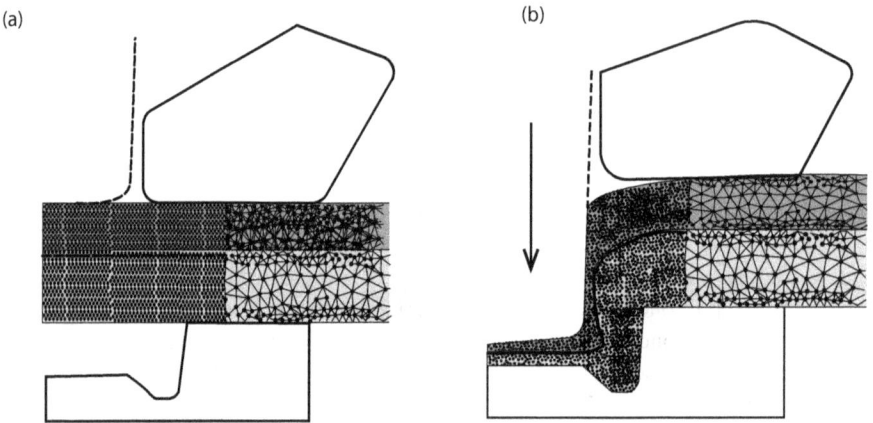

FIGURE 10.30   Refinement of mesh: (a) initial state and (b) final state (Roux and Bouchard, 2013).

10.7.3   Remeshing Method

Nonlinear issues such as material plasticity, tool shape, and contact interactions appear insurmountable during the FEM of clinching. Hence, numerous tried-and-true numerical approaches were widely used. The simulation accuracy suffers from extreme distortion throughout the clinching process. The remeshing process is frequently used to address this issue. A mapping technique to allocate key variables among the old and new meshes, a remeshing criteria, and an algorithm for node relocation often make up a remeshing technique. To prevent grid distortion brought on by significant deformation, Hamel et al. (2000) suggested an autonomous remeshing method based on the Lagrangian formulation. This remeshing approach helps to get a result with excellent precision, as evidenced by comparing simulations with and without it.

Coppieters et al. (2011) examined the effects of the remeshing technique with three different smoothing methods (volume smoothing, Laplacian smoothing, and equipotential smoothing). Such a method places each node depending on the nodes or elements around it. They discovered that the force-displacement curves of clinching are mostly unaffected by the smoothing strategy used and that the volume and Laplacian smoothing methods handle the meshes better in the interlock region than the equipotential approach does. Using the adaptive remeshing approach in ABAQUS, Oudjene and Ben-Ayed (2008) discovered that except for the folding region of the top material (shown in Figure 10.31), the clinching joints created by simulation with and without remeshing had almost comparable forms. Without remesh, however, certain simulation mesh distortions still happened. The maximal plastic strain obtained by modeling without remeshing is greater than that obtained with remeshing. It may be due to the elements' distortion, according to the equivalent plastic strain distribution in both situations. According to Jónás et al. (2019), the interlock and neck thickness are significantly impacted by the remeshing process. It also affects the anticipated contour and quality of clinched connections (Figure 10.32).

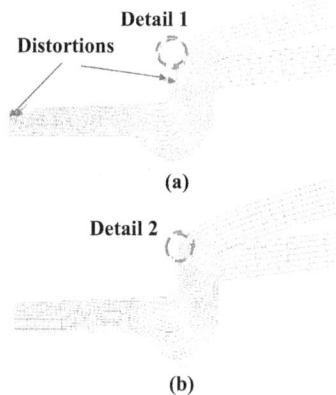

FIGURE 10.31   Deformed mesh (a) without remeshing and (b) with meshing (Oudjene and Ben Ayed, 2008).

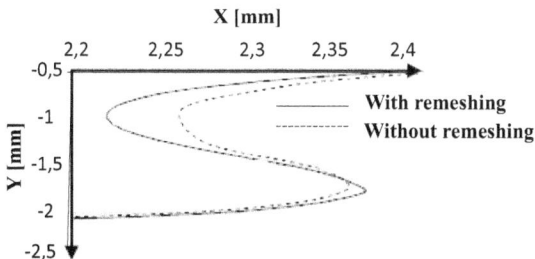

FIGURE 10.32   Comparison of meshing techniques (Jónás et al., 2019).

The Arbitrary Lagrangian-Eulerian (ALE) approach to adaptive meshing is typically used when employing the ABAQUS software for simulation. The ALE approach might enhance the features of the grids in the one-stage solution method without altering the topological structure of the original grid (the number of elements, nodes, and connections do not alter). The grid of clinched joints with and without the ALE approach was compared by Shi et al. (2018). The usefulness of the ALE approach for addressing mesh distortion during clinching has also been extensively proven by the work carried out by other researchers (Atia and Jain, 2018; Jónás et al., 2019; Köhler et al., 2020; Shi et al., 2018).

10.7.4 CONTACT MODELING

Due to the very nonlinear frictional phenomena, the contact modeling in the clinching process mostly consists of tool-work material contact and work-work contact. The friction conditions may also be affected by several other variables, such as the normal stress at the juncture, the greasing state, relative velocity, temperature, roughness, and the work or tool material properties. By controlling the material flow of joining partners, friction substantially influences the formation and quality of clinched junctions. Consequently, the friction state is essential for precise modeling of clinching. Although it is exceedingly challenging to characterize the real friction properties using standard friction testing and friction concepts. The method used most frequently to assess friction during plastic forming is the ring compression test. Regarding the clinching process, writers have generally utilized the traditional Coulomb friction law, which states that "frictional force is proportional to normal force". Kaščák et al. (2016) used the ABAQUS pure penalty contact approach derived from the Coulomb model. The Newton-Raphson method-based iterative solution effectively finds the stiffness matrix. The friction conditions were modeled by Lambiase and Di Ilio (2016), Roux and Bouchard (2013), and Coppieters et al. (2013, 2017) using a modified Coulomb friction model with a shear stress limit ($\tau_{max} = \dfrac{\sigma_y}{3}$). Cumin et al. (2018) also used the Coulomb friction model using the arctangent approach.

10.8 CONCLUSION AND FUTURE SCOPE

This chapter on mechanical clinching presents a basic comprehension of clinching's growth and inception. This is particularly beneficial for welding researchers and the automotive industry, as clinching has been widely used to join car body parts. It may also serve as a sound basis for interested researchers to broaden their fundamental concepts of clinching. The present body of knowledge on clinching is organized around numerous categories: flat clinching, hole clinching, LAC, friction stir clinching, clinch-adhesive joining, etc. The findings indicate that, while the literature on clinching is presently limited, the publication trend suggests that this subject of study is quickly expanding. On the other hand, clinching offers a lot of potential for future research. Future scholars may focus on the research gap to further develop the process.

- Clinching has traditionally been employed to join two thin worksheets; however, research on joining three sheets using clinching is limited. Clinching technology should be extended to join three sheets, given the prevalence of multi-material models in the automotive industry.
- Researchers have conducted numerous FEM simulations and numerical modeling to optimize the tool's shape and dimensions. However, FEM modeling in hybrid clinch-adhesive joining, hybrid clinch-resistance welding, etc., is somewhat limited.
- Clinching should also be extended to join novel, similar, and dissimilar materials.
- The clinching joint has less strength than other welding or joining processes. Therefore, researchers should try to increase the strength of clinched connections through adhesive, reshaping, or other novel methods.
- The material surface circumstances and coating effect on the clinching joint must be thoroughly investigated.
- Till now, mechanical properties have been studied by destructive testing only. Researchers have a minimal investigation of non-destructive testing of clinching joints. Non-destructive testing of clinching joints is another prospective, unexplored research area.

REFERENCES

Abe, Y., Ishihata, S., Maeda, T., & Mori, K. I. (2018a). Mechanical clinching process using preforming of lower sheet for improvement of joinability. *Procedia Manufacturing, 15*, 1360–1367.

Abe, Y., Kishimoto, M., Kato, T., & Mori, K. (2009). Joining of hot-dip coated steel sheets by mechanical clinching. *International Journal of Material Forming, 2*, 291–294.

Abe, Y., Saito, T., Nakagawa, K., & Mori, K. I. (2018b). Rectangular shear clinching for joining of ultra-high strength steel sheets. *Procedia Manufacturing, 15*, 1354–1359.

Abibe, A. B., Amancio-Filho, S. T., Dos Santos, J. F., & Hage Jr, E. (2013). Mechanical and failure behaviour of hybrid polymer-metal staked joints. *Materials & Design, 46*, 338–347.

Abibe, A. B., Sônego, M., dos Santos, J. F., Canto, L. B., & Amancio-Filho, S. T. (2016). On the feasibility of a friction-based staking joining method for polymer-metal hybrid structures. *Materials & Design, 92*, 632–642.

Atia, M. K. S., & Jain, M. K. (2018). A parametric study of FE modeling of die-less clinching of AA7075 aluminum sheets. *Thin-Walled Structures, 132*, 717–728.

Babalo, V., Fazli, A., & Soltanpour, M. (2018). Electro-Hydraulic Clinching: A novel high speed joining process. *Journal of Manufacturing Processes, 35*, 559–569.

Bach, F. W., Beniyash, A., Lau, K., & Versemann, R. (2005). Joining of steel-aluminium hybrid structures with electron beam on atmosphere. *Advanced Materials Research, 6*, 143–150.

Balawender, T., Kneć, M., & Sadowski, T. (2011). Technological problems and experimental investigation of hybrid: Clinched-adhesively bonded joint. *Archives of Metallurgy and Materials, 56*, 439–446.

Benabderrahmane, B., & Ali, B. (2013). Finite element analysis of the parameters affect the mechanical strength of a point clinched. *International Journal of Engineering Research and Technology, 2*, 795–799.

Busse, S., Merklein, M., Roll, K., Ruther, M., & Zürn, M. (2010). Development of a mechanical joining process for automotive body-in-white production. *International Journal of Material Forming, 3*, 1059–1062.

Carboni, M., Beretta, S., & Monno, M. (2006). Fatigue behaviour of tensile-shear loaded clinched joints. *Engineering Fracture Mechanics, 73*(2), 178–190.

Chastel, Y., & Passemard, L. (2014). Joining technologies for future automobile multi-material modules. *Procedia Engineering, 81*, 2104–2110.

Chen, C., Zhao, S., Han, X., Wang, Y., & Zhao, X. (2017). Investigation of flat clinching process combined with material forming technology for aluminum alloy. *Materials, 10*(12), 1433.

Coppieters, S., Cooreman, S., Lava, P., Sol, H., Houtte, P. V., & Debruyne, D. (2011). Reproducing the experimental pull-out and shear strength of clinched sheet metal connections using FEA. *International Journal of Material Forming, 4*, 429–440.

Coppieters, S., Lava, P., Hecke, R. V., Cooreman, S., Sol, H., Houtte, P. V., & Debruyne, D. (2013). Numerical and experimental study of the multi-axial quasi-static strength of clinched connections. *International Journal of Material Forming, 6*, 437–451.

Coppieters, S., Zhang, H., Xu, F., Vandermeiren, N., Breda, A., & Debruyne, D. (2017). Process-induced bottom defects in clinch forming: Simulation and effect on the structural integrity of single shear lap specimens. *Materials & Design, 130*, 336–348.

Cumin, J., Samardžić, I., & Dunđer, M. (2018). Mechanical clinching process stress and strain in the clinching of EN-AW5754 (AlMg3), and EN AW-5019 (AlMg5) metal plates. *Metalurgija, 57*(1-2), 107–110.

Eshtayeh, M. M., Hrairi, M., & Mohiuddin, A. K. M. (2016). Clinching process for joining dissimilar materials: State of the art. *The International Journal of Advanced Manufacturing Technology, 82*, 179–195.

Gerstmann, T., & Awiszus, B. (2014). Recent developments in flat-clinching. *Computational Materials Science, 81*, 39–44.

Hamel, V., Roelandt, J. M., Gacel, J. N., & Schmit, F. (2000). Finite element modeling of clinch forming with automatic remeshing. *Computers & Structures, 77*(2), 185–200.

He, X., Liu, F., Xing, B., Yang, H., Wang, Y., Gu, F., & Ball, A. (2014). Numerical and experimental investigations of extensible die clinching. *The International Journal of Advanced Manufacturing Technology, 74*, 1229–1236.

Hiller, M., & Volk, W. (2015). Joining aluminium alloy and mild steel sheets by roller clinching. *Applied Mechanics and Materials, 794*, 295–303.

Jäckel, M., Grimm, T., & Falk, T. (2017, November). Process development for mechanical joining of 7xxx series aluminum alloys. In *European Aluminium Congress*. Düsseldorf, Germany.

Jónás, S., Tisza, M., Felhős, D., & Kovács, P. Z. (2019, July). Experimental and numerical study of dissimilar sheet metal clinching. In *AIP Conference Proceedings* (Vol. 2113, No. 1, p. 050021). AIP Publishing, Vitoria-Gasteiz, Spain.

He, X. (2017). Clinching for sheet materials. *Science and Technology of Advanced Materials, 18*(1), 381–405.

Kaščák, L., Spišák, E., Kubik, R., & Mucha, J. (2016). FEM analysis of clinching tool load in a joint of dual-phase steels. *Strength of Materials, 48*, 533–539.

Köhler, D., Kupfer, R., & Gude, M. (2020). Clinching in in-situ CT-A numerical study on suitable tool materials. *Journal of Advanced Joining Processes, 2*, 100034.

Lambiase, F. (2015a). Mechanical behaviour of polymer-metal hybrid joints produced by clinching using different tools. *Materials & Design, 87*, 606–618.

Lambiase, F. (2015b). Clinch joining of heat-treatable aluminum AA6082-T6 alloy under warm conditions. *Journal of Materials Processing Technology, 225*, 421–432.

Lambiase, F., & Di Ilio, A. (2016). Damage analysis in mechanical clinching: Experimental and numerical study. *Journal of Materials Processing Technology*, *230*, 109–120.

Lee, C. J., Kim, J. Y., Lee, S. K., Ko, D. C., & Kim, B. M. (2010). Parametric study on mechanical clinching process for joining aluminum alloy and high-strength steel sheets. *Journal of Mechanical Science and Technology*, *24*, 123–126.

Lee, C. J., Lee, S. H., Lee, J. M., Kim, B. H., Kim, B. M., & Ko, D. C. (2014). Design of hole-clinching process for joining CFRP and aluminum alloy sheet. *International Journal of Precision Engineering and Manufacturing*, *15*, 1151–1157.

Lin, P. C., & Lo, S. (2016). Development of friction stir clinching process for alclad 2024-T3 aluminum sheets. *SAE International Journal of Materials and Manufacturing*, *9*(3), 756–763.

Mayyas, A., Qattawi, A., Omar, M., & Shan, D. (2012). Design for sustainability in automotive industry: A comprehensive review. *Renewable and Sustainable Energy Reviews*, *16*(4), 1845–1862.

Mehta, K. (2017). Advanced joining and welding techniques: An overview. In: Gupta, K. (ed.) *Advanced Manufacturing Technologies. Materials Forming, Machining and Tribology*. Springer, Cham.

Messler, R. W. (2000). Trends in key joining technologies for the twenty-first century. *Assembly Automation*, *20*(2), 118–128.

Mucha, J. (2017). Clinching technology in the automotive industry. *Archiwum Motoryzacji*, *76*(2), 75–94.

Neugebauer, R., Mauermann, R., & Grützner, R. (2005). Combination of hydroforming and joining. *Steel Research International*, *76*(12), 939–944.

Neugebauer, R., Mauermann, R., & Grützner, R. (2008). Hydrojoining. *International Journal of Material Forming*, *1*, 1303–1306.

Neugebauer, R., Todtermuschke, M., Mauermann, R., & Riedel, F. (2008). Overview on the state of development and the application potential of dieless mechanical joining processes. *Archives of Civil and Mechanical Engineering*, *8*(4), 51–60.

Oudjene, M., & Ben-Ayed, L. (2008). On the parametrical study of clinch joining of metallic sheets using the Taguchi method. *Engineering Structures*, *30*(6), 1782–1788.

Peng, H., Chen, C., Ren, X., & Wu, J. (2022). Development of clinching process for various materials. *The International Journal of Advanced Manufacturing Technology*, *119*(1–2), 99–117.

Peng, H., Chen, C., Zhang, H., & Ran, X. (2020). Recent development of improved clinching process. *The International Journal of Advanced Manufacturing Technology*, *110*, 3169–3199.

Reich, M., Osten, J., Milkereit, B., Kalich, J., Füssel, U., & Kessler, O. (2014). Short-time heat treatment of press hardened steel for laser assisted clinching. *Materials Science and Technology*, *30*(11), 1287–1296.

Roux, E., & Bouchard, P. O. (2013). Kriging metamodel global optimization of clinching joining processes accounting for ductile damage. *Journal of Materials Processing Technology*, *213*(7), 1038–1047.

Sadowski, T., Balawender, T., Golewski, P., Sadowski, T., Balawender, T., & Golewski, P. A. (2015). *Technological Aspects of Manufacturing and Numerical Modelling of Clinch-Adhesive Joints* (pp. 1–59). SpringerBriefs in Applied Sciences and Technology. Springer, Cham.

Shi, B., Zhang, Z., Yang, M., & Zhong, J. (2018). Simulation and optimization analysis of clinching joint performance based on mould parameters. *DEStech Transactions on Computer Science and Engineering (ICMSIE)*, 537–584. ISBN: 978-1-60595-516-2

Varis, J. P. (2002). The suitability of round clinching tools for high strength structural steel. *Thin-Walled Structures*, *40*(3), 225–238.

Wang, X., Li, C., Ma, Y., Shen, Z., Sun, X., Sha, C., Gao, S., Li, L., & Liu, H. (2016). An experimental study on micro clinching of metal foils with cutting by laser shock forming. *Materials*, *9*(7), 571.

Wiesenmayer, S., & Merklein, M. (2021). Potential of shear-clinching technology for joining of three sheets. *Journal of Advanced Joining Processes*, *3*, 100043.

Yuce, C., Karpat, F., Yavuz, N., & Sendeniz, G. (2014). A case study: Designing for sustainability and reliability in an automotive seat structure. *Sustainability*, *6*(7), 4608–4631.

Zhang, Y., Shan, H., Li, Y., Guo, J., Luo, Z., & Ma, C. Y. (2017). Joining aluminum alloy 5052 sheets via novel hybrid resistance spot clinching process. *Materials & Design*, *118*, 36–43.

Zhao, L., He, X. C., & Lu, Y. (2014). Research of mechanical behavior for rounded and rectangular clinched joint. *Advanced Materials Research*, *1035*, 144–148.

11 Systematic Study of Digital Twins for Welding Processes

Vivek Warke, Sameer Sayyad, Satish Kumar, Arunkumar Bongale, and Suresh R.

11.1 INTRODUCTION

The welding process has become increasingly important in fabrication of parts, structures, and components for a wide range of industrial applications. In order to keep up with the ever-increasing demands placed on manufacturers in terms of efficiency, quality, and reliability, the conventional welding manufacturing process needs to be developed into an intelligent welding manufacturing process (Edwin & Jenkins, 2012). In real time, the operating status of the welding process is monitored, and control techniques are implemented to make adaptive adjustments to the welding parameters to keep the welding state within an acceptable range (Benyounis & Olabi, 2008).

In order to solve these concerns, researchers implemented the digital twin (DT) framework for monitoring, controlling, and visualizing the welding process (Söderberg et al., 2017). The term "digital twin" refers to a more generic framework for digitalizing and representing physical entities in the virtual realm. Not only do the developed DTs possess the same elements as their physical counterparts, but they also follow the same operating dynamics and regulations. DT is regarded as an enabler for the implementation of Industry 4.0 in this technological era, and conscientious efforts are being made to enact the concepts of Industry 4.0 in the industrial environment. In addition, meticulous efforts are being made to implement the concepts of Industry 4.0 in the modern tech era. The obstacles, such as the retesting of physical prototypes, are removed as a result of the DT idea, which also makes it possible for employees to acquire new skills in an online setting. The challenges experienced by the operator can be promptly remedied, and there is no requirement to retrace the steps or rebuild prototypes in order to do additional testing. In addition, DT systems can be used to inspect the process, enabling visualization of modifications' impact. Implementing DT in welding operations results from integrating real-time monitoring and real-time modeling into the process (Tao et al., 2019). For the purpose of data communication, developments in information technology such as the Internet-of-Things (IoT) and Augmented Reality (AR) can be utilized to create a link between the actual system and its digital counterpart. Real-time data collection, data analysis, and physics-based simulation are the fundamental stages of defect detection (Paritala et al., 2017). These phases aim to improve the quality of cutting-edge technology by reducing the occurrence of flaws and preventing their occurrence. As a result, a comprehensive analysis of monitoring technologies for the welding process using DT technology is presented in this chapter.

11.2 LITERATURE REVIEW

11.2.1 WELDING PROCESS

The term "welding" refers to a fabrication process in which two or more components are joined using heat, pressure, or even a combination to establish a joint while the parts cool. Most often, welding is performed on metals, thermoplastics, and sometimes wood. A weldment is a term used for a finished welded junction. The assemblies produced by joining processes such as bolt assembly, riveting, brazing, seaming, and soldering are all momentary joints. The only method that can permanently join two metals is welding. Higher corrosion resistance, higher fatigue strength, tensile strength, toughness, hardness, ductility, and better efficiency are the significant characteristics of welding, due to which it is preferred over the other joining processes. Primarily, welded joints are classified based on methods of joint formation, such as a fusion method with and without pressure and a non-fusion plan (Swapna Sai et al., 2020). Metals of the same or different types can be fused without pressure in a process known as non-pressure fusion welding. This may be done with or without filler metal, but it must be done without applying pressure. Arc, gas, and chemical reactions are familiar heat sources used in fusion welding without pressure. On the other hand, pressure welding is a type in which similar metals are joined without the need for filler metal by first melting them to a plastic or partially molten state and then pushing or pounding the metals together. Electric resistance (resistance welding), forge, or friction are the familiar heat sources used in pressure fusion welding. Non-fusion welding, on the other hand, is a technique for welding dissimilar metals that uses a filler metal with a low melting point, and no pressure is applied to avoid melting the ends of the base metal. As heat is the primary source to fuse the metal, welding processes are categorized as indicated in Figure 11.1, depending on different heat sources.

The most prevalent materials that may be joined using various welding procedures are ferrous, non-ferrous, and alloys. The chemical properties of the materials, such as corrosion, oxidation, and reduction, are involved in the chemical effect and impact weld quality. Also, when metals are subjected to heat created during welding, their physical characteristics, likewise grain growth, melting point, thermal conductivity, and thermal expansion, are affected. Further, mechanical properties like brittleness, ductility, hardness, tensile strength, and toughness determine the behavior of the metal during welding and under load conditions. In addition, the various types of weld joints, including bead welds, butt or groove welds, fillet welds, and plug or slot welds, all play a vital part in producing the dimensions and attributes of joints that are required.

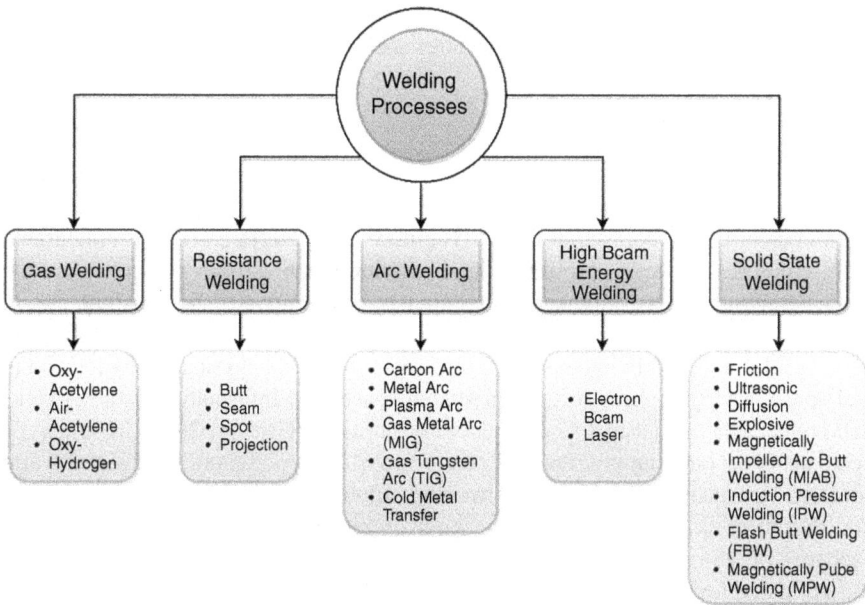

FIGURE 11.1 Classification of welding processes.

The various welding process parameters affect the part quality and performance of the joint during the application. Improper selection or sudden changes in various process parameters, like arc voltage, electrode size, feed, gas flow rate, gas pressure orientation, polarity, shielding gas composition, welding current, and welding speed, result in defective welded joints. Generally, welding defects are classified as external, visual, internal, or hidden. The external defects that occur externally at the weld surface can be observed as cracking, porosity, undercut, overlap, lack of penetration, excessive penetration, and spatter. At the same time, internal defects are those that occur internally and are characterized as slag inclusion, lack of fusion, incomplete fusion, shrinkage cavities, porosity, and internal stresses. There are numerous destructive and non-destructive testing (NDT) techniques to determine the defect in the weldment. The destructive testing involves acid etch, bend, break, and tensile strength tests. However, NDT comprises an acoustic emission test, a dye test, eddy current test, a gamma ray test, a hardness test, a magnetic particle test, and an X-ray test. Although these techniques are used to test the weld quality, there are a few drawbacks to them, which limit their use in real-time monitoring, control, and inspection. Testing the weld quality using these methods requires trained personnel, part cleaning, and, most importantly, the review is only carried out after the completion of the process.

Thus, advancements in sensing, communication, and computational technologies are used to overcome the limitations of destructive testing and NDT in the welding process. Furthermore, in the epoch of Industry 4.0, the use of DT for control, monitoring, and optimization of the welding process has become widespread owing to developments in ICT (information and communication technology). Therefore, a brief summary of DT and its use in various welding procedures is provided in the next section.

11.2.2 DIGITAL TWIN

DT has been described in a number of ways by a variety of organizations and academics. Different industrial research groups and academics have proposed their own definitions of DT. According to NASA (Bigliardi et al., 2021), a DT is "an integrated multi-physics, multiscale, probabilistic simulation of an as-built vehicle or system that leverages the best available physical models, sensor updates, fleet history, etc., to replicate the life of its corresponding flying twin." According to Grieves, "the core paradigm for a digital twin consists of three components: a physical entity located in real space, a virtual entity located in virtual space, and information data networks that connect the actual and virtual places or entities" (Grieves, 2015). To track the behavior of an ongoing process, DT creates a virtual representation of a real-world situation that includes condition monitoring, anomaly detection, and trend prediction (Singh et al., 2021). Hence, a six-layer architecture is proposed to enable the exchange of data and information between digital and physical twins and the rest of the world and for the proper synchronization of physical and virtual equipment in the cyber-physical realm. A six-layer architecture comprises the following: a physical layer, a layer for data transport and collection, a layer for data storage and processing, a layer for communication, a layer for cloud computing and storage, and a layer for virtualization (Tao & Zhang, 2017). Additionally, the real-time implementation of DT is made possible due to the recent advancements in ICT and computational methodologies.

A real-time DT is the virtual reproduction of a physical thing or procedure functioning in a physical space. Therefore, designing and implementing a DT for a physical entity or process undergoes a sequence of steps, as shown in Figure 11.2.

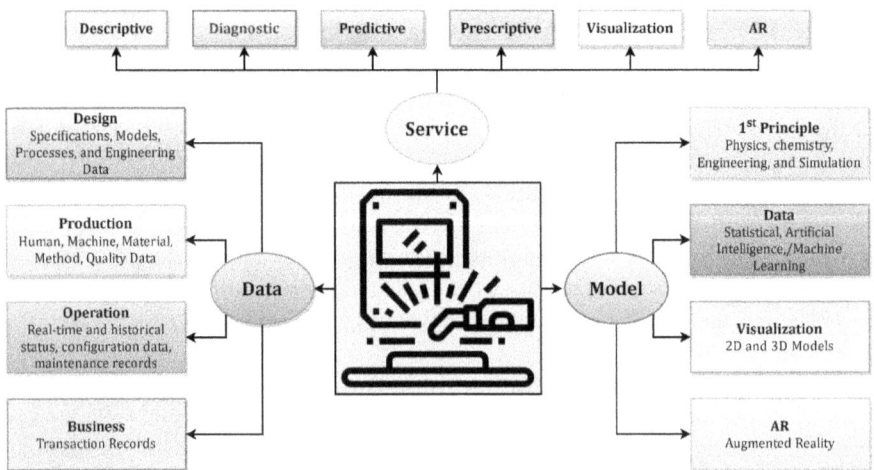

FIGURE 11.2   Design and implementation of DT.

It consists of computational models, data, and services. A DT comprises data for models to describe and comprehend the states and behaviors of the real-world entity it represents. In the case of equipment, it may contain data from the design, manufacture, operation, and end-of-life phases (decommissioning and disposal data) (Warke et al., 2021). It might also include corporate data such as transaction records.

In addition, a DT comprises computational or analytic models used to describe, grasp, and forecast the operational states and behaviors of a physical item, as well as models used to prescribe actions based on the objectives and business logic connected with the physical entity (K. J. Wang et al., 2020). These models might be established on the fundamentals of physics or chemistry; they could also be engineering or simulation models; data models founded on machine learning, statistics, and artificial intelligence; or they could be models for data (AI) (Sayyad et al., 2021, 2022). It may also include AR models and 3D visuals to help people better understand how things work in the actual world. In addition to this, a DT will have its own set of service interfaces, which will allow other DTs as well as industrial applications to access its data and activate its capabilities (Jordan & Mitchell, 2015; Negri et al., 2017). Thus, with these abilities and enabling technologies such as the IoT, cyber-physical systems, machine learning, cloud computing, and AR/virtual reality, the DT for various welding processes can be constructed to optimize, control, and monitor the process (Alcácer & Cruz-Machado, 2019; Botkina et al., 2018; Farsi et al., 2020). Even though DT is a fundamental enabling technology of Industry 4.0, which allows for the testing of new systems before they are manufactured, the improvement of efficiency and productivity, the forecasting of the future behaviors of the system, and the provision of improved service, there are still issues with the application of these models to real-world scenarios. The development of methodologies for applying DT models to industrial domains, notably manufacturing, predictive maintenance, and after-sales services, is still in its early stages. Real-time changes throughout the machining process are difficult to recognize and simulate, making it difficult to design, create, interpret, and manage the machine with an accurate multiscale DT model of work-in-process (Melesse et al., 2020). Hence, the following section explains the different approaches to implementing DT for different welding processes to overcome a few challenges with the help of case studies.

11.3   DIGITAL TWIN IN WELDING

The researchers used the DT framework to overcome and monitor the different problems that occurred in welding processes. A few case studies related to the application of DTs in welding are discussed here as follows.

11.3.1   Weld Joint Expansion and Penetration Regulation of Gas Tungsten Arc Welding Process Using a Digital Twin

This work reports on how a DT-based graphical user interface (GUI) could be used to track, manage, and display data about a gas tungsten arc welding operation (Q. Wang et al., 2020). During welding, the quality of the joint can be determined in one of the three stages based on the size of the weld pool: partial fusion, adequate fusion, or burn-through. An effective join depends on the materials used, the heat used, and the surface tension present. So, constant monitoring of the procedure is required. Because of the required width of the backside bead, the quality of the weld connection is guaranteed to be less than the backside bead width. It would be arduous to monitor this comparison using the conventional method; as a result, the DT, as shown in Figure 11.3, is being utilized to simplify the process.

It is possible to collect the precise feature, which correlates significantly with process output, by processing the raw data gathered from the sensors. Using a watershed segmentation-based technique to locate the weld pool boundary in the picture captured by an industrial camera, the TSBW width of the weld pool can be determined. This algorithm and discovered border can then be used to locate the weld pool's peak, from which the pool's width can be calculated. Pre-processing is performed on the raw camera data to remove noise and locate the region of interest, and then the filtered images are utilized for training a convolutional neural network algorithm to locate the backside bead width. Finally, a GUI-based DT model is developed, displaying data on the welding process in four dimensions: the geometry of the weld joints, imagery of the welds, the growth dynamics of the weld joints, and arc information. Therefore, the quality of the weld joint and the depth of penetration may both be monitored and controlled with the aid of a developed DT model for the gas tungsten arc welding process.

11.3.2 DIGITAL TWIN-BASED PROCESS MONITORING IN LASER-WELDED BLANKS OF LIGHT METAL BLANKS

An important study reports that by utilizing a DT technique, laser welding machine operators might improve their capacity to anticipate structural breakdowns and their ability to schedule maintenance tasks. This approach will decrease welding-related downtime and maintenance expenses (Aminzadeh et al., 2022).

In laser welding, they are cleaning the surface of aluminum sheets before welding is done. The clamping process is then modified to exert uniform pressure across the plates. In order to accomplish this, alignment sensors are utilized to determine the fault, and the findings are then uploaded to the cloud in order to facilitate the future course of action, as shown in Figure 11.4. In the event that the alignment is deemed satisfactory, the automatic laser welding process will begin, and real-time monitoring will be carried out. Both digital platforms and physical activities can transmit data to one another simultaneously by utilizing sensors and artificial intelligence. Big data should be able to be processed and refined on a cloud platform, with the ability to do so based on pass/fail production criteria. In the end, a permit will be provided to evaluate each production's overall level of quality. It is important to note that this model may be utilized in mass production and on remote manufacturing platforms in any region worldwide. Both of these applications are possible thanks to advancements in technology. As a result, this model is the initial step toward cloud manufacturing and linked production, which will eventually be accessed remotely at any time.

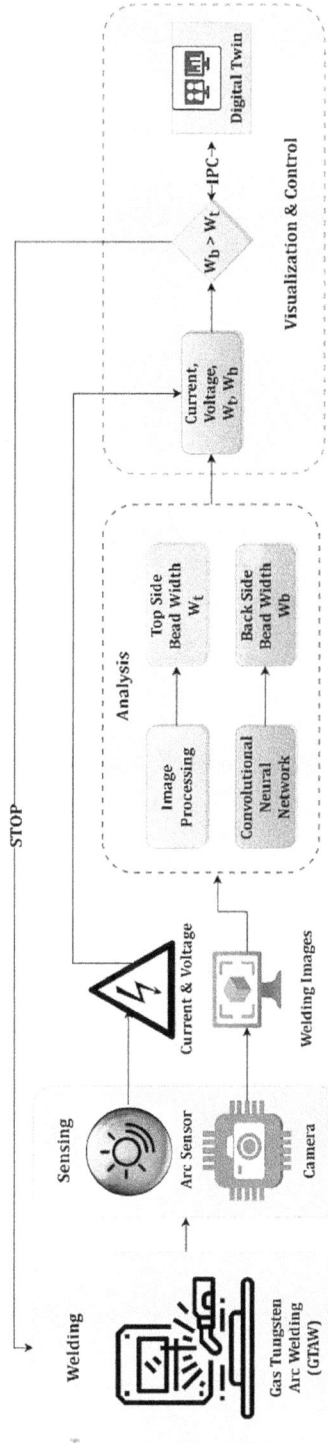

FIGURE 11.3 DT framework for weld joint expansion and penetration control.

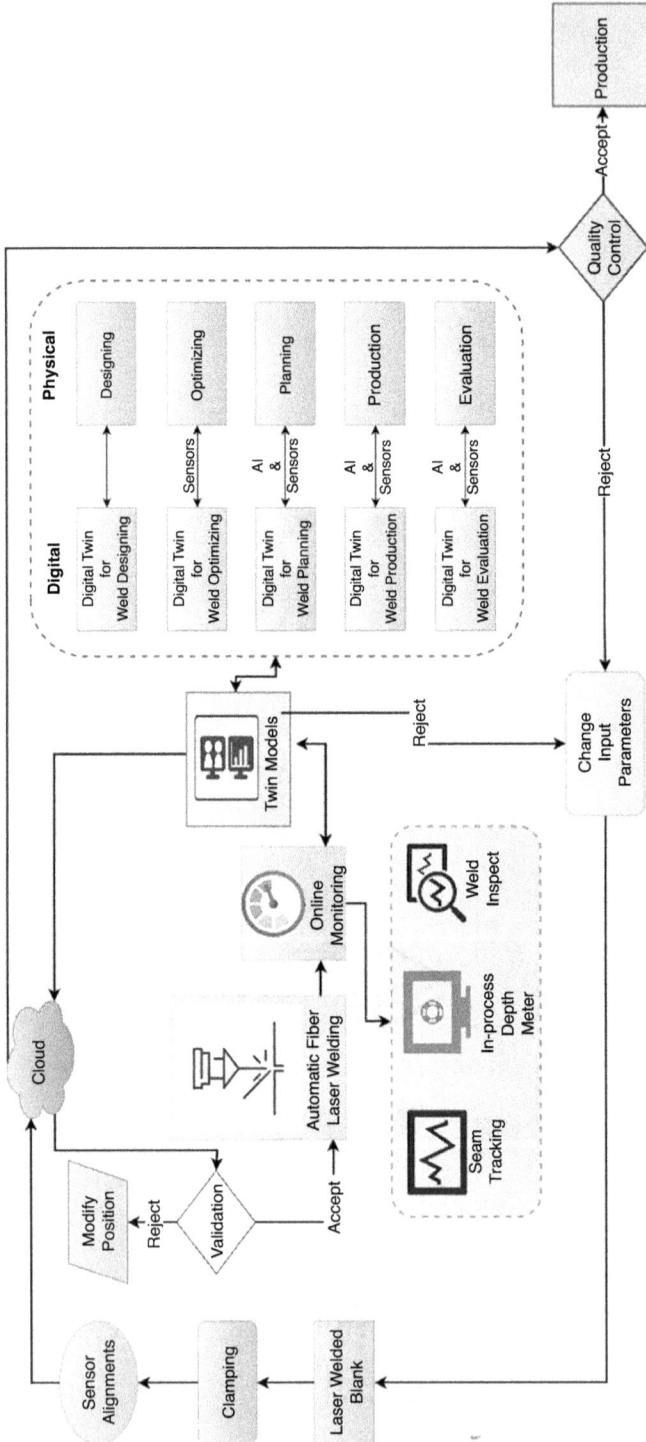

FIGURE 11.4 DT-based process monitoring of laser welding.

11.3.3 Sequence Optimization of Spot Welding for Geometry Assurance Digital Twin

A DT for geometry assurance is comprised of a number of different studies that are carried out in order to guide the actual production in the direction of ensuring the geometries of the final assembly. The goal of geometry assurance is to reduce the impact of geometrical variance across the whole product realization process. Spot welding is commonly used to secure components in sheet metal assemblies. The welding sequence has a significant impact on the geometrical output of the final assembly. In this work (Tabar et al., 2020), the geometrical consequence of spot welding in a sequence is retrieved by the application of non-rigid variation simulation, which takes into consideration the variations introduced on the parts. The fundamental procedures for simulating non-rigid sheet metal assembly involve clamping, connecting, and releasing the components from the fixture when spring back occurs. Each process is carried out with contact modeling in place to prevent any neighboring part penetration. Thus, to find the optimal spot-welding sequence to minimize the total geometrical deformations in a manufacturing assembly, the DT approach is used. The overall goal of using DT is to evaluate the sequence of operations that corresponds to the minimal geometrical deviations. Therefore, the proposed DT approach consists of two steps, viz., contact displacement minimization and assembly deviation minimization. In contact displacement minimization, a DT approach is used to generate the sequence of contact points for the minimum contact displacements. Thus, the different combinations of contact points are generated using the neural networks, and the proposed contact sequences are simulated using the CAT tools RD&T. A modified GA is presented to extract the best sequence in terms of assembly deviation. The suggested GA makes use of the results of the contact displacement minimization phase and makes a call to the CAT tool RD&T in order to analyze the assembly deviation of the produced populations in order to accomplish this sequence, as shown in the figure. Therefore, in real time, the sequence of operations and geometrical deformations are observed, and information is stored in the database. Based on the data, DT will again take the corrective action as explained above and minimize the geometrical deformations in spot welding. The proposed approach, as shown in Figure 11.5, has been proven to minimize calculation time for the chosen reference assemblies while optimizing both subassemblies. The proposed approach demonstrates that the influence of the sequences on the geometrical output may be eliminated, decreasing 86%–99% of the overall geometrical outcome.

11.3.4 Digital Twin-Based Simulation and Optimization of Robotic Arc Welding Station

In Industry 4.0, the manufacturing industry continuously evolves and adapts itself toward intelligence and flexibility; hence, the industrial robots are widely used in various applications such as grinding, welding, handling, etc. Traditional welding robots are often designed to optimize the path based on manual expertise, making it impossible to assure welding quality and efficiency. Thus, with the advancements in new ICT technologies, DT approaches are used to simulate and optimize the robotic

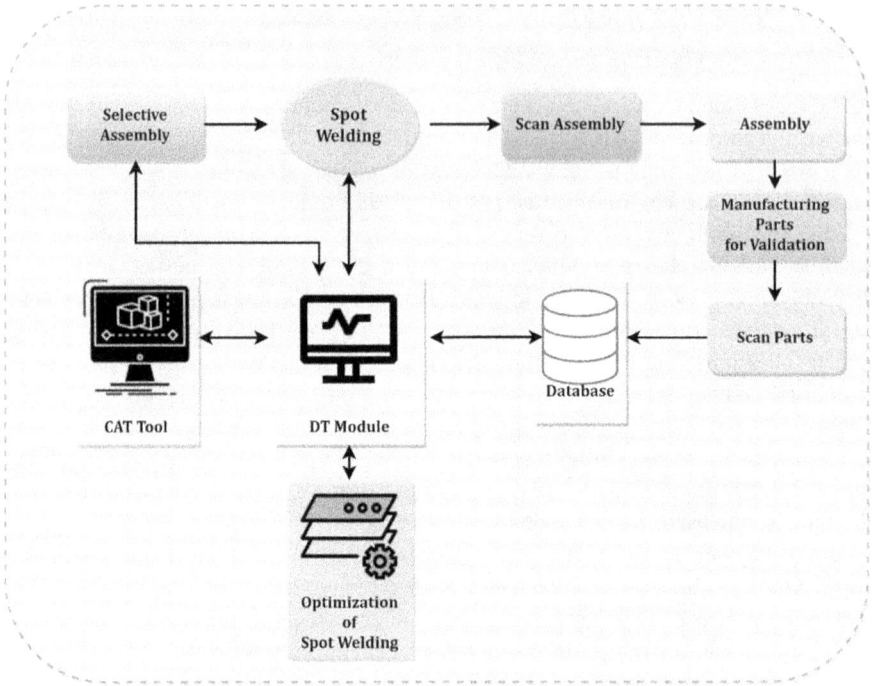

FIGURE 11.5   DT approach for sequence optimization in spot welding.

welding process. Hence, the work described in Zhang et al. (2023) proposes the DT-based simulation and optimization of a robotic arc welding station.

The most crucial part of the manufacturing line for welding is the robot welding cell. A standard arc welding robot workstation includes a robot system, a positioner system, a gantry system, a welding system, and other components. In accordance with the requirements of the welded workpiece, the robot is outfitted with the appropriate welding torch end-effector. Because of the complexities of arc welding and the severe welding environment, the robot welding workstation has a lengthy cycle time. DT welding workstations are key to achieving a breakthrough in the existing technological bottleneck. The goal is to create a virtual simulation environment that provides high-fidelity mapping of real-world physical behavior as well as a modular, universal DT welding system. Thus, the proposed DT approach for the robotic arc welding workstation is shown in Figure 11.6, which primarily includes physical robotic arc welding, virtual entities, data-driven DT, and service entities.

The actual physical unit is a welding production system made up of a welding robot and related production tools, namely physical objects like guidance mechanisms, controllers, moving arm workpieces, position changers, sensors, welding guns, welding robots, wire shearing and gun clearing devices, etc. Moreover, position-finding status, position, space planning, size, workpiece welding process, welding current, welding arc status detection, and other data information are also included. The virtual entities comprise the digital model of a welding robot, which is primarily built at

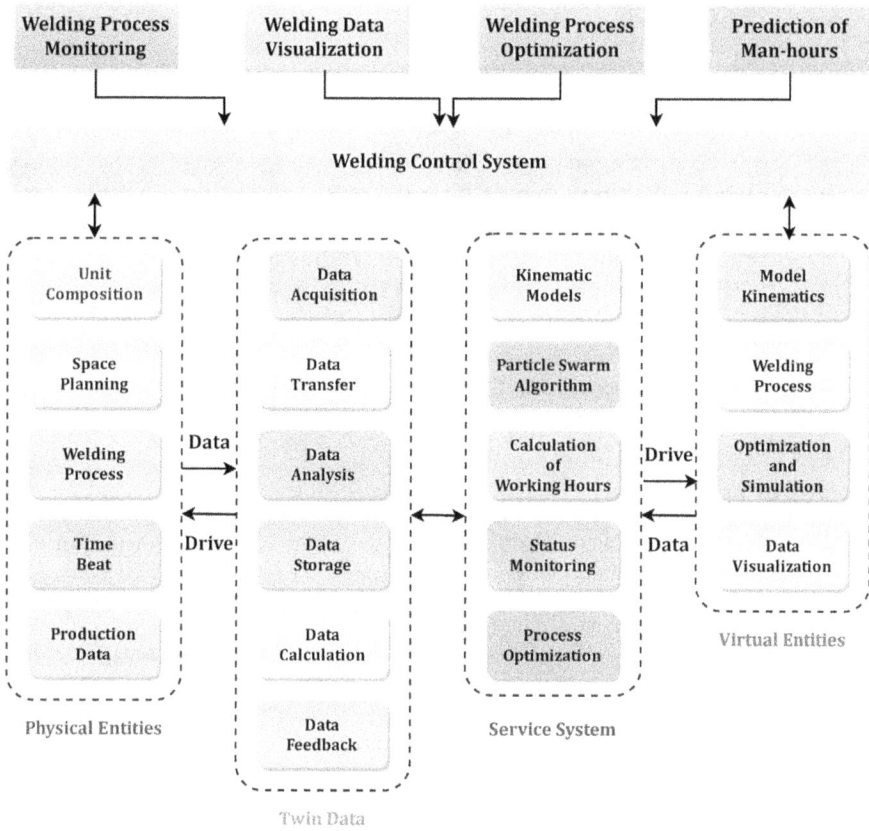

FIGURE 11.6 DT-based optimization of robotic arc welding workstation.

three different levels, such as elements, behaviors, and rules. The elements are consists mainly of physical equipment, workstation layout, and auxiliary production elements. However, linkages, welding behaviors, and other process behaviors form the behavioral part of the virtual model of a robotic arc welding workstation. The welding process optimization and other evolutionary algorithms jointly form the rules for DT-based workstations. The core of the service system is twin data, which provides services like motion control and logic driving for the DT, analysis, and optimization of the welding process, such as time beat and path of robotic arc welding, and mapping it to the virtual entity cell to perform motion simulation of the welding process. Furthermore, the twin data is made up of three different types of information: data pertaining to physical units, data pertaining to virtual units, and data pertaining to the service system itself. This data offers relative analysis, verification, and decision information for the service system by means of data transmission, interaction, and updating between the different layers. Thus, the proposed approach realizes effective collaboration between the positioner and robot during welding, which helps enhance the efficiency of the welding process and optimize the idle time.

11.4 CONCLUSION

This chapter has broadly covered the application of DT in different welding techniques. The DT is mainly used in welding to control, monitor, visualize, and optimize the joint growth and quality of the weld. The various aspects of designing a DT, such as data, service, and model, are implemented for the DT of the welding process to optimize their performance through various emerging technologies, such as machine learning, AR, sensors, and IoT. It reveals the application of deep learning and image processing in the DT of welding. The welding industry experiences a few challenges while implementing DT for the welding process. This primarily includes the design and development of DT within a stipulated time, risk while adopting a DT, modeling of a process, effective information exchange, and simultaneous integration of different domains. Therefore, a future direction for study may be to develop a framework capable of resolving these challenges.

REFERENCES

Alcácer, V., & Cruz-Machado, V. (2019). Scanning the Industry 4.0: A literature review on technologies for manufacturing systems. *Engineering Science and Technology, an International Journal*, 22(3), 899–919. https://doi.org/10.1016/j.jestch.2019.01.006.

Aminzadeh, A., Karganroudi, S. S., Meiabadi, M. S., Mohan, D. G., & Ba, K. (2022). A survey of process monitoring using computer-aided inspection in laser-welded blanks of light metals based on the digital twins concept. *Quantum Beam Science*, 6(2), 1–12. https://doi.org/10.3390/qubs6020019.

Benyounis, K. Y., & Olabi, A. G. (2008). Optimization of different welding processes using statistical and numerical approaches - A reference guide. *Advances in Engineering Software*, 39(6), 483–496. https://doi.org/10.1016/J.ADVENGSOFT.2007.03.012.

Bigliardi, B., Casella, G., & Bottani, E. (2021). Industry 4.0 in the logistics field: A bibliometric analysis. *IET Collaborative Intelligent Manufacturing*, 3(1), 4–12. https://doi.org/10.1049/cim2.12015.

Botkina, D., Hedlind, M., Olsson, B., Henser, J., & Lundholm, T. (2018). Digital twin of a cutting tool. *Procedia CIRP*, 72, 215–218. https://doi.org/10.1016/j.procir.2018.03.178.

Edwin, R. D. J., & Jenkins, H. D. S. (2012). A review on optimization of welding process. *Procedia Engineering*, 38, 544–554. https://doi.org/10.1016/J.PROENG.2012.06.068.

Farsi, M., Daneshkhah, A., Hosseinian-Far, A., & Jahankhani, H. (2020). *Internet of Things Digital Twin Technologies and Smart Cities*. Berlin/Heidelberg, Germany: Springer. https://www.springer.com/series/11636.

Grieves, M. (2015). Digital Twin : Manufacturing Excellence through Virtual Factory Replication. This paper introduces the concept of a Whitepaper by Dr. Michael Grieves. *White Paper, March*. https://www.researchgate.net/publication/275211047_Digital_Twin_Manufacturing_Excellence_through_Virtual_Factory_Replication.

Jordan, M. I., & Mitchell, T. M. (2015). Machine learning: Trends, perspectives, and prospects. *Science*, 349(6245), 255–260.

Melesse, T. Y., Di Pasquale, V., & Riemma, S. (2020). Digital twin models in industrial operations: A systematic literature review. *Procedia Manufacturing*, 42(2019), 267–272. https://doi.org/10.1016/j.promfg.2020.02.084.

Negri, E., Fumagalli, L., & Macchi, M. (2017). A review of the roles of digital twin in cps-based production systems. *Procedia Manufacturing*, 11(June), 939–948. https://doi.org/10.1016/j.promfg.2017.07.198.

Paritala, P., Manchikatla, S., & Yarlagadda, P. (2017). Digital manufacturing-applications past, current, and future trends. *Procedia Engineering*, 174, 982–991. https://doi.org/doi:10.1016/j.proeng2017.01.250.

Sayyad, S., Kumar, S., Bongale, A., Kamat, P., Patil, S., & Kotecha, K. (2021). Data-driven remaining useful life estimation for milling process: Sensors, algorithms, datasets, and future directions. *IEEE Access*, 9(July), 1–1. https://doi.org/10.1109/access.2021.3101284.

Sayyad, S., Kumar, S., Bongale, A., Kotecha, K., Selvachandran, G., & Suganthan, P. N. (2022). Tool wear prediction using long short-term memory variants and hybrid feature selection techniques. *The International Journal of Advanced Manufacturing Technology*, 121(9), 6611–6633. https://doi.org/10.1007/s00170-022-09784-y.

Singh, M., Fuenmayor, E., Hinchy, E. P., Qiao, Y., Murray, N., & Devine, D. (2021). Digital twin: Origin to future. *Applied System Innovation*, 4(2), 1–19. https://doi.org/10.3390/asi4020036.

Söderberg, R., Wärmefjord, K., Carlson, J. S., & Lindkvist, L. (2017). Toward a Digital Twin for real-time geometry assurance in individualized production. *CIRP Annals*, 66(1), 137–140. https://doi.org/10.1016/J.CIRP.2017.04.038.

Swapna Sai, M., Dhinakaran, V., Manoj Kumar, K. P., Rajkumar, V., Stalin, B., & Sathish, T. (2020). A systematic review of effect of different welding process on mechanical properties of grade 5 titanium alloy. *Materials Today: Proceedings*, 21, 948–953. https://doi.org/10.1016/J.MATPR.2019.09.027.

Tabar, R. S., Warmefjord, K., Soderberg, R., & Lindkvist, L. (2020). Efficient spot welding sequence optimization in a geometry assurance digital twin. *Journal of Mechanical Design, Transactions of the ASME*, 142(10). https://doi.org/10.1115/1.4046436.

Tao, F., Qi, Q., Wang, L., & Nee, A. Y. C. (2019). Digital twins and cyber-physical systems toward smart manufacturing and Industry 4.0: Correlation and comparison. *Engineering*, 5(4), 653–661. https://doi.org/10.1016/j.eng.2019.01.014.

Tao, F., & Zhang, M. (2017). Digital twin shop-floor: A new shop-floor paradigm towards smart manufacturing. *IEEE Access*, 5, 20418–20427. https://doi.org/10.1109/ACCESS.2017.2756069.

Wang, K. J., Lee, Y. H., & Angelica, S. (2020). Digital twin design for real-time monitoring-a case study of die cutting machine. *International Journal of Production Research*. https://doi.org/10.1080/00207543.2020.1817999.

Wang, Q., Jiao, W., & Zhang, Y. M. (2020). Deep learning-empowered digital twin for visualized weld joint growth monitoring and penetration control. *Journal of Manufacturing Systems*, 57, 429–439. https://doi.org/10.1016/j.jmsy.2020.10.002.

Warke, V., Kumar, S., Bongale, A., & Kotecha, K. (2021). Sustainable development of smart manufacturing driven by the digital twin framework: A statistical analysis. *Sustainability (Switzerland)*, 13(18). https://doi.org/10.3390/su131810139.

Zhang, Q., Xiao, R., Liu, Z., Duan, J., & Qin, J. (2023). Process simulation and optimization of arc welding robot workstation based on digital twin. *Machines*, 11(1), 53.

12 Application of Machine Learning Techniques for Fault Detection in Friction Stir-Based Advanced Joining Techniques

Pragya Saxena, Arunkumar Bongale,
Satish Kumar, and Suresh R.

12.1 INTRODUCTION

Friction stir processing (FSP) is an advanced solid-state processing technique that is recently being used to form surface composites of lightweight alloys such as aluminum, magnesium, and copper alloys, generally embedded with varying reinforcement particles. The process has gained importance as the temperature at which it operates is below the melting point of the parent metal, resulting in a lesser heat-affected zone. Also, since the recrystallization occurs in a semi-solid state, the grains are refined with improved microstructure, and minimal surface finishing operations are required after the preparation of surface composites. Some reinforcements (ceramic materials generally) may be added by the ex-situ method, which involves no intermetallic bonding with the matrix material, so that they add their characteristics to the composite without hindering parent metal properties. On the other hand, some reinforcements are added to the matrix material by the in-situ method, which involves the formation of intermetallic compounds with the matrix and hence improves the strength and other physical characteristics of the composite. Recently, many researchers have incorporated a hybrid approach that includes two or more nanoparticles as reinforcements added in-situ as well as ex-situ by the FSP method to form surface hybrid nanocomposites.

FSP, as derived from the friction stir welding (FSW) technique, utilizes the heat produced by the friction because of the rubbing of the tool and workpiece surfaces. It consists of a tool shoulder with an integrated pin so that the shoulder comes into contact with the composite surface and the pin is protruded into the surface.

DOI: 10.1201/9781003327769-12

So, the material exchange occurs in solid phase between the tool and the composite surface caused by the pin stirring the plastically deformed material around it. With the traverse motion of the rotating tool, the pin stirs the material around it and this method causes recrystallization of the plastically deformed material and hence grain size refining of the metal. FSP refines the microstructure of the composite surface (aluminum, magnesium, and copper alloys, generally) during metal matrix composite fabrication and hence improves their physical properties such as hardness, stiffness, strength-to-weight ratio, wear resistance, etc. They are in high demand for manufacturing components in the automobile, aerospace, and industrial sectors owing to their exceptional strength, corrosion resistance, and electrical conductivity.

In order to compete in commercial industries today, FSP needs to be explored with Industry 4.0 techniques and artificial intelligence. In this context, it is desired to obtain controlled size and uniformity in the dispersion of reinforcement particles on the composite surface. Also, it is crucial to detect the defects occurring during the process and attempt to prevent them by modifying the process parameters. Although some recent studies have introduced the development of various strategies to measure different parameters during FSW and FSP using sensors to automate and control the process, the monitoring and control of FSP process parameters are rarely found in the literature. So, there is a wide scope for research to explore FSP for multiple process parameters and provide an accurate control strategy. Numerous industrial-grade sensors can be implemented, such as temperature sensors, vibration sensors, current sensors, dynamometers, sound sensors, etc., to measure respective parameters during the process. Hence, the process information such as heat generated, movement of the tool, current produced, tool rotation, feed, and forces generated, along with the machine information such as machine vibration and power input can be analyzed and interpreted to identify the possible defects in the part as well as the anomaly in the machine during FSP. The online data generated can be used for further improvement of the process and integration with similar types of machines.

In this work, it is proposed to detect and classify the defects occurring in the aluminum-based surface hybrid composites during fabrication by FSP using multiple sensor data analysis with the help of machine learning algorithms. However, it is highly challenging to synchronize and analyze the multiple sensor data simultaneously and understand the influence of each parameter individually and collectively. Similarly, the occurrence of defects during the FSP of composites is crucial to be identified and correlated with the associated causes.

The multiple sensor data is collected during the FSP of the composites with the help of a data acquisition system and is stored as multiple files for each process. This data is pre-processed and analyzed using various machine learning algorithms. Further modification of parameters to minimize the defects and obtain a uniform microstructure and improved physical characteristics of the prepared composites is attempted. For other machining processes, a lot of work has been attempted for condition monitoring and defect prediction with the help of multiple sensor data analyses. This leads to the promotion of real-time monitoring systems and defect prediction systems.

12.2   ARTIFICIAL INTELLIGENCE IN FSW AND FSP

In the manufacturing sector, the control and monitoring of machine and tool conditions are emerging as advanced technological developments in this area. Real-time monitoring of the process parameters during machining with the help of suitable sensors is crucial in the path of Industry 4.0. Various machine learning techniques are implemented during FSP to obtain a controlled process. The abnormalities during machine operations are identified by changes in operational conditions such as temperature, vibration, force, power consumption, etc. Hence, the online monitoring of the tool and machine condition is crucial for fault detection and preventing failures in the machining processes. As reviewed from the relevant literature, defect detection and condition monitoring have been performed recently for online break condition and defect identification in bearings (Jegadeeshwaran and Sugumaran 2015; Sakthivel, Sugumaran, and Nair 2010), centrifugal pumps, and gear boxes of automobiles (Amarnath et al. 2015). However, condition monitoring and fault detection during FSP of surface composites using the Industry 4.0 approach is quite rare and needs further exploration. Some of the relevant research on control methods with a machine learning approach for improvement in FSW techniques is hereby discussed in brief. Mishra et al. (2020) developed a real-time monitoring system consisting of a feedback-based system to monitor the FSW process by collecting and analyzing signals from multiple sensors with the help of a machine learning approach to predict and prevent the occurrence of defects in the produced welds. The same group of researchers has attempted to implement the industrial 4.0 approach in the process by utilizing various methods and techniques for sensor data analysis for an efficient data exchange feature in FSW (Mishra et al. 2018).

Balachandar, Jegadeeshwaran, and Gandhikumar (2020) collected real-time vibration data collected from vibration sensors for normal as well as defective welding operation for the process monitoring of FSW. They utilized a machine learning classifier technique to detect and classify and, hence, prevent the occurrence of defects during the process. Another group of researchers developed an online defect prediction system for monitoring and controlling the FSW of AA2219 with the help of various machine learning algorithms for feature extraction, selection, classification, and prediction of defects (Liao et al. 2019). Some review research for various controlling methods deals with welding issues such as the back support issues required for weld, thinning of weld, and other surface defects occurring during the FSW (Meng et al. 2021). In order to minimize the defects, it is crucial to analyze the individual effects of all the parameters on the physical characteristics and the technical modifications required for them. Some researchers have attempted to measure the thrust and torque of the rotating tool to develop an automation system in FSP (Ammouri and Hamade 2018). Another group of researchers reviewed various strategies to control force and torque on the FSP tool during the process on the basis of varying tool rotation speeds, feed, and tool tilt angle (Busu et al. 2015).

A group of researchers (Sigl et al. 2020) developed a digital twin model of FSW for online temperature monitoring during the process with the help of torque signals obtained from the tool spindle. This feedback control system is verified by the experimental results and leads to improved weld quality. However, the control system exhibited some fluctuations for a few weld parameters and the geometry of the work-pieces. Bagheri, Abbasi, and Givi (2019) studied the influence of vibration in friction stir spot welding on the microstructure and thermal characteristics of the aluminum alloy experimentally and using FEA and then comparing the results. It was found that the temperature increased and the size of the grain in the microstructure decreased due to vibration in FSW. In another similar study, it was attempted to investigate the main effects of various welding parameters on the dynamic forces generated during FSW. It is concluded that the tool position influences the axial force, while the rotation speed and traverse feed affect the lateral and traverse forces during the welding.

Iqbal et al. (2020) examined the effects of the tool rotational speed and the plunge depth on the axial forces-generated torque, temperature, and power output during the FSW process of aluminum pipes. It was found that the grain size of the particles increases during the process due to high temperature, which results in increased hardness of the joints. Sahu et al. (2020) studied the effects of tool geometry on the force and torque signals generated during the FSW. Also, the study involves an attempt at predicting joint strength using sensor data analysis. Another study attempts to prevent welding defects by using a heat and force thermocouple and dynamometer. The study evaluates the temperature distribution, material displacement, etc., hence verifying the predictions (Hussein et al. 2014). One recent study involves a closed-loop feedback control system developed with FSW so that the tool plunge depth and temperature are controlled by varying the tool speed and position (Fehrenbacher et al. 2014).

It is reviewed that the FSP or FSW machine setup is designed for sensor data collection, management, and processing of signals from multiple sensors. Vita, Bruneo, and Das (2020) developed and applied a defect detection algorithm so that the data from multiple heterogeneous sensors could be used to monitor the working condition of the process. The model developed helps predict the defects by classifying the normal and abnormal range of the readings with the help of edge computing for each sensor (Yin, Li, and Yin 2020). Another group of researchers attempted to detect the faults in the FSW process by analyzing the clustering of the correlated data (Yoo 2020), so that the interrelationships among the data were obtained based on measuring the random distances. One of the recent studies involving the defect prediction in sensor data readings from a surveillance network attempted to identify and classify the faults (Marzat, Piet-Lahanier, and Bertrand 2018). Also, modification of the parameters is attempted in order to minimize the defects. Another work (Du, Mukherjee, and DebRoy 2019) attempts to find faults in the energy data and develops a statistical model that uses domain data knowledge (virtual sensors). The sensor data collected is compared with domain knowledge data using suitable machine learning algorithms. Another work by Cauteruccio et al. (2021) involves the study of defects and their detection and classification in the path of Industry 4.0. The study includes the process parameters' effects on the network's strength and the data exchange feasibility from the sensors. Rabe, Schiebahn, and Reisgen (2022) utilized deep learning techniques to examine the three-dimensional forces and torques in order to detect the defects that occurred inside the weld seam. The force and torque data are classified over a wide parameter range across the cross-section of the weld area with the help of feature extraction and classification algorithms. The authors found the Convolutional Neural Networks algorithm outperformed other techniques for the classification of defects. Another study (Effertz et al. 2022) optimized the plunge depth, time, and rotational speed of the tool during friction stir spot welding of aluminum plates to obtain the maximum shear force between the joints by utilizing machine learning algorithms. They also verified the predicted values with experimental results. An application of a novel unsupervised learning technique, i.e., self-organizing map, in determining the fracture areas in the friction stir-welded aluminum alloy plates and verifying its performance accuracy in comparison with other machine learning techniques such as decision trees, logistic classification, random forests, and Ada Boost has been attempted recently (Mishra and Dasgupta 2022). A recent attempt at the feature engineering approach (Camps et al. 2022) to correlate the FSW experimental data with the quality inspection data for obtaining effective predictive quality measures for the process. The authors predicted the defects that occurred in the weld, ensuring good-quality products with reduced inspection efforts.

12.3 FAULT DETECTION APPROACH IN FSW OR FSP USING ARTIFICIAL INTELLIGENCE

The idea of applying artificial intelligence to the FSP of surface composites for detecting surface defects is quite rare in this field. Some case studies related to the fault detection approach in friction stir-welded composites are hereby described in brief in the following sub-sections:

12.3.1 DIGITAL TWIN MODELING OF FSW PROCESS

A recent study by Roy et al. (2020) on various manufacturing processes includes a case study of developing a digital twin model for the physical FSW process. The raw sensor data is collected from various industrial-grade sensors, which are input to the model. This data is further processed and analyzed for online monitoring of the process and machine conditions. The digital twin inspects the selected characteristics during the FSW process and sends information regarding the maintenance work required. Figure 12.1 shows a flowchart for the workings of the digital twin model. There are four different working phases of the model, as shown in the figure, namely, data acquisition, fault

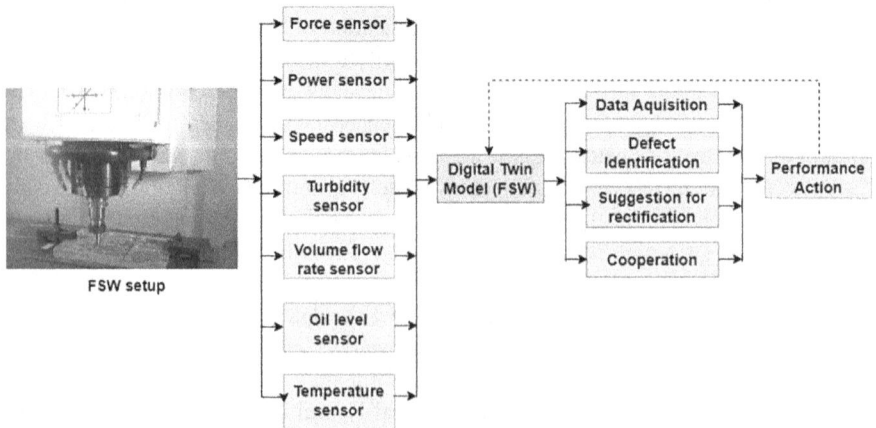

FIGURE 12.1   Flowchart for the digital twin model for defect detection in FSW.

detection, suggestion, and cooperation. In the data acquisition stage, the sensor data is acquired from various sensors: force, power, rotational speed, traverse speed, turbidity, flow rate, oil level, and temperature. In the fault detection phase, various machine learning techniques are employed to identify possible defects in the weld and machine condition. In the suggestion phase, the improvement measures are given to the digital twin model according to the present condition observed in the model by modifying the respective parameters for better condition of the machine and smooth functioning of the process. The rectification techniques are suggested for the faults that occurred during the welding process using a graphical interface system developed by software. In the cooperation phase, the condition of the machine and process is determined with the help of the information gathered from the sensor data. Using optimization techniques, the amendments in the workflow are implemented to minimize defects and improve the machine condition of the FSW setup. Hence, the workload assigned to a particular machine depends on its health condition.

12.3.2 SURFACE DEFECT DETECTION IN FSW JOINTS USING A MACHINE LEARNING APPROACH

One of the recent studies in the field has been attempted by Ranjan et al. (2016), who utilized machine learning techniques to detect and categorize the defects occurring in the AA1100 alloy joints during the FSW process. The surface of the welded joints is subjected to various kinds of defects, such as cracks, grooves, flashes, keyholes, and porosity. In this work, the accurate location of each defect is identified using image processing techniques. First, the grayscale image is filtered using a median filter. Then data compression and image resolution expansion are accomplished with the help of an image pyramid using pattern information on different scales. The reduction of image leads to the removal of diminishing texture variations or cracks so that significant defects such as porosity, pinholes, grooves, etc., are easily visible. These defects are identified by edge detection near the defect regions. Some morphological operations are performed to spot these defects easily. After identifying the major defects, image reconstruction processing using a suitable machine learning algorithm is utilized to identify the minor undetected cracks and edges by removing insignificant texture defects. Then, the Hough transform algorithm is used to detect flash defects along the length of the retreating side weld and in the middle region of the advancing and retreating sides of the weld. After fragmentation of various kinds of defects using image processing, the defects are classified as keyholes, porosity, cracks, grooves, flashes, etc. The graphs for the surface area of all the defects recognized are obtained, and all are combined to get the final location and surface area of the defects identified. The researchers obtained the images of the weld with accurate locations and magnitudes of various defects on the advanced as well as retreating sides superimposed successfully. The intensity of each defect is displayed with the help of the number of white pixels in the area plots obtained. Figure 12.2 shows the working methodology adopted in this study to utilize image processing in identifying surface defects in FSW joints.

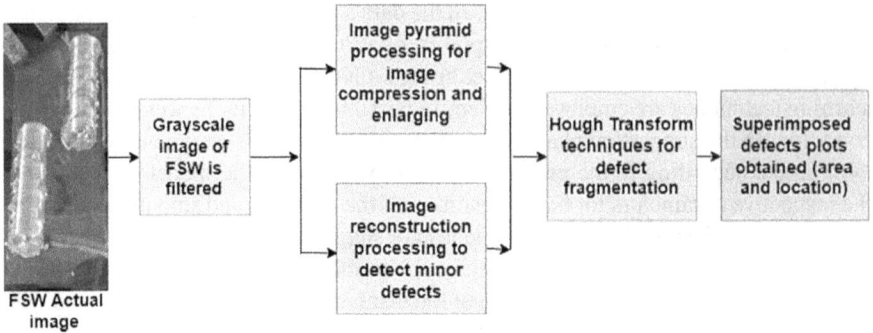

FIGURE 12.2   Image processing techniques for the classification of defects in FSW.

12.3.3   Artificial Intelligence for Detecting Surface Defects in Friction Stir-Welded Joints

Another approach by a group of researchers (Mishra and Dutta 2020) includes the use of image processing methods to detect the irregularities occurring at the joints while welding Al6060 plates by FSW using the H13 tool in a slightly different way from that discussed in the previous case study. The image pyramid is constructed to identify the porosity, pin holes, and cracks on the surface, which smoothens and resamples the signal repeatedly. However, the image reconstruction method is utilized to detect flashes and irregular welds that occurred due to improper surface contact between the tool and the metal plates. It reconstructs the image and obtains desired features using morphological techniques such as dilation, erosion, etc. In image processing, first the images are converted to grayscale data, and then the path of each file is defined. Then the noise is removed by using a two-dimensional median filter. After reading the images, the essential information is extracted from the image data with the help of data processing. This step is further categorized into image pyramid design, image reconstruction, and image fusion. Image pyramid step consists of reducing the size of the image and identifying the horizontal and vertical edges in order to locate the porosity, pin holes, and cracks with the help of the Sobel edge detection technique. In image reconstruction, the eroded mask images are reformed using the repeated dilation process of the marker image until its contour is fitted with the mask image. Again, the images are filtered, and the edges are detected with the help of the Sobel detection method to identify the flash and rough textures on the retreating side of the weld. Finally, image fusion is applied to merge the output binary images obtained from the image pyramid and image reconstruction in order to achieve a single binary image showing all the defects, as illustrated in Figure 12.3. The authors suggest the use of a Convolutional Neural Network for accurate detection of various kinds of defects during FSW processes.

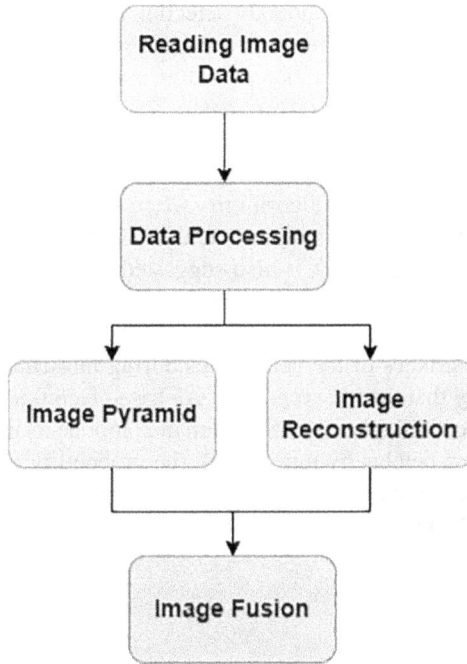

FIGURE 12.3   Methodology adopted in image processing for fault identification in FSW joints.

12.4   SUMMARY

This study provides a review of recent developments in FSW and processing and its scope in the context of artificial intelligence for prediction of defects in composites. From the case studies, it is evident that fault detection in friction stir-welded or processed composites is successfully implemented by both techniques, i.e., digital twin modeling and image processing techniques. The digital twin modeling includes the study of real-time data collected from various sensors during the actual FSW process. This method is useful for real-time condition monitoring and prediction of faults with feedback to prevent their occurrence. However, the image processing techniques involve the study of images captured of the welded or processed region after the process. This helps to identify and classify the faults that occurred with accurate location, intensity, area, etc. So, this technique focuses on the detection of faults both qualitatively and quantitatively in detail.

In view of observations from the discussion made in this work based on the available literature, research project is proposed to detect and classify defects in the composite surface with the help of data collection by multiple sensors. The FSP tool will be designed, fabricated, and tested using a vertical CNC milling machine. The data from various sensors, such as vibration, temperature, sound, current sensor, dynamometer, etc., mounted on the machine with FSP setup, is collected during fabrication with the help of a data acquisition system for different composites. Thus, the collected data can be used for real-time optimization of the process parameters and for condition monitoring and anomaly detection in sensor data, further improving the process to achieve the desired surface microstructure in hybrid composites. This data, which consists of information regarding normal and defected composites, is analyzed further to identify the type of defect. This sensor data is pre-processed and classified using machine learning classifiers (fault classification), and a range of parameters (associated with the respective sensor data) can be predicted using various machine learning prediction algorithms. After the fabrication of composites, their mechanical properties like hardness, tensile strengths, etc. can be obtained using conventional experiments. It is also suggested that the experimental results obtained can be correlated with the obtained sensor data, which may be utilized for development of an online prediction system to detect faults and predict the probable physical characteristics of the composites during fabrication using the sensor data. Thus, it is likely that the intervention of AI-based techniques like digital twin, image processing, etc., will take the fabrication of components using FSP-type joining techniques to new heights by minimizing the probability of the occurrence of surface defects. This will certainly help in improving the physical quality of products and, hence, reducing the overall cost, time, and energy consumption of the production system.

REFERENCES

Amarnath, Muniyappa V.V., Jain, Deepak K., Sugumaran, V., and Kumar, Hemantha. 2015. "Fault Diagnosis of Helical Gear Box Using Naïve Bayes and Bayes Net." *International Journal of Decision Support Systems* 1 (1): 4. https://doi.org/10.1504/ijdss.2015.067252.

Ali Ammouri, and Ramsey F. Hamade. 2018. "Toward an Affordable Automation Scheme of Friction Stir Processing," 1–7.

Bagheri, Behrouz, Abbasi, Mahmoud, and Besharati Givi, Mohammad Kazem. 2019. "Effects of Vibration on Microstructure and Thermal Properties of Friction Stir Spot Welded (FSSW) Aluminum Alloy (Al5083)." *International Journal of Precision Engineering and Manufacturing* 20 (7): 1219–27. https://doi.org/10.1007/s12541-019-00134-9.

Balachandar, K., Jegadeeshwaran, R., and Gandhikumar, D. 2020. "Condition Monitoring of FSW Tool Using Vibration Analysis-A Machine Learning Approach." *Materials Today: Proceedings* 27: 2970–74. https://doi.org/10.1016/j.matpr.2020.04.903.

Busu, Nur'Amirah, M. Shamil Jaffarullah, Cheng Yee Low, M. Saiful Bahari Shaari, Armansyah, and Ahmed Jaffar. 2015. "A Review of Force Control Techniques in Friction Stir Process." *Procedia Computer Science* 76 (Iris): 528–33. https://doi.org/10.1016/j.procs.2015.12.331.

Camps, Marta, Maddi Etxegarai, Francesc Bonada, William Lacheny, Sylvain Pauleau, and Xavier Domingo. 2022. "Feature Engineering and Machine Learning Predictive Quality Models for Friction Stir Welding Defect Prediction in Aerospace Applications" 4–7. https://doi.org/10.3233/faia220330.

Cauteruccio, Francesco, Luca Cinelli, Enrico Corradini, Giorgio Terracina, Domenico Ursino, Luca Virgili, Claudio Savaglio, Antonio Liotta, and Giancarlo Fortino. 2021. "A Framework for Anomaly Detection and Classification in Multiple IoT Scenarios." *Future Generation Computer Systems* 114: 322–35. https://doi.org/10.1016/j.future.2020.08.010.

Du, Yang, Tuhin Mukherjee, and Tarasankar DebRoy. 2019. "Conditions for Void Formation in Friction Stir Welding from Machine Learning." *NPJ Computational Materials* 5 (1): 1–8. https://doi.org/10.1038/s41524-019-0207-y.

Pedro Effertz, Willian Sales De Carvalho, Guimarães, R. P. M., G. Saria, and Sergio Amancio-Filho. 2022. "Optimization of Refill Friction Stir Spot Welded AA2024-T3 Using Machine Learning." *Frontiers in Materials* 9 (April): 1–9. https://doi.org/10.3389/fmats.2022.864187.

Fehrenbacher, Axel, Christopher B. Smith, Neil A. Duffie, Nicola J. Ferrier, Frank E. Pfefferkorn, and Michael R. Zinn. 2014. "Combined Temperature and Force Control for Robotic Friction Stir Welding." *Journal of Manufacturing Science and Engineering, Transactions of the ASME* 136 (2): 1–15. https://doi.org/10.1115/1.4025912.

Hussein, Sadiq Aziz, S. Thiru, Raja Izamshah, and Abd Salam Md Tahir. 2014. "Unstable Temperature Distribution in Friction Stir Welding." *Advances in Materials Science and Engineering.* https://doi.org/10.1155/2014/980636.

Iqbal, Md Perwej, Ranjan Kumar Vishwakarma, Surjya K. Pal, and Parthasarathi Mandal. 2020. "Influence of Plunge Depth during Friction Stir Welding of Aluminum Pipes." *Proceedings of the Institution of Mechanical Engineers, Part B: Journal of Engineering Manufacture.* https://doi.org/10.1177/0954405420949754.

Jegadeeshwaran, R., and Sugumaran, V. 2015. "Fault Diagnosis of Automobile Hydraulic Brake System Using Statistical Features and Support Vector Machines." *Mechanical Systems and Signal Processing* 52–53 (1): 436–46. https://doi.org/10.1016/j.ymssp.2014.08.007.

Liao, Thunshun Warren, Roberts, J., Wahab, Muhammad A., and Okeil, Ayman M. 2019. "Building a Multi-Signal Based Defect Prediction System for a Friction Stir Welding Process." *Procedia Manufacturing* 38: 1775–91. https://doi.org/10.1016/j.promfg.2020.01.089.

Marzat, Julien, Hélène Piet-Lahanier, and Sylvain Bertrand. 2018. "Cooperative Fault Detection and Isolation in a Surveillance Sensor Network: A Case Study." *IFAC-PapersOnLine* 51 (24): 790–97. https://doi.org/10.1016/j.ifacol.2018.09.665.

Meng, Xiangchen, Yongxian Huang, Jian Cao, Junjun Shen, and Jorge F. dos Santos. 2021. "Recent Progress on Control Strategies for Inherent Issues in Friction Stir Welding." *Progress in Materials Science* 115: 100706. https://doi.org/10.1016/j.pmatsci.2020.100706.

Mishra, Akshansh, and Anish Dasgupta. 2022. "Supervised and Unsupervised Machine Learning Algorithms for Forecasting the Fracture Location in Dissimilar Friction-Stir-Welded Joints." *Forecasting* 4 (4): 787–97. https://doi.org/10.3390/forecast4040043.

Mishra, Akshansh, and Saloni Bhatia Dutta. 2020. "Detection of Surface Defects in Friction Stir Welded Joints by Using a Novel Machine Learning Approach." *Applied Engineering Letters* 5 (1): 16–21. https://doi.org/10.18485/aeletters.2020.5.1.3.

Mishra, Debasish, Abhinav Gupta, Pranav Raj, Aman Kumar, Saad Anwer, Surjya K. Pal, Debashish Chakravarty, et al. 2020. "Real Time Monitoring and Control of Friction Stir Welding Process Using Multiple Sensors." *CIRP Journal of Manufacturing Science and Technology* 30: 1–11. https://doi.org/10.1016/j.cirpj.2020.03.004.

Mishra, Debasish, Rohan Basu Roy, Samik Dutta, Surjya K. Pal, and Debashish Chakravarty. 2018. "A Review on Sensor Based Monitoring and Control of Friction Stir Welding Process and a Roadmap to Industry 4.0." *Journal of Manufacturing Processes* 36 (May): 373–97. https://doi.org/10.1016/j.jmapro.2018.10.016.

Rabe, P., Schiebahn, Alexander, and Reisgen, Uwe. 2022. "Deep Learning Approaches for Force Feedback Based Void Defect Detection in Friction Stir Welding." *Journal of Advanced Joining Processes* 5: 100087. https://doi.org/10.1016/j.jajp.2021.100087.

Ranjan, Ravi, Aaquib Reza Khan, Chirag Parikh, Rahul Jain, Raju Prasad Mahto, Srikanta Pal, Surjya K. Pal, and Debashish Chakravarty. 2016. "Classification and Identification of Surface Defects in Friction Stir Welding: An Image Processing Approach." *Journal of Manufacturing Processes* 22 (April): 237–53. https://doi.org/10.1016/j.jmapro.2016.03.009.

Roy, Rohan Basu, Debasish Mishra, Surjya K. Pal, Tapas Chakravarty, Satanik Panda, M. Girish Chandra, Arpan Pal, Prateep Misra, Debashish Chakravarty, and Sudip Misra. 2020. "Digital Twin: Current Scenario and a Case Study on a Manufacturing Process." *International Journal of Advanced Manufacturing Technology* 107 (9–10): 3691–3714. https://doi.org/10.1007/s00170-020-05306-w.

Sahu, Santosh K., Debasish Mishra, Kamal Pal, and Surjya K. Pal. 2020. "Multi Sensor Based Strategies for Accurate Prediction of Friction Stir Welding of Polycarbonate Sheets." *Proceedings of the Institution of Mechanical Engineers, Part C: Journal of Mechanical Engineering Science*. https://doi.org/10.1177/0954406220960772.

Sakthivel, N. R., V. Sugumaran, and Nair, Binoy B. 2010. "Application of Support Vector Machine (SVM) and Proximal Support Vector Machine (PSVM) for Fault Classification of Monoblock Centrifugal Pump." *International Journal of Data Analysis Techniques and Strategies* 2 (1): 38–61. https://doi.org/10.1504/IJDATS.2010.030010.

Sigl, Martina E., Andreas Bachmann, Thomas Mair, and Michael F. Zaeh. 2020. "Torque-Based Temperature Control in Friction Stir Welding by Using a Digital Twin." *Metals* 10 (7): 1–18. https://doi.org/10.3390/met10070914.

Vita, Fabrizio De, Dario Bruneo, and Sajal K. Das. 2020. "On the Use of a Full Stack Hardware/Software Infrastructure for Sensor Data Fusion and Fault Prediction in Industry 4.0." *Pattern Recognition Letters* 138: 30–37. https://doi.org/10.1016/j.patrec.2020.06.028.

Yin, Chunyong, Bo Li, and Zhichao Yin. 2020. "A Distributed Sensing Data Anomaly Detection Scheme." *Computers and Security* 97. https://doi.org/10.1016/j.cose.2020.101960.

Yoo, Young. June 2020. "Data-Driven Fault Detection Process Using Correlation Based Clustering." *Computers in Industry* 122: 103279. https://doi.org/10.1016/j.compind.2020.103279.

13 Friction Stir Welding Characteristics of Dissimilar/Similar Ti-6Al-4V-Based Alloy and Its Machine Learning Techniques

P. S. Samuel Ratna Kumar and P. M. Mashinini

13.1 INTRODUCTION

In general, the Ti-6Al-4V double-phase titanium alloy is regarded as a cornerstone by businesses that create flying vehicles, including aviation, defense, and aerospace (Thomas 1991). In addition to having a significant mechanical property at ambient temperature and being resistant to corrosive environments, fracture propagation under cyclic loading, and deformation, Ti-6Al-4V also meets all other material performance criteria. Massive constructions and components with several distinct features and faces must be built using materials that can be linked with ease. It is undeniably true that titanium and its alloys are weldable (Mishra et al. 1999, Banerjee and Williams 2013, Liu and Shin 2019). There are several welding techniques that connect titanium by completely melting the adjacent area. Tensile residual tension was created as a result of melting and subsequent solidification. A fraction of the gas that was absorbed by molten metal and often confined caused undesirable pores. Distortion was a frequent problem with thin layers of sheet. The joint's elasticity was either significantly or partially lost. The aforesaid unwanted characteristics in the weld zone can be eliminated if the welding technique does not melt joint areas. In rotary or linear friction welding methods, the heat generated by the severe rubbing of the butting edges is used to combine titanium (Kaur and Singh 2019). Despite impressive, combined performance, only a small number of combinations can be connected. The metals that are flashed out require additional machining processes to be removed. Friction stir welding (FSW), which was created and developed at TWI thirty years ago, quickly gained popularity as a welding

DOI: 10.1201/9781003327769-13

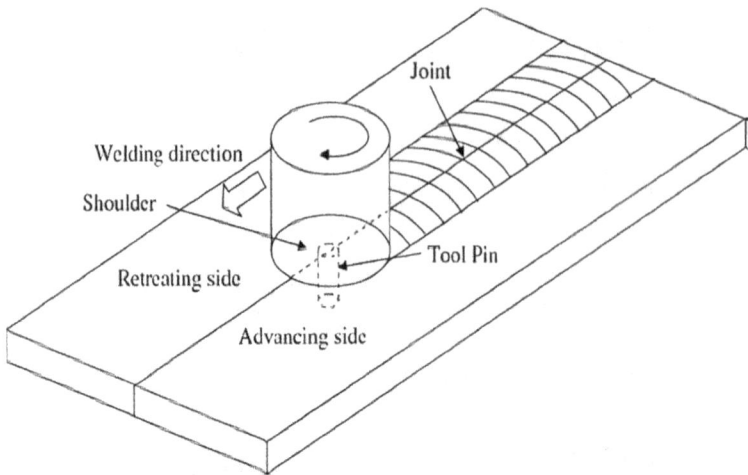

FIGURE 13.1   Schematic diagram of friction stir welding process.

technology for joining metals in their plastic limit without getting them to a molten condition (Singha et al. 2017). At first, a compressive force plasticizes and forges the adjacent edges together. Weldments are of superior ingredients, while joining via fusion has undesirable microstructural characteristics that can be reduced. Air from the atmosphere won't be drawn into the joint area to create pores. Capacity change brought on by the heat cycle is too small to significantly increase tensile residual stress (Singh et al. 2020, Jiang et al. 2019, Gao et al. 2020). Compared to a conventionally solidified structure, the grains have been greatly purified and smaller in dimension. For plasticization to occur, frictional heat must be produced by an external tool, as shown in Figure 13.1. As the crystallization and the ability of the substance to be connected increase, the ease of plasticization is significantly reduced. Maximum temperature and stress increase the tool's breakdown. Thus, further, FSW to be utilized to combine metals like steel, titanium, and nickel has been shown to be a challenging process (Gibson et al. 2014, Liu et al. 2018, Mironov et al. 2018, Buffa et al. 2013, and Wu et al. 2019). The impact of various welding process parameters and conditions on the physical qualities and microstructure of FSWed joints is illustrated in Figure 13.2. In regard to the welding process variables, the thermomechanical characteristics of the welded sheets are very important in choosing the appropriate tool material and tool geometry (Kumara et al. 2019, Auwal et al. 2018). To produce sound FSWed joints, the tool geometry's many interrelated components should be properly chosen. In particular, altering the tool shoulder diameter may have an impact on the microstructure of the welded connections because applying the ideal shoulder diameter to pin diameter ratio may result in improved grain growth. Notwithstanding ongoing difficulties, the FSW technique was used to combine highly purified Ti and Ti-6Al-4V pieces that ranged in thicknesses between 5.9 mm. The tool materials were typically focused on tungsten, either in its raw state or mixed with several other elements, including Re, La, and Co (Dhinakaran et al. 2017, Qina et al. 2019, and McAndrew et al. 2018). Every inquiry made some joints efficiently and concentrated on learning

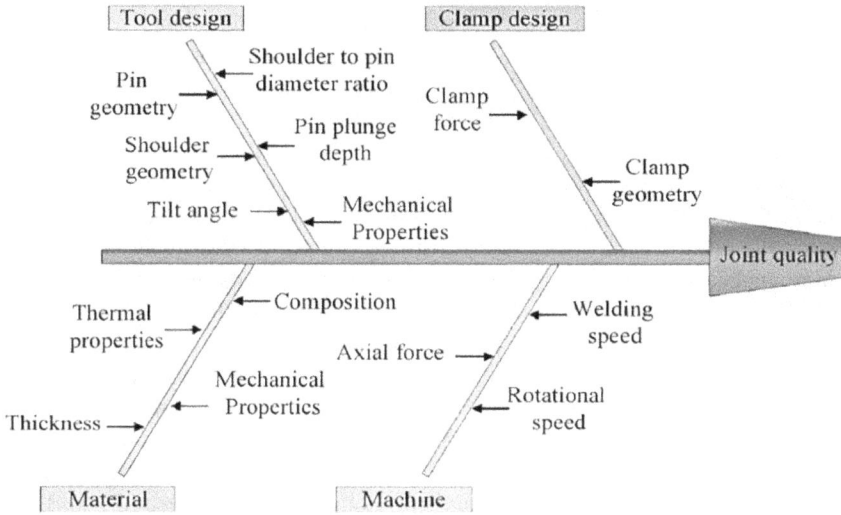

FIGURE 13.2 Friction stir welding process parameters.

about the stir zone's microstructural development. Ti-6Al-4V titanium alloy additive manufacturing has grown dramatically in the last several years, principally as a result of its widespread use as an alternative to the traditional manufacturing technique for complicated and relatively close net shape fabrication (Haghshenas and Gerlich 2018, Gangwar and Ramulu 2018). The parts require an appropriate joining procedure because printing large, complicated structures can be challenging. FSW with a tool rotation speed of 650 rpm or above should improve joint strength and effectiveness. All the welds' nearly identical average grain sizes demonstrate grain refinement after the process of FSW. The welding microstructure is made up of a fully developed lamellar phase and very small, equiaxed grains that have undergone transformation and are surrounded by a refined phase. This suggests that the heat generated during FSW is higher (Singh et al. 2020).

Machine learning (ML) techniques have lately been used, including the FSW technique. In FSW, numerous ML models and neural networks successfully identified flaws and evaluated their properties (Baraka et al., 2015). For additional investigation, some of the noteworthy readings and data were looked at. As an outcome, a variety of algorithmic sets, such as categorization, clustering, and regression models, have almost constrained reliability. Principle component analysis, random forest techniques, K-nearest neighbor and multi-layer perceptron are evaluated in the FSW environment (Elsheikh 2023). The ML model accepts inputs such as rotational speed, forging speed, travel speed, transverse and longitudinal stresses, torque, and energy densities. In order to match input restrictions and output variables, several ML algorithms are employed to investigate the necessary output values. Among the most common and regularly used methods for predicting or estimating is regression analysis. It is employed to establish the relationship between dependent and independent variables. Interactions between many input variables and various output variables are established via regression models. Machin Learning Regression is a suitable tool for determining actual system connections among variables used as inputs and outputs that can be employed to determine the kind of joints (Verma et al. 2018, Hartl et al. 2020, Van Otterlo and Wiering 2012, Kaelbling et al. 1996, Yang et al. 2019, Du et al. 2020).

An ML branch known as artificial neural network (ANN) is where deep learning got its start. Since most deep learning techniques rely on neural network design, the two terms deep neural networks are sometimes used interchangeably. Deep learning attempts to understand the information from a hierarchy representation by using numerous non-linear processing layers, either supervised or unsupervised. Deep learning applications are present across all sectors, from automation to healthcare equipment. Also, ML algorithms that are supervised and unsupervised were separated. Among the earliest ML techniques, ANN was created in the 1940s based on the neuronal network of the human nervous system. Due to its capability of eliminating complicated and erratic interactions between the features of diverse systems, it found its first application later in the 1980s and is now employed in many application areas (Elsheikh 2023, Elsheikh et al., 2019). However, ANN can only produce accurate findings whenever a large amount of training data is employed (Figure 13.3). Sometimes, it offers relatively weak analysis capabilities and effects on a specific optimum solution rather than the most appropriate global response. The biggest weakness of ANN models is their inability to define the weight values that go along with them. Moreover, ANN can be used to address these problems employing a hybrid artificial technique known as a neurofuzzy system, which denotes the fusion of fuzzy theory and ANN. The adaptive neurofuzzy inference system (ANFIS) incorporates both neural networks and fuzzy systems, combining the learning and connectionist structures of neural networks with the human-like reasoning style of fuzzy systems. By giving each function a weight in the fuzzy inference system, the major drawback of the ANN model is eliminated in ANFIS. As a result, the ANFIS system's capacity to handle weight functions makes it superior to others, as research demonstrates (Elaziz et al., 2019). A supervised ML method known as support vector machine (SVM) was established in the early 1990s as a non-linear approach for both regression and classification functions. SVM was discovered by researchers to be a superior

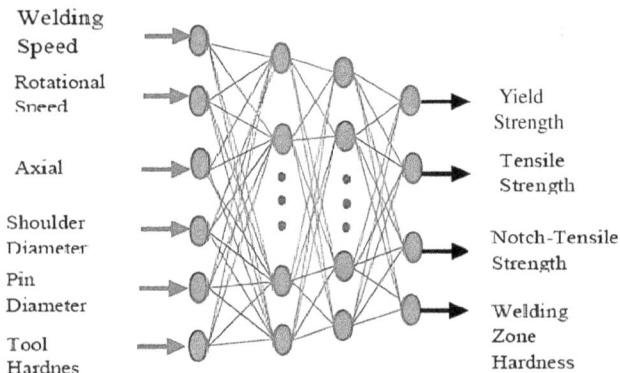

FIGURE 13.3   A concept of ANN model structure.

forecasting tool in several domains. SVM provides superior analytical capabilities, an exceptional structure that is unmatched worldwide, and fast information retraining capabilities. These attributes increase the sturdiness, suitability, and reliability of SVM. Scientists are using ML models in practical applications, while they are widely used in a variety of industrial operations. Because most manufacturing applications use labeled data, SVM has been found to be effective in these scenarios. The most frequently employed algorithm in ML with supervision for manufacturing industries is the SVM (Suthaharan 2016). A strong tool, ML will become much more valuable in the days to come. Each discipline of study is discovering uses for ML, including facial recognition, image analysis, the manufacturing industry, and healthcare, among many other fields.

13.2 FRICTION STIR WELDING

As a potential solution to problems with traditional fusion welding processes like large distortions, solidification cracking, macro and micro segregation, coarse dendritic structures, gas and shrinkage porosities, solid inclusions, surface oxidation or discoloration, the development of brittle intermetallics between dissimilar materials, with respect to the broad heat-impacted zone, and many more (Nandan et al. 2008, Mishra and Ma 2005). Therefore, it is crucial to comprehend the connection among FSW circumstances and structural characteristics in order to better tune weld qualities due to the capacity to regulate morphology and get crystal structures that are difficult to obtain by other methods. Moreover, friction-based stir processing (FSP) has the ability to produce fairly close crystal structures that are difficult to get in conventional ways. Both friction-based welding and processing are, however, getting utilized in a variety of industrial uses. According to the basic principles of FSW, a rotary tool is placed at the intersection of two workpiece material and moved along the weld zone direction. Typically, the FSW/FSP tool consists of a shoulder and a circular pin. A solid-state weld is created as a result of the deformation and heat that the rotating tool causes. FSP, on the other hand, can be used to either change the surface microstructure or add a second stage. In the last scenario, the particles from the second stage are introduced into the material's surface by drilling holes or cutting grooves, which allows the rotating tool to integrate the particle into the matrix surface and produce a surface-layered material. Many FSW/FSP factors have an impact on the final crystal structure of the recast layer. The weldment substance endures severe thermomechanical excursions because of the special characteristics of the FSW process, which cause grain growth and restoration reactions. Moreover, a significant portion (about 79.9%) of the effort associated with elastic deformation is transformed into heat, leading to localized adiabatic heat. Due to the geometry of the tool, the rotary and traverse speeds, the applied load of the shoulder on the specimen, the pin tilt, and the characteristics of the different materials, such as the thermal conductivity, the temperature-dependent flow stress, etc., the thermal field, the imposed strain, and the strain rate are consequently highly non-uniform. They also vary with distance from the rotating pin and shoulder (Heidarzadeh et al. 2021).

The titanium-grade 5 alloy is the most frequently used material in the titanium alloy group. For this reason, researchers have conducted numerous FSW studies on this material, focusing primarily on the FSW parameters with respect to the surface and physical properties of joint surfaces. Ramirez and Juhas (2003) documented the emergence of small, lamellar-shaped communities in the friction stir zone and also emphasized the necessity of FSW parameter optimization. Zhang et al. (2008) and Su et al. (2013) reported that the friction stir zone formed a fully lamellar structure, and they noted that the mechanical and physical properties of the joints improved as the tool traverse speed increased and the tool rotation rate decreased. Mashinini et al. (2022) used advanced FSW equipment (Figure 13.4) and changed the welding speed in increments of 40 mm/min from 40 to 200 mm/min. Using a tool made of lanthanide tungsten, sound joints were produced with no macroscopic flaws. The joints' microstructure, hardness, and residual stress were assessed. The base metal's

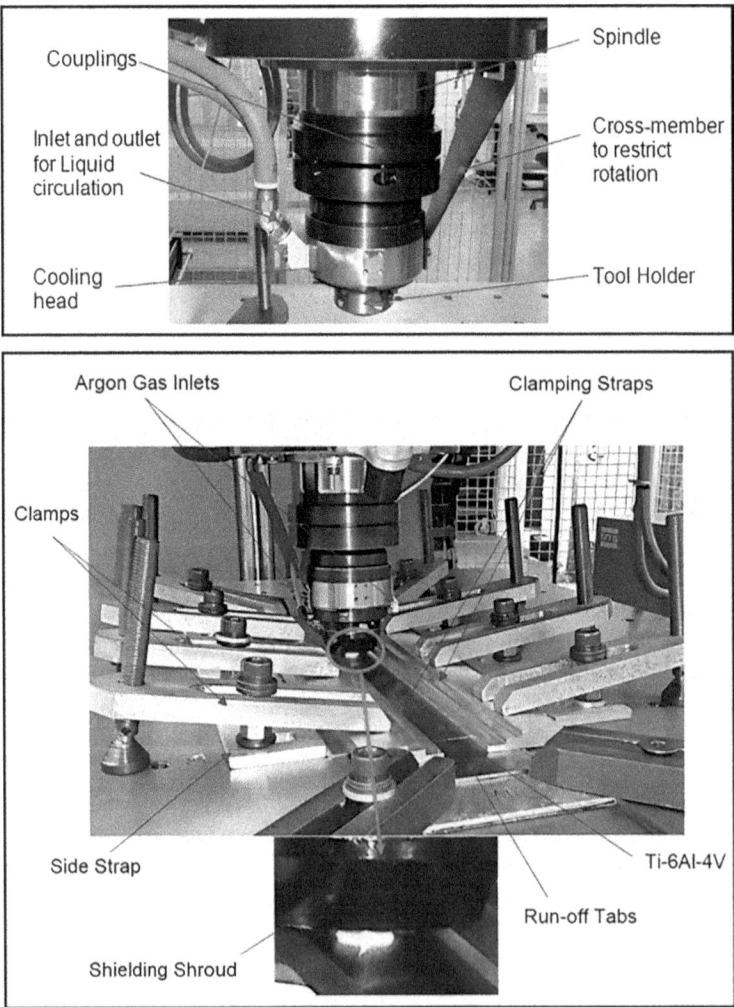

FIGURE 13.4 Advanced friction stir welding system with cooling and shielding configuration.

microstructure underwent phase transformation in the stir zone and changed into a lamellar structure mimicking a widmanstätten pattern. After FSW, it showed a slight increase in hardness value. The transverse path of the weld seam recorded residual stresses of the tensile type. At a stress ratio of 0.1, fatigue tests were performed in the welded and polished condition. Regardless of the quality of the interface or the welding speed, all samples failed at the stir zone. For welded specimens, shoulder-produced striation marks functioned as crack starting sites. The fatigue strength was somewhat influenced by the microstructural heterogeneity. The fatigue strength decreased as the welding speed increased. Shehabeldeen et al. (2021) used the right process parameter, and it was possible to successfully weld the dissimilar friction stir lap joints made of the metals AA6061-T6 and Ti6Al4V. The impact of tool rotational speed on mechanical properties, morphology, and fracturing mechanisms was investigated. The quantity of heat generation and viscoelasticity at the Al-Ti interaction, in addition to the production of intermetallic phases and various brittle phases that were generated at the Al-Ti combination, were found to be greatly influenced by the tool rotating speed. A lower tool rotational speed (600 rpm) and a peak hardness of 384 HV were used to reach the maximum shear strength of 3.5 KN, while a high rotational motion (1000 rpm) and a minimum hardness of 343.2 HV were used to obtain the minimum shear strength of 2.5 KN. While brittle failure mechanisms were seen at high rates of rotation because of the concentration of brittle intermetallic phases, hybrid fracture surfaces with both brittle and ductile features were generated at lower rotational speeds. As increased heat intake results in cracking and holes and reduces weld quality, TiAl and TiAl3 phases are present at the interface. Researchers also explained that although the mechanical properties of the joints improved with lower heat input, their transverse tensile qualities lagged behind those of the matrix material. Recrystallization is a crucial metallurgy phenomenon for the evolution of the microstructure, and the FSW can be viewed as one of the plastic processes. On the other hand, because Ti-6Al-4V is classified as a "+" alloy, the development of the resulting microstructure should be significantly influenced by the phase change. The quantity of different crystalline lattice distortions, such as grain boundaries, has an impact on both recrystallization and phase change. If the FSW process is applied commercially to titanium alloys, there are various issues that need to be resolved. The link among FSW process parameters that influences the temperature field, macrostructure, microstructure, and mechanical characteristics is the main issue and requires a thorough investigation via optimization tools.

13.3 CONVENTIONAL OPTIMIZATION TECHNIQUES

Welds can be produced using a variety of optimal welding settings; hence, a mixture of optimized welding parameters has not been attained. By defining the relationship for both the input and output variables, various optimization and modeling techniques can be used to find the optimized output parameters and avoid this issue. A number of conventional optimization techniques are available for accurately predicting and defining the relationship between the input and output variables of the process parameters without requiring a lot of time, resources, or personnel (Figure 13.5). Such conventional optimization techniques and trail-and-error

Input variable or Independent	Process	Output

FIGURE 13.5 Schematic diagram of the process to be optimized.

methods were used to identify the FSW process parameters for the dissimilar and similar titanium alloys.

(Prasomthong et al. 2022) FSW was used to examine the shear force of different welding processes between Ti-4V-6Al titanium alloys and Al5052 aluminum alloy. The experimental design made use of the Taguchi L9 orthogonal. The parameters for the welding process include rotation speed, feeds, and dwell time. To identify the ideal values for statistically significant components, an ANOVA analysis of the S/N ratio of shear force was performed. Also, consider the reaction and degree of the indispensability factor while analyzing the process parameters, as well as the forecast and regression model for the ideal tensile strength. Thus, because both variables had a P-value of less than 0.05, the experiment revealed that rotation speed and dwell time had a significant impact on the shear force. Thus, at a 95% confidence level, the rotation speed and dwell duration parameters are the most important ones in the process. However, a detailed analysis shows a negligible shear force on the weld because of a P-value greater than 0.05 (Patel et al. 2020). By employing Taguchi-based gray relational analysis, multi-response optimization of FSW of dissimilar Al 6061-Titanium alloy. The gray relational factors for every output feature are averaged to get the gray relational rating. For such gray relational rating, the experimental data of the multi-response characteristics are assessed. The process parameter level again with the greatest gray relational rating is the ideal level. According to ANOVA analysis that took into account the gray relational grade as a factor, feed rate is the most important processing parameter determining a variety of output responses (Shehabeldeen et al. 2021). By using a proper process parameter, it was possible to successfully weld the dissimilar friction stir lap joints made of the metals AA6061-T$_6$ and Ti6Al4V (Figure 13.6). The impact of tool rotational speed on the mechanical properties, crystalline lattice, and crack pathways was investigated. The quantity of heat generation and ductility at the aluminum/Titanium interface, along with the production of various brittle and intermetallic phases that were generated at the aluminum/Titanium mixture, were found to be greatly influenced by the tool rotating speed. A maximum shear strength was able to be attained at the tool rotation speed of 600 rpm, and a minimum shear strength with an increased hardness value was attained at 1000 rpm. While brittle fracture mechanisms were seen at high speeds of rotation because of the dense distribution of brittle intermetallic phases like TiAl and TiAl$_3$ phases at the interface due to the increased heat input that mostly induces porosities and fractures and subsequently deteriorates joint strength, hybrid fracture surfaces were produced at lower rotational speeds that had both brittle and ductile properties (Figure 13.7).

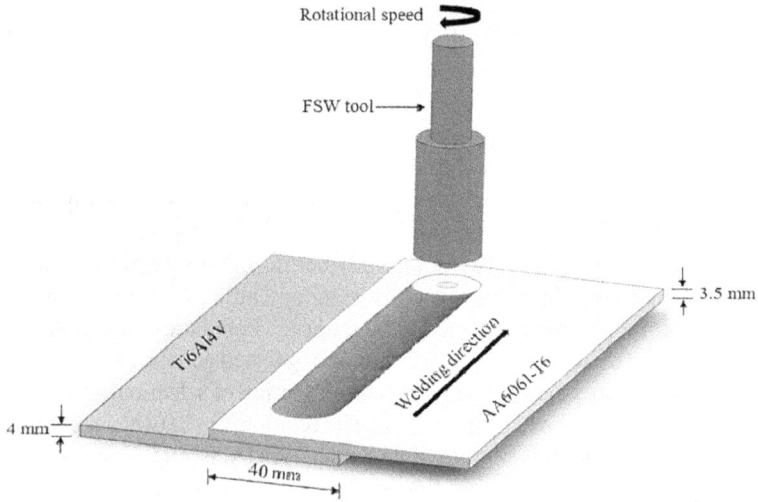

FIGURE 13.6 AA6061 T6-Ti6Al4V dissimilar material friction stir welding.

Joint appearance	Cross-section	Rotational speed (rpm)
		600
		800
		1000

FIGURE 13.7 Surface morphological images of the FSW-processed dissimilar material.

(Nirmal and Jagadesh 2021) For the FSW process, a finite element simulation has been created in the ABAQUS program to understand the process mechanics and forecast the mechanical properties of FSW-processed dual-phase titanium alloy components. Changes in the process parameters were made to the modeling, and the observations, including stress strain graphs, fraction deformation, yielding, and the final tensile strength, were calibrated and validated against experimental data. To improve FS weld quality, the welding parameters were genetically improved. Regression analysis is used to assess how well the genetic algorithm's fitness function works. Figure 13.8 shows the example structure of how neural network works along with the genetic algorithm for optimization. The FSW process optimization takes into account both upper and lower boundaries. Population types and sizes are fifty with a double vector, accordingly. With the suggested genetic algorithm, the size of the tournament is two, and it is regarded as a function of selection. The function of creation is assumed to be a feasible sample. The relationship between a predictor and responder variable, which serves as a fitness value for GA, is studied using the regression approach. For the fundamental linear regression, X and Y serve as the predictor

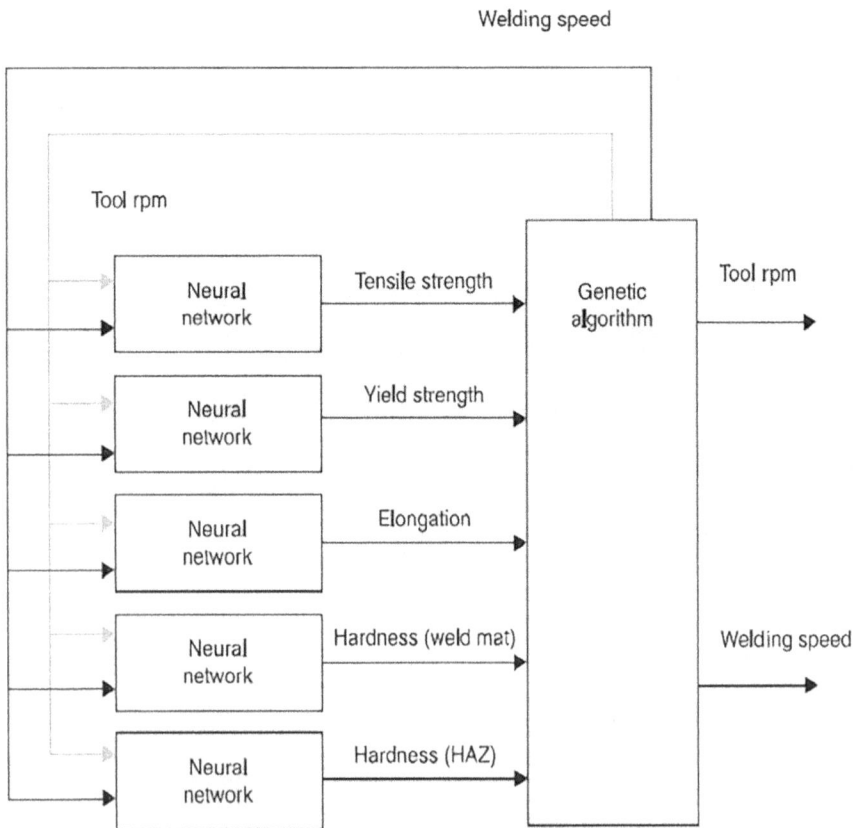

FIGURE 13.8 Neural network structure for genetic algorithm.

and response variables, respectively. A polynomial function of X is taken into consideration when one approaches a non-linear calculation. The confidence interval value was within 90%, and the prediction values are accurate with the experimental value. A better-optimized weld quality was obtained with the help of GA, with fewer defects and increased mechanical properties.

13.4   MACHINE LEARNING

The modeling of the FSW process is a challenging issue due to the wide variety of FSW technologies, numerous welding parameters and circumstances, and complex thermomechanical behavior of the welding process. In recognition of their outstanding ability to determine the relationship between the control factors and responses of the FSW process during its training stage, ML approaches have been suggested to model FSW. Due to their ability to generalize, once the ML methods have been taught, they may be utilized to ascertain the process responses under welding conditions that haven't been used in training stages. Applications of ML in several engineering fields, like mechanical behavior, nano, machining, FSP, and so on. Process of managing the welding process is a crucial component of ML method applications in FSW processes. The necessity for creating effective continuous approaches to continuously monitor the welding process has increased due to the requirement to manufacture high-quality FSWed joints at affordable production costs (Elsheikh 2023). It has been claimed that an effective method for monitoring the welding process involves the use of various signals measured throughout the welding process. To control the welding process and ensure the desired product quality, these signals could be gathered using multiple sensors on the welding machine and then analyzed using ML algorithms. Nonetheless, ML methods can be used to evaluate the FSW tool lifespan, which can be forecast. Due to the changing compressive and tensile loads that FSW tools are subjected to at high temperatures. Unwanted fatigue failure could be the outcome of this fluctuating loading. Due to the tool material's drastically reduced yield stress at high temperatures, plastic deformation of the tool may also happen when welding at high temperatures and varying loads. Additionally, owing to the increased welding temperature, alloying elements may be transferred from the welding joint to the tool. Such dispersed components produce tensile substances, reducing the tool's capacity. Due to the excessive heat developed during the friction, a high tool wear rate will occur during the FSW process. Consequently, a number of interrelated mechanisms may interact to produce the failure of FSW techniques. As a result, determining the tool life using traditional mechanical methods is a challenging challenge, and ML approaches may be employed as efficient replacements for traditional modeling methods. Thus, the application of ML to forecast the metallurgical and mechanical characteristics of FSWed joints, where ML techniques are employed to determine the correlation among weld process control parameters and processing actions. The combination of ML with finite element methods (FEM), where ML techniques are used to forecast the adjustable parameters that are utilized to adapt FEM to real-world FSW conditions or to function as a virtual machine to produce enough forecasted datasets for additional data analysis. ML techniques are used to regulate the process parameters in real time and enhance the effectiveness of the

FSWeld joints (Shojaeefard et al. 2014). Miaoquan et al. (2002), with experimental findings for Ti-6Al-4V titanium alloy with homogenous deformation under various process and technological factors, developed an adaptive model of grain size with the aid of fuzzy neural networks. The microstructural parameters, such as the volume percentage of the alpha-phase and the typical grain diameter, have been discovered. The discrepancy between the experimental and predicted results demonstrates the current fuzzy neural network model's high generalizability. This algorithm demonstrated good learning precision and good generalization when it was applied to the isothermal forging of Ti-6Al-4V titanium alloy. For processing noisy data or data with significant non-linear relationships, the neural network method's outputs are in good agreement with experimental data (Buffa et al. 2012). Engineers may choose to construct numerical simulations of the process in order to pursue cost reductions and a time-efficient solution. To anticipate both the microhardness and the grain structure of the welded butt joints at altering the key process parameters, an ANN was correctly trained and connected to an existing 3D FEM model for the FSW of Ti-6Al-4V alloy. For the microhardness prediction, a satisfactory agreement was discovered; however, for the microstructure, an excellent network prediction capability was attained. The established parametric simulation can be seen as a useful resource for an efficient and time- and money-saving process design, given the strong correlation between the microhardness and microstructure of the welded joints and their mechanical performances. In order to forecast tensile shear strength, several ML algorithms have been applied to dissimilar AA 7075-T651/Ti-6Al-4V alloys under various process circumstances to weld joints. We looked at how weld parameters like dwell time and rotational speed affected the mechanical and microstructural properties of weld junctions. SEM and optical microscopy were used to perform microstructural studies. At a rotating speed of 1000 rpm and a time duration of 10 s, the highest tensile shear strength of 3457.2 N was reached. The tensile shear strength of weld joints is significantly impacted by time. During prolonged dwell times and high revolving speeds, tensile shear strength was shown to rapidly decrease. At reduced stay times and fast rotating speeds, tensile shear strength also showed a significant improvement of 53.38%. In order to predict the tensile shear strength of welded joints at particular welding parameters, the researchers used the most significant ML data-driven methods used in welding, such as ANN, ANFIS, SVM, and regression model. All models' results are evaluated in the training and testing phases and contrasted with the experimental observations. The two independent conventional measures of forecast error% and root mean squared error were used to assess how well the generated models performed. Regression analysis, ANN, ANFIS, and SVM performance were examined, and it was discovered that SVM regression analysis performed much better than ANN and ANFIS in predicting the tensile shear strength of weld joints (Asmael et al. 2022).

13.5 ADVANTAGES OF MACHINE LEARNING IN FSW

A branch of data analysis known as ML consists of a set of methods that allow computers to learn to forecast future events using historic information. Programmers create if-else conditions based on their knowledge, as opposed to expert systems.

It accomplishes this without having been specifically instructed too. Whenever the connection between the information being provided and the result is too hazy or complex to spend time constructing the system and determining the formulas and regulations that will guide the method, ML techniques are applied. By examining prior information to construct a framework of what a connection would entail in the future, algorithms that use ML can learn to anticipate interactions. Algorithms are usually 100% accurate, unlike expert systems, although an experienced ML engineer can decrease the accuracy gap by adjusting model input variables, extraction of features, and so on. ML is used in FSW to enhance the procedure in a variety of ways. Quality control was one of the MLs in FSW applications that was most frequently used. In predicting the weld width, penetration phase, or tensile strength, various models, including convoluted neural networks, SVMs, and random forests, were given data from cameras (top and side views of the weld pool) or numerical data related to the probe (both rotational and translational speeds). These results facilitate and secure the process of quality assurance and verification. The findings can be applied to increase factory quality and output, or they can be parsed and inserted into a controlling system that makes quick modifications to create an improved welding. The variables that have the biggest effects on the FSW process were also discovered using ML. The agent was programmed to adjust the translation and rotation speeds of the probe in order to get the best weld in terms of time and strength. This has been done via a supervised learning algorithm. Further iterations were carried out using the RSM (response surface technique) to identify the factors that affected the FSW joint the most. Dimensionality reduction was yet another application. Prior to being transferred through a control system, the multimodal or multidimensional sensor outputs are required to be suppressed. Several neural networks perform this by default. In FSW, ML is largely implemented to accomplish two goals: enhanced quality control and integration. Neural networks were primarily used for quality control, working with images of the weld zone to guarantee there were no flaws. The type of material or the rotational and translational speeds of the FSW pin were nominal and numerical data that the neural networks also ingested. Utilizing self-learning or exploration algorithms, such as RL and Bayesian estimation, can solve one of the key issues, namely the lack of data. As was, such algorithms are more effective since they can learn the relationship between the input and the output without the need for data. For the finest weld in the shortest amount of time, RL has been utilized to set parameters like the FSW pin's rotational and translational speeds. Such algorithms also have very high accuracy levels that can be relied upon because they thoroughly investigate the surroundings in an effort to masterfully govern the intermediary. The one and only drawback of RL or other self-learning algorithms is their high computational and time requirements. The lack of temperature information in most models is yet another drawback. The materials must be preheated to the appropriate temperatures so that they remain in a plastic condition and do not melt in order to be cast, in order to achieve the fewest distortions and losses in their final tensile strength. Hence, temperature management is essential in FSW. Current research on ML has demonstrated that it serves as a useful tool in FSW and is growing in popularity. As a result of the input parameters, researchers can now predict the output response. Thus, ML is an increasingly prominent technique that has been shown to be helpful in FSW by a recent study. Investigators can now estimate the output response based on the input characteristics.

By properly adjusting welding process parameters like heat, torque, and shear forces, one can prevent the creation of FSW voids. By taking into account the process parameters relating to tool pins like heat, torque, and maximum shear stress, one could obtain a higher accuracy around 96% by expanding the dataset size and preprocessing with proper handling of class imbalance. As a result, there will be 4% or less faulty welds overall. The expense and duration of FSW have been greatly reduced by recent developments in ANN and methods for image processing.

13.6   CONCLUSION

The uses of ML techniques in the field of FSW are summarized in this chapter. The fundamentals of the widely used ML techniques are described. The scientific metrics used to assess the effectiveness of the ML approaches are provided.

- Employing ML techniques to quantify the connection between the controlled process parameters and the responses of the welding process, one may forecast the attributes of FSW Ti-6Al-4V joints.
- By merging ML and simulation to provide enough anticipated information for additional data analysis and optimization or acting as a virtual machine to forecast the simulation parameters needed to adapt the model to actual FSW circumstances.
- ML techniques are used to regulate the FSW process in real time and produce better-quality FSW Ti-6Al-4V joints.
- Predicting various FSW processes for various welding circumstances and similar/dissimilar titanium materials has shown promise when using ML approaches.
- In modeling the FSW process, ML methods fared better than traditional statistical techniques like trail-and-error, Taguchi, genetic algorithms, and gray relational analysis.
- When compared to statistical methodologies, the fundamental disadvantage of ML algorithms is their slightly longer computational time. There isn't a single ML model structure that works for all welding operations and controlled factors.

REFERENCES

Asmael M, Nasir T, Zeeshan Q. et al. (2022) Prediction of properties of friction stir spot welded joints of AA7075-T651/Ti-6Al-4V alloy using machine learning algorithms. *Archiv Civ Mech.Eng* 22:94. https://doi.org/10.1007/s43452-022-00411-x.
Auwal ST, Ramesh S, Yusof F, Manladan SM (2018) A review on laser beam welding of titanium alloys. *Int J Adv Manuf Technol* 97:1071–1098.
Banerjee D, Williams JC (2013) Perspectives on titanium science and technology. *Acta Mater* 61:844–879.

Baraka A, Panoutsos G, Cater S (2015) A real-time quality monitoring framework for steel friction stir welding using computational intelligence. *J Manuf Process* 20:137–148. https://dx.doi.org/10.1016/j.jmapro.2015.09.001.

Buffa G, Ducato A, Fratini L (2013) FEM based prediction of phase transformations during friction stir welding of Ti6Al4V titanium alloy. *Mater Sci Eng A* 581:56–65.

Buffa G, Fratini L, Micari F (2012) Mechanical and microstructural properties prediction by artificial neural networks in FSW processes of dual phase titanium alloys. *J Manuf Proc* 12. https://doi.org/10.1016/j.jmapro.2011.10.007.

Dhinakaran V, Siva Shanmugam N, Sankaranarayanasamy K (2017) Experimental investigation and numerical simulation of weld bead geometry and temperature distribution during plasma arc welding of thin Ti-6Al-4V sheets. *J Strain Anal Eng Des* 52:30–44.

Du Y, Mukherjee T, Mitra P, DebRoy T (2020) Machine learning based hierarchy of causative variables for tool failure in friction stir welding. *Acta Materialia* 192:67–77.

Elaziz MA, Elsheikh AH, Sharshir SW (2019) Improved prediction of oscillatory heat transfer coefficient for a thermoacoustic heat exchanger using modified adaptive neuro-fuzzy inference system. *Int J Refrig* 102:47–54. https://dx.doi.org/10.1016/j.ijrefrig.2019.03.009.

Elsheikh AH (2023) Applications of machine learning in friction stir welding: prediction of joint properties, real-time control and tool failure diagnosis. *Eng Appl Artif Intell* 121. https://doi.org/10.1016/j.engappai.2023.105961.

Elsheikh AH, Abd Elaziz M (2019) Review on applications of particle swarm optimization in solar energy systems. *Int J Environ Sci Technol*. 16. http://dx.doi.org/10.1007/s13762-018-1970-x.

Gangwar K, Ramulu M (2018) Friction stir welding of titanium alloys: a review. *Mater Des* 141:230–255.

Gao FY, Guo YF, Qiu SW, Yu Y, Yu W (2020) Fracture toughness of friction stir welded TA5 titanium alloy joint. *Mater Sci Eng A* 776:138962.

Gibson BT, Lammlein DH, Prater TJ, Longhurst WR, Cox CD, Ballun MC, Dharmaraj KJ, Cook GE, Strauss AM (2014) Friction stir welding: process, automation, and control. *J Manuf Process* 16:56–73.

Haghshenas M, Gerlich AP (2018) Joining of automotive sheet materials by friction-based welding methods: a review. *Eng Sci Technol Int J* 21:130–148.

Hartl R, Hansjakob J, Zaeh MF (2020) Improving the surface quality of friction stir welds using reinforcement learning and Bayesian optimization. *Int J Adv Manuf Technol* 110(11–12):3145–3167.

Heidarzadeh A, Mironov S, Kaibyshev R, Çam G, Simar A, Gerlich A, Khodabakhshi F, Mostafaei A, Field DP, Robson JD, Deschamps A, Withers JP (2021) Friction stir welding/processing of metals and alloys: a comprehensive review on microstructural evolution. *Prog Mater Sci* 117:100752.

Jiang LY, Huang WJ, Liu CL, Chai LJ, Yanga XS, Xu QL (2019) Microstructure, texture evolution and mechanical properties of pure Ti by friction stir processing with slow rotation speed. *Mater Charact* 148:1–8.

Kaelbling LP, Littman ML, Moore AW (1996) Reinforcement learning: a survey. *J Artif Intell Res* 4:237–285.

Kar J, Chakrabarti D, Roy SK, Roy GG (2019) Beam oscillation, porosity formation and fatigue properties of electron beam welded Ti-6Al-4V alloy. *J Mater Process Technol* 266:165–172.

Kaur M, Singh K (2019) Review on titanium and titanium based alloys as biomaterials fororthopaedic applications. *Mater Sci Eng C* 102:844–862.

Kumara K, Masantaa M, Sahoo SK (2019) Microstructure evolution and metallurgical characteristic of bead-on-plate TIG welding of Ti-6Al-4V alloy. *J Mater Process Technol* 265:34–43.

Liu FC, Hovanski Y, Miles MP, Sorensena CD, Nelson TW (2018) A review of friction stir welding of steels: tool, material flow, microstructure, and properties. *J Mater Sci Technol* 34:39–57.

Liu SY, Shin YC (2019) Additive manufacturing of Ti6Al4V alloy: a review. *Mater Des* 164:107552.

Mashinini PM, Dinaharan I, Hattingh DG et al. (2022) Microstructure evolution and high cycle fatigue failure behavior of friction stir-welded Ti-6Al-4V at varying welding speeds. *Int J Adv Manuf Technol* 122:4041–4054. https://doi.org/10.1007/s00170-022-10161-y.

McAndrew AR, Colegrove PA, Bühr C, Flipo BCD, Vairis A (2018) A literature review of Ti-6Al-4V linear friction welding. *Prog Mater Sci* 92:225–257.

Miaoquan L, Dunjun C, Aiming X, Li L (2002) An adaptive prediction model of grain size for the forging of Ti-6Al-4V alloy based on fuzzy neural networks. *J Mate Process Technol* 123. https://doi.org/10.1016/S0924-0136(02)00040-7.

Mironov S, Sato YS, Kokawa H (2018) Friction-stir welding and processing of Ti-6Al-4V titanium alloy: a review. *J Mater Sci Technol* 34:58–72.

Mishra RS, Ma ZY (2005) Friction stir welding and processing. *Mater Sci Eng: R: Reports* 50:1–78.

Mishra RS, Mahoney MW, McFadden SX, Mara NA, Mukherjee, AK (1999) High strain rate superplasticity in a friction stir processed 7075 Al alloy. *Scripta Mater* 41:163–168. https://dx.doi.org/10.1016/S1359-6462(99)00329-2.

Nandan R, DebRoy T, Bhadeshia HKDH (2008) Recent advances in friction-stir welding-process, weldment structure and properties. *Prog Mater Sci* 53:980–1023.

Nirmal K, Jagadesh T (2021) Numerical simulations of friction stir welding of dual phase titanium alloy for aerospace applications *Mater Today: Proc* 46. https://doi.org/10.1016/j.matpr.2020.10.300.

Patel S, Fuse K, Gangvekar K, Badheka V (2020) Multi-response optimization of dissimilar Al-Ti alloy FSW using Taguchi - Grey relational analysis. *Key Eng Mater* 833. https://doi.org/10.4028/www.scientific.net/KEM.833.35.

Prasomthong S, Kaewchaloon A, Charoenrat S (2022) Optimization of friction stir spot welding between aluminium alloys and titanium alloy by the Taguchi method. *SNRU J Sci Technol* 14. https://doi.org/10.55674/snrujst.v14i2.245169.

Qina PT, Damodaram R, Maity T, Zhang WW, Yang C, Wang Z, Prashanth KG (2019) Friction welding of electron beam melted Ti-6Al-4V. *Mater Sci Eng A* 761:138045.

Ramirez AJ, Juhas MC (2003) Microstructural evolution in Ti-6Al-4V friction stir welds. *Mater Sci Forum* 426:2999–3004.

Shehabeldeen TA, Yin Y, Ji X, Shen X, Zhang Z, Zhou J (2021) Investigation of the microstructure, mechanical properties and fracture mechanisms of dissimilar friction stir welded aluminium/titanium joints. *J Mater Res Technol* 11:507–518. https://dx.doi.org/10.1016/j.jmrt.2021.01.026.

Shojaeefard MH, Akbari M, Asadi P (2014) Multi objective optimization of friction stir welding parameters using FEM and neural network. *Int J Precis Eng Manuf* 15(11):2351–2356.

Singh AK, Kumar B, Jha K, Astarita A, Squillace A, Franchitti S, Arora A (2020) Friction stir welding of additively manufactured Ti-6Al-4V: microstructure and mechanical properties. *J Mater Process Technol* 277:116433.

Singha P, Pungotra H, Kalsi NS (2017) On the characteristics of titanium alloys for the aircraft applications. *Mater Today: Proc* 4:8971–8982.

Su J, Wang J, Mishra RS, Xu R, Baumann JA (2013) Microstructure and mechanical properties of a friction stir processed Ti-6Al-4V alloy. *Mater Sci Eng A* 573:67–74.

Suthaharan S (2016) Support vector machine. In: Shan Suthaharan (ed.), *Machine Learning Models and Algorithms for Big Data Classification*. Springer, pp. 207–235. https://doi.org/10.1007/978-1-4899-7641-3_9

Thomas WM (1991) Friction stir butt welding. Int. Patent № PCT/GB92/02203.

Van Otterlo M, Wiering M (2012) Reinforcement learning and Markov decision processes. In: Marco Wiering, Martijn Otterlo (eds.), Reinforcement Learning. Springer, Berlin, Heidelberg.

Verma S, Gupta M, Misra JP (2018) Performance evaluation of friction stir welding using machine learning approaches. *MethodsX* 5:1048–1058. https://dx.doi.org/10.1016/j.mex.2018.09.002.

Wu LH, Jia CL, Han SC, Li N, Ni DR, Xiao BL, Ma ZY, Fu MJ, Wang YQ, Zeng YS (2019) Superplastic deformation behavior of lamellar microstructure in ahydrogenated friction stir-welded Ti-6Al-4V joint. *J Alloy Compd* 787:1320–1326.

Yang D, Mukherjee T, DebRoy T (2019) Conditions for void formation in friction stir welding from machine learning. NPJ *Comput Mater* 5(1):1–8.

Zhang Y, Sato YS, Kokawa H, Park SHC, Hirano S (2008) Microstructural characteristics and mechanical properties of Ti-6Al-4V friction stir welds. *Mater Sci Eng A* 485:448–455.

Index

Note: **Bold** page numbers refer to tables; *italic* page numbers refer to figures.

For Product Safety Concerns and Information please contact our EU
representative GPSR@taylorandfrancis.com
Taylor & Francis Verlag GmbH, Kaufingerstraße 24, 80331 München, Germany